Protein Stability and Folding

Methods in Molecular Biology™

John M. Walker, Series Editor

40. **Protein Stability and Folding,** edited by ***Bret A. Shirley, 1995***
39. **Baculovirus Expression Protocols,** edited by ***Christopher D. Richardson, 1995***
38. **Cryopreservation and Freeze-Drying Protocols,** edited by ***John G. Day and Mark R. McLellan, 1995***
37. **In Vitro Transcription and Translation Protocols,** edited by ***Martin J. Tymms, 1995***
36. **Peptide Analysis Protocols,** edited by ***Michael W. Pennington and Ben M. Dunn, 1994***
35. **Peptide Synthesis Protocols,** edited by ***Ben M. Dunn and Michael W. Pennington, 1994***
34. **Immunocytochemical Methods and Protocols,** edited by ***Lorette C. Javois, 1994***
33. ***In Situ* Hybridization Protocols,** edited by ***K. H. Andy Choo, 1994***
32. **Basic Protein and Peptide Protocols,** edited by ***John M. Walker, 1994***
31. **Protocols for Gene Analysis,** edited by ***Adrian J. Harwood, 1994***
30. **DNA–Protein Interactions,** edited by ***G. Geoff Kneale, 1994***
29. **Chromosome Analysis Protocols,** edited by ***John R. Gosden, 1994***
28. **Protocols for Nucleic Acid Analysis by Nonradioactive Probes,** edited by ***Peter G. Isaac, 1994***
27. **Biomembrane Protocols:** *II. Architecture and Function,* edited by ***John M. Graham and Joan A. Higgins, 1994***
26. **Protocols for Oligonucleotide Conjugates,** edited by ***Sudhir Agrawal, 1994***
25. **Computer Analysis of Sequence Data:** *Part II,* edited by ***Annette M. Griffin and Hugh G. Griffin, 1994***
24. **Computer Analysis of Sequence Data:** *Part I,* edited by ***Annette M. Griffin and Hugh G. Griffin, 1994***
23. **DNA Sequencing Protocols,** edited by ***Hugh G. Griffin and Annette M. Griffin, 1993***
22. **Optical Spectroscopy, Microscopy, and Macroscopic Techniques,** edited by ***Christopher Jones, Barbara Mulloy, and Adrian H. Thomas, 1994***
21. **Protocols in Molecular Parasitology,** edited by ***John E. Hyde, 1993***
20. **Protocols for Oligonucleotides and Analogs,** edited by ***Sudhir Agrawal, 1993***
19. **Biomembrane Protocols:** *I. Isolation and Analysis,* edited by ***John M. Graham and Joan A. Higgins, 1993***
18. **Transgenesis Techniques,** edited by ***David Murphy and David A. Carter, 1993***
17. **Spectroscopic Methods and Analyses,** edited by ***Christopher Jones, Barbara Mulloy, and Adrian H. Thomas, 1993***
16. **Enzymes of Molecular Biology,** edited by ***Michael M. Burrell, 1993***

Earlier volumes are still available. Contact Humana for details.

Methods in Molecular Biology™ • 40

Protein Stability and Folding

Theory and Practice

Edited by

Bret A. Shirley

Boehringer Ingelheim Pharmaceuticals, Inc., Ridgefield, CT

Humana Press **Totowa, New Jersey**

999 Riverview Drive, Suite 208
Totowa, New Jersey 07512

This publication is printed on acid-free paper. ∞
ANSI Z39.48-1984 (American National Standards Institute)
Permanence of Paper for Printed Library Materials.

Printed in the United States of America. 10 9 8 7 6 5 4 3 2 1

Library of Congress Cataloging in Publication Data

Main entry under title:

Methods in molecular biology™.

Protein stability and folding: theory and practice/edited by Bret A. Shirley.
p. cm.—(Methods in molecular biology™; 40)
Includes index.
ISBN 0-89603-301-5
1. Protein folding. 2. Proteins—Conformation. I. Shirley, Bret A. II. Series: Methods in molecular biology™ (Totowa, N.J.); 40.
QP551.P697636 1995
574.19'245—dc20 94-39996
CIP

Preface

The intent of this work is to bring together in a single volume the techniques that are most widely used in the study of protein stability and protein folding. Over the last decade our understanding of how proteins fold and what makes the folded conformation stable has advanced rapidly. The development of recombinant DNA techniques has made possible the production of large quantities of virtually any protein, as well as the production of proteins with altered amino acid sequence. Improvements in instrumentation, and the development and refinement of new techniques for studying these recombinant proteins, has been central to the progress made in this field.

To give the reader adequate background information about the subject, the first two chapters of this book review two different, yet related, aspects of protein stability. The first chapter presents a review of our current understanding of the forces involved in determining the conformational stability of proteins as well as their three-dimensional folds. The second chapter deals with the chemical stability of proteins and the pathways by which their covalent structure can degrade. The remainder of the book is devoted to techniques used in the study of these two major areas of protein stability, as well as several areas of active research. Although some techniques, such as X-ray crystallography and mass spectroscopy, are used in the study of protein stability, they are beyond the scope of this book and will not be covered extensively.

I would like to thank everyone whose time and effort made this work possible. I would especially like to thank all of the authors for their forbearance and understanding about the exceedingly painstaking process of getting a book into its final form. Finally, I would like to thank my wife, Francine, for the time that she gave up to allow me to complete this task.

Bret A. Shirley

Contents

Contributors

PHILIP N. BRYAN • *Center for Advanced Research in Biotechnology, Biotechnology Institute, University of Maryland, Rockville, MD*

CARL J. BURKE • *Department of Pharmaceutical Research, Merck Research Laboratories, West Point, PA*

THOMAS E. CREIGHTON • *European Molecular Biology Laboratory, Heidelberg, Germany*

NIGEL DARBY • *European Molecular Biology Laboratory, Heidelberg, Germany*

ANTHONY L. FINK • *Department of Chemistry and Biochemistry, The University of California, Santa Cruz, CA*

ERNESTO FREIRE • *Departments of Biology and Biophysics and the Biocalorimetry Center, The Johns Hopkins University, Baltimore, MD*

PAUL M. HOROWITZ • *Department of Biochemistry, Health Science Center, University of Texas, San Antonio, TX*

THOMAS KIEFHABER • *Department of Biochemistry, Beckman Center, Stanford University Medical Center, Stanford, CA*

KUNIHIRO KUWAJIMA • *Department of Physics, School of Science, University of Tokyo, Tokyo, Japan*

HENRYK MACH • *Department of Pharmaceutical Research, Merck Research Laboratories, West Point, PA*

C. RUSSELL MIDDAUGH • *Department of Pharmaceutical Research, Merck Research Laboratories, West Point, PA*

KENNETH P. MURPHY • *Department of Biochemistry, University of Iowa, Iowa City, IA*

ANDREW D. ROBERTSON • *Department of Biochemistry, University of Iowa, Iowa City, IA*

CATHERINE A. ROYER • *School of Pharmacy, University of Wisconsin, Madison, WI*

James A. Ryan • *Department of Pharmaceutical Research, Merck Research Laboratories, West Point, PA*

Gautam Sanyal • *Department of Pharmaceutical Research, Merck Research Laboratories, West Point, PA*

J. Martin Scholtz • *Department of Medical Biochemistry and Genetics, Texas A & M University, College Station, TX*

Bret A. Shirley • *Boehringer Ingelheim Pharmaceuticals Inc., Ridgefield, CT*

Kathryn L. Stone • *W. M. Keck Foundation Biotechnology Resource Laboratory and Howard Hughes Medical Institute Biopolymer Facility, Yale University, New Haven, CT*

Serge N. Timasheff • *Graduate Department of Biochemistry, Brandeis University, Waltham, MA*

David B. Volkin • *Department of Pharmaceutical Research, Merck Research Laboratories, West Point, PA*

Kenneth R. Williams • *W. M. Keck Foundation Biotechnology Resource Laboratory and Howard Hughes Medical Institute Biopolymer Facility, Yale University, New Haven, CT*

CHAPTER 1

Noncovalent Forces Important to the Conformational Stability of Protein Structures

Kenneth P. Murphy

1. Introduction

Proteins are the workhorses of life, responsible for catalyzing the numerous chemical reactions necessary to living cells as well as being important structural molecules. The functionality of proteins requires the correct spatial placement of amino acid side chains and prosthetic groups that, in turn, requires a defined three-dimensional structure of the protein chain. The three-dimensional structure is obtained through the folding of the polypeptide chain from an ensemble of fairly loose, disordered configurations, collectively referred to as the unfolded state, into a well-defined compact configuration, known as the native state.

There are three related questions that can be asked regarding a protein's native structure:

1. *What* is it?
2. *How* is it attained? and
3. *Why* is the native structure the one observed?

The first question is primarily addressed through physical techniques, such as X-ray crystallography or multidimensional NMR spectroscopy. The second question is primarily addressed by kinetic studies aimed at assessing the pathway of folding. The third question, with which this chapter is concerned, is primarily addressed through structural thermodynamics, the study of the connection between a protein's structure and the forces that maintain it.

From: *Methods in Molecular Biology, Vol. 40: Protein Stability and Folding: Theory and Practice*
Edited by: B. A. Shirley

Most of us have been taught in undergraduate biology and biochemistry that proteins are stable because of the "hydrophobic effect," a force that is entirely entropic and results from the restructuring of water around apolar surfaces. Recently it has become increasingly clear that this simple view of protein stability is inadequate to account for the experimental data. The emerging view is not yet completely clear and there seems to be considerable disagreement within the literature. Nevertheless, the points of agreement between various investigators, although not emphasized in the literature, are many and will be discussed in this chapter.

This chapter will address the following questions:

1. What are the primary forces responsible for the stabilization of globular proteins?
2. How can quantitative estimates of these forces be obtained? and
3. Can quantitative estimates of the various forces be related to protein structures in a predictive manner?

In the latter part of the chapter, we will discuss an empirical approach that we have recently applied to estimating the thermodynamics of folding/unfolding transitions and binding reactions from high resolution structures. This approach illustrates that even with the current, incomplete understanding of the noncovalent interactions in proteins, accurate predictions of the thermodynamics of processes involving proteins are possible and the future for a structural energetic understanding of proteins is promising.

2. Energetic Description of Protein Folding/Unfolding Transitions

The stability of a protein molecule customarily is described by the difference in free energy between the unfolded and native states, ΔG. Using the standard convention, a positive value of ΔG indicates that the native state is energetically favored over the unfolded state.

The free energy difference, ΔG, is defined in terms of the enthalpy, ΔH, and entropy, ΔS, differences as:

$$\Delta G = \Delta H - T\,\Delta S \quad (1)$$

For the case of protein denaturation, ΔH and ΔS are dependent on temperature through the heat capacity difference, ΔC_p, as:

$$\Delta H = \Delta H_R + \Delta C_p\,(T - T_R) \quad (2)$$

and

$$\Delta S = \Delta S_R + \Delta C_p \ln\,(T / T_R) \quad (3)$$

where ΔH_R and ΔS_R are the enthalpy and entropy changes at the reference temperature, T_R, and T is the absolute temperature in K.

In the native state many of the protein functional groups are buried in the core, from which solvent water is excluded. Additionally, the amino acid side chains in the core are essentially fixed by specific interactions and tight packing. The peptide backbone is also relatively fixed throughout the protein. The amino acid side chains on the surface of the protein are also likely to have their motion somewhat restricted because of their close proximity to each other, although this restriction should be much less than that of the interior side chains. On unfolding, the interactions between protein groups within the core are disrupted and replaced with interactions with the solvent. Also, the backbone and side chains become much more mobile and, as discussed later, this gain in configurational entropy strongly favors the unfolded state.

3. Primary Forces Involved in Protein Stability

Since this chapter will discuss our current, largely empirical, understanding of the role of various forces in stabilizing globular protein structures, a detailed physical description of those interactions will not be given here. A brief overview is, nevertheless, appropriate. There are many ways in which one can envision a partitioning of the forces involved in protein stability. Here we consider van der Waals interactions, hydrogen bonds and electrostatic interactions, configurational entropy, and the unique role played by water.

3.1. Van der Waals Interactions

Van der Waals interactions, also known as London dispersion forces, result from transient dipoles that nonbonded atoms induce in each other. Such interactions, although ubiquitous, are fairly weak and short range. Because of the strong distance dependence of van der Waals interactions, the packing of atoms in the protein core, relative to their interaction with solvent, is important in determining whether they will stabilize or destabilize the native state.

3.2. Hydrogen Bonds

Hydrogen bonds are noncovalent interactions that arise from the partial sharing of a hydrogen atom between a hydrogen bond donor group, such as a hydroxyl –OH or an amino –NH, and a hydrogen bond acceptor atom, such as oxygen or nitrogen. Many potential hydrogen bond donor and acceptor groups are present in proteins, namely the peptide backbone groups and the polar amino acid side chains. A hydrogen bond is essentially a dipole–dipole interaction, although it is often modeled as a charge–charge interaction between partial charges. Because hydrogen bonds involve permanent dipoles, the interaction also has an angular dependence. A thorough review of hydrogen bonding in proteins has been given by Baker and Hubbard *(1)*, and the distribution of hydrogen bond patterns and their implications for folding also have been recently discussed by Stickle et al. *(2)*.

3.3. Electrostatic Interactions

Electrostatic interactions occur between charges on protein groups. Such charges are present at the amino- and carboxy- termini and on many ionizable side chains. Charges buried in the protein interior will interact strongly since the protein interior is considered a low-dielectric medium, but it has been suggested that such interactions may only play a minor role in protein stability since they are not overly common *(3)*. Electrostatic repulsion may be more important, not only in destabilizing the native state but also in terms of its effect on the degree of extension of the unfolded state *(4–6)*.

3.4. Configurational Entropy

Whereas the interactions discussed above tend to stabilize the native protein structure, configurational entropy destabilizes it. The gain in configurational entropy relates to the increased degrees of freedom available to the protein chain in the unfolded state relative to the native state. This gain comes from both the side chains and the backbone. Although the peptide backbone of most residues in a globular protein is relatively fixed (i.e., has low entropy), those residues that are most buried within the core of the protein have even fewer backbone degrees of freedom. The entropic effect of burying side chains is more pronounced since they have considerable flexibility on the protein surface. As larger proteins bury more of their side chains, they will have an overall larger configurational entropy change per residue. This effect may help to set a limit on the size of a globular folding domain.

The amino acid composition also affects the configurational entropy. For example, proteins containing a large proportion of proline residues will have a lower entropy in the unfolded state and thus will be more stable. The opposite will be true for proteins containing a large proportion of glycine.

3.5. The Role of Water

Water plays a crucial role in the stabilization of proteins. The small molecular size of water relative to other liquids, along with its complex hydrogen-bonded structure, make it an excellent solvent for many functional groups. These same features also give rise to the hydrophobic effect. The complex role of water in protein stability has been long recognized, but defining that role in quantitative terms is a topic of intense discussion more than 30 yr after Kauzmann's seminal review *(3)*.

One of the difficulties in discussing the role of water in stabilizing proteins is simply the definition of terms. Many phrases, such as "hydration," "solvation," and "the hydrophobic effect," are in common use but are not clearly defined (*see*, for example, refs. *7–9*). In particular, the term "hydrophobic effect" has been used to refer to:

1. The transfer of a compound from an organic liquid into water *(9)*;
2. The transfer of apolar surface from any initial phase into water *(10,11)*; and
3. Any transfer into water accompanied by a large ΔC_p *(7)*.

In this chapter, "hydration" refers to the transfer of any group from the gas phase into water and "hydrophobic effect" refers to the transfer of apolar groups from any initial phase into water. It is important to bear in mind that this usage is purely operational. The hydrophobic effect does not refer only to the properties of water or imply a "purely entropic" effect. In this sense, the hydrophobic effect will be different for gases, liquids, and solids. Considering protein stability, the hydrophobic effect refers to the energetic consequences of removing apolar groups from the protein interior and exposing them to water.

4. Experimental Evaluation of Forces

In attempting to assess experimentally the magnitude of the various contributions to protein stability, there are two basic approaches that can be taken: to study the stability of proteins as a function of various environmental variables, such as temperature, denaturant concentration, site-specific mutations, and so on; and to study model systems. In the study of model systems one attempts to mimic the unfolding process, but with a system simpler than a protein, which, hopefully, will be easier to interpret. Since a comprehensive review of the literature of these areas is beyond the scope of the present work, we will present here only those results necessary to understanding the various, current viewpoints on the partitioning of energetic contributions.

4.1. Protein Studies

In studying proteins the goal is to measure stability as a function of an environmental perturbant. Perhaps the most fundamental studies of protein stability involve temperature as the environmental variable. Differential scanning calorimetry (DSC), in which the excess heat capacity of a protein solution is determined as a function of temperature, can provide all the thermodynamic parameters that specify the stability of the protein as a function of temperature: ΔH, ΔS, and ΔC_p (for reviews *see* Chapter 9; refs. *12,13*).

4.1.1. Differential Scanning Calorimetry Studies

Numerous DSC studies have revealed several regular features regarding the thermal denaturation of globular proteins. Denaturation is always accompanied by a large positive ΔC_p. The specific heat capacity change normalized to the number of residues in the protein, $\Delta\overline{C}_p$, is directly proportional to the number of apolar contacts in the protein interior *(13)*. (Overscored values below are always normalized to the number of residues). Likewise, $\Delta\overline{C}_p$ is proportional to the amount of apolar surface area buried in the protein. Although it is now clear that the exposure of polar surface makes a significant, negative, contribution to $\Delta\overline{C}_p$ *(10,14–16)*, globular proteins generally bury a constant amount of polar surface area per residue *(11,17)*, and $\Delta\overline{C}_p$ can be considered to some degree as a measure of the "hydrophobicity" of the protein.

For a large number of globular proteins, the specific enthalpy changes, $\Delta\overline{H}$, when plotted as a function of temperature and taking $\Delta\overline{C}_p$ to be independent of temperature, converge to a common value, $\Delta\overline{H}^*$, at a specific temperature, T^*_H *(13)*. This behavior is revealed by plotting $\Delta\overline{H}$ at some convenient temperature, usually 298 K, vs $\Delta\overline{C}_p$ for a set of proteins *(18)* in what has been called an MPG enthalpy plot *(19)*. The slope of this plot is $(298 - T^*_H)$ and the intercept is $\Delta\overline{H}^*$. Analysis of the set of globular proteins shown in Fig. 1 indicates a value of 373 ± 6 K for T^*_H and a value of 1.35 ± 0.12 kcal/(mol-res) for $\Delta\overline{H}^*$ *(10)*. These values are independent of whether or not $\Delta\overline{C}_p$ is constant with temperature; they are

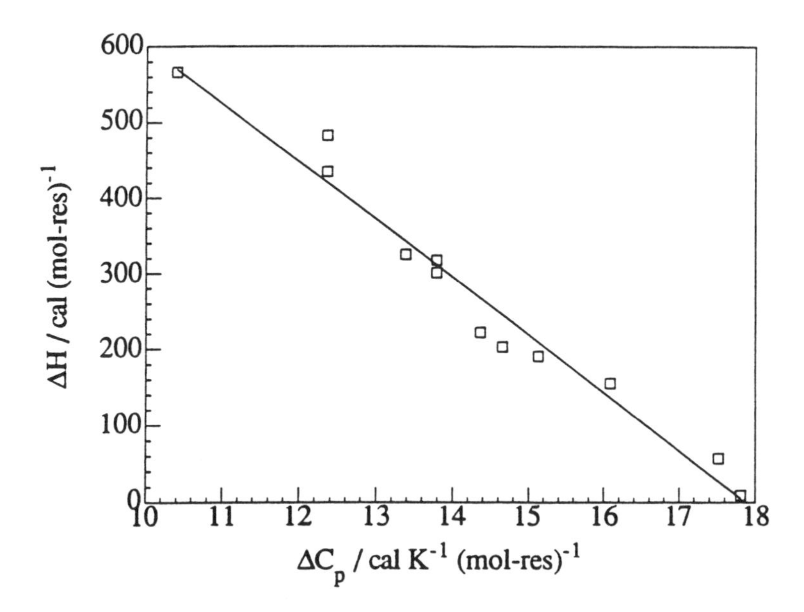

Fig. 1. Plot of $\Delta\overline{H}$ vs $\Delta\overline{C}_p$ for globular proteins (adapted from ref. *10*). The slope of the line ($298.15 - T_H$*) gives a T_H* value of 373 K. The intercept gives $\Delta\overline{H}$* equal to 1.35 kcal/mol. It should be noted that $\Delta\overline{H}$ decreases with increasing $\Delta\overline{C}_p$, suggesting that the hydrophobic groups make a negative contribution to the ΔH at 298 K. Data are from ref. *13,* Table 1, with the correction that $\Delta\overline{C}_p$ for parvalbumin is 13.4 cal/K/(mol-res) and $\Delta\overline{H}$ for pepsinogen is positive.

simply reference values useful for describing the enthalpic behavior of proteins from ~0–80°C. As discussed below, the convergence of the enthalpy terms is of fundamental importance in permitting an empirical calculation of energetic terms.

Inspection of Fig. 1 leads to the somewhat surprising conclusion that $\Delta\overline{H}$ *decreases* with increasing hydrophobicity. This is a key observation in interpreting the partitioning of the forces that contribute to protein stability, as will be seen later.

Just as with the specific enthalpy changes, it is observed that the specific entropy changes also converge to a defined value, $\Delta\overline{S}$*, at a specific temperature, T_S^*. This convergence is readily seen in an MPG entropy plot (Fig. 2) in which $\Delta\overline{S}$ at 298 K is plotted against $\Delta\overline{C}_p$ *(18)*. Here, the slope is ln ($298/T_S$*) and the intercept, $\Delta\overline{S}$*. T_S^* has the same value for all processes involving the transfer of apolar groups into water *(18,20)*,

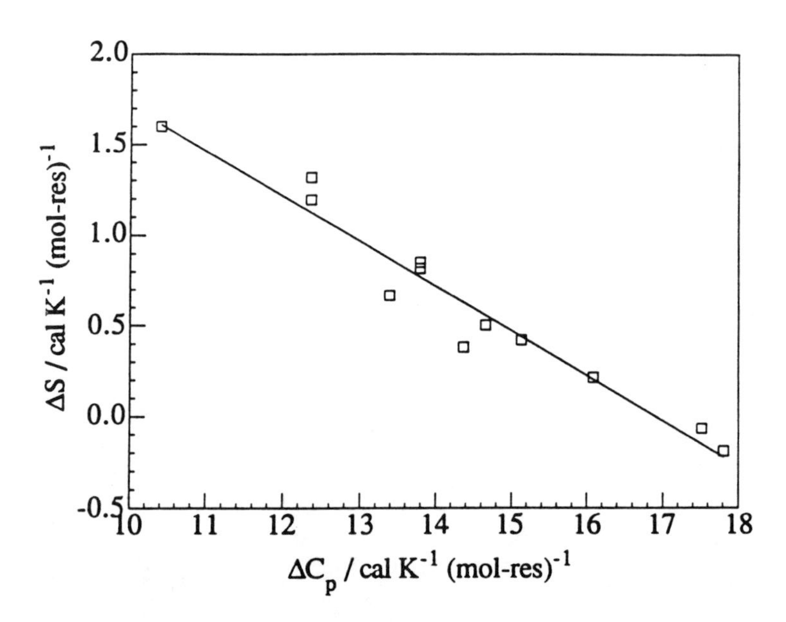

Fig. 2. Plot of $\Delta\overline{S}$ vs $\Delta\overline{C}_p$ for globular proteins (adapted from ref. *10*). The slope of the line (ln [298/T_S*]) gives a T_S* value of 385 K. The intercept gives $\Delta\overline{S}$* equal to 4.3 cal K/(mol-res). The negative slope is indicative of the negative ΔS associated with the exposure of apolar surface on unfolding. Data are from ref. *13,* Table 1.

385 ± 1 K. For globular proteins, the value of $\Delta\overline{S}$* is 4.3 ± 0.1/cal K/(mol-res) *(13,18)*.

As with $\Delta\overline{H}$, it is seen that $\Delta\overline{S}$ *decreases* with increasing hydrophobicity of the protein. This is to be expected, since the exposure of apolar groups to water has long been known to be accompanied by a negative ΔS *(21)*. Furthermore, Baldwin *(20)* observed that the transfer of apolar liquids into water is accompanied by a ΔS of zero at T^*_S; hence, $\Delta\overline{S}$* can be considered as the configurational entropy change accompanying the unfolding of globular proteins. The ΔS attributable to solvent restructuring, i.e., primarily the hydrophobic effect, is thus given as $\Delta\overline{C}_p$ ln (T/T^*_S).

Interestingly, a plot of $\Delta\overline{G}$ at 298 K vs $\Delta\overline{C}_p$ (not shown) shows essentially no correlation. In other words, the slope is zero and the intercept is just the average $\Delta\overline{G}$. This indicates that $\Delta\overline{G}$ is practically independent of the hydrophobicity of the protein at 298 K.

The convergence behavior discussed above presents an intriguing puzzle in understanding the forces stabilizing proteins and has been addressed by several investigators *(10,13,18–20,22–25)*. Why do the specific enthalpy and entropy converge at some high temperature, and why do the specific values at 298 K decrease with increasing "hydrophobicity?" Various interpretations of these phenomena will be discussed later.

4.1.2. Mutational Studies

The ability to make single amino acid changes has provided another means by which investigators can probe the stabilization of proteins *(26)*. The basic rationale is that the contribution of a specific interaction can be determined by adding or removing a side chain involved in that interaction and comparing the mutant stability to that of the wild-type protein.

For example, a hydrogen-bonded serine can be mutated to an alanine, thus removing a hydrogen bond. The difference between the wild-type stability and the mutant stability, $\Delta\Delta G$, is typically taken as the hydrogen bond contribution to the free energy. Similarly, values of $\Delta\Delta H$, $\Delta\Delta S$, and $\Delta\Delta C_p$ can be obtained. Shirley et al. have studied a number of hydrogen bond mutants of RNase T1 *(27)*. On average, they find a $\Delta\Delta G$ of 1.5 kcal/mol for a hydrogen bond mutant. However, one must be cautious in the interpretation of such results, as can be seen by considering the effect of a mutation on the unfolding process.

The overall hydrogen bond contribution to the energetics of unfolding can be represented in the following way. First, the hydrogen bond between the donor, *D*, and the acceptor, *A*, is broken in the interior of the protein:

$$A \bullet D \rightarrow A + D \tag{4}$$

Then both A and D are transferred to solvent:

$$A \rightarrow A_{\mathrm{aq}} \tag{5a}$$

$$D \rightarrow D_{\mathrm{aq}} \tag{5b}$$

where the subscript $_{\mathrm{aq}}$ indicates that the group is solvated in an aqueous environment. In the hydrogen bond mutant described above, the donor has been deleted and the remaining contribution of the hydrogen bond groups is represented by Eq. (5a). The difference between the wild-type and mutant reactions corresponds to ($A \bullet D \rightarrow A + D_{aq}$) rather than to ($A \bullet D \rightarrow A_{aq} + D_{aq}$) the desired overall reaction. Furthermore, this simple analysis assumes that the mutation has no other effects, such as structural rearrangements or bond strains.

Shirley et al. *(27)* have pointed out that if a water molecule replaces the deleted donor, then the measured difference will be closer to the desired contribution. Even so, it cannot be assumed that the reaction $A \bullet H_2O \rightarrow A_{aq}$ makes no contribution to the energetics. The energetic effects of site-specific mutations may be complex, even in the simple case where no structural rearrangements occur in the protein.

Numerous mutation studies have also been directed at determining the contribution of the hydrophobic effect by, for instance, replacing large hydrophobic side chains with Ala or Gly *(28–32)*. Interpreting the results of such studies is again difficult, which can be seen by replacing the hydrogen bond in Eq. (4) with a van der Waals contact. Matthews and coworkers *(32)* have recognized this complication and have included a "cavity" term to account for it. However, the flexibility of proteins can allow for repacking in response to a deletion, further complicating the problem. Lee *(33)* recently analyzed this problem in detail, showing that the maximum destabilization will occur for a rigid structure in which no repacking occurs.

Most studies of the effects of mutations investigate only the changes in ΔG, and not in ΔH, ΔS, and ΔC_p. Although ΔG directly reflects the overall stability of the protein, because of the usually compensating effects of ΔH and ΔS *(34,35)* it provides scant information on the actual contributors to the stability and may, in fact, be misleading. As an example, Connelly et al. *(36)* studied a series of T4 lysozyme mutations in which seven different residues were substituted for Thr 157. The results showed that even though the changes in ΔG are relatively small and regular, the changes in ΔH (and hence ΔS) are large and reveal no trends. If a trend in ΔG arises from some specific interaction, for example the hydrophobic effect, then all other thermodynamic functions should be correlated with that same interaction.

It is also important to note that mutations may affect both the native and unfolded states of a protein. This possibility was addressed early in mutational studies *(37,38)* and is the subject of a recent review by Dill and Shortle *(39)*.

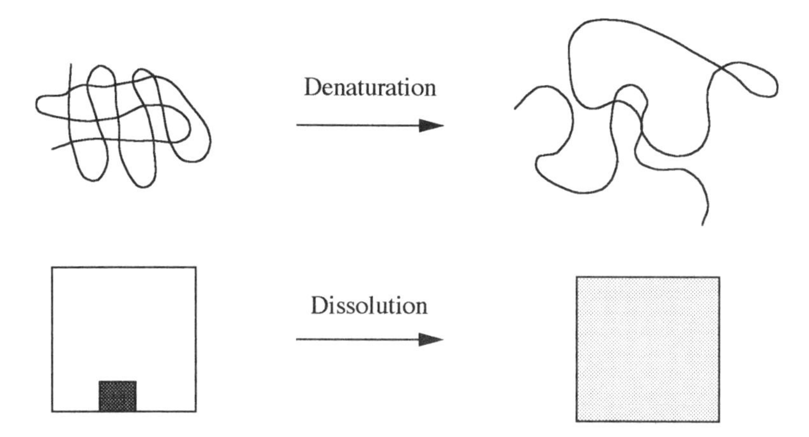

Fig. 3. Schematic representation of the similarities between protein unfolding and the aqueous dissolution of model compounds. In both cases interactions between functional groups are disrupted and new interactions are formed between those groups and water.

4.2. Model Compound Studies

The process of unfolding a globular protein involves the disruption of interactions within the protein and the transfer of the formerly interacting groups into water. An analogous process, illustrated in Fig. 3, is the dissolution of a solute in which the interactions of that solute with some initial phase are disrupted and the solute is subsequently transferred into water.

In attempting to model the denaturation of a globular protein, the initial phase in the dissolution process must mimic the interior of the protein as much as possible. The protein interior is characterized by a specific spatial arrangement of atoms that can be thought of as an aperiodic crystal. Further evidence for the solid-like nature of the protein interior is the observation that amino acid side chains in proteins occupy the same volume as they do in amino acid crystals *(40,41)*. Indeed, the average packing density of both proteins and organic crystals is around 0.75 *(40,41)* leading to the idea that the protein interior is nearly as close packed as it can be. In spite of the solid-like nature of the protein interior, model compound studies involving transfer from a liquid have been more common, largely because of the comparative ease with which they can be studied.

4.2.1. Liquid Transfer Studies

Most model compound studies focused on transfer of various compounds from organic liquids into water. For example, Nozaki and Tanford *(42)* studied the transfer of amino acids from ethanol to water, whereas Fauchere and Pliska *(43)* studied the transfer of *N*-methylamino-acid-amides from octanol to water. Such studies helped to establish hydrophobicity scales for amino acids, but since only ΔG values were determined, these results cannot be used in trying to interpret the detailed thermodynamics available from the protein studies cited above.

Direct calorimetric studies of the transfer of apolar liquids into water were performed with four aromatic and three aliphatic compounds *(44,45)*. Several important conclusions could be drawn from these results *(46)*. First, all the thermodynamic parameters are proportional to the surface area of the compound, as had been shown previously for the ΔG of transfer of apolar compounds *(47)*. Second, the ΔH of transfer near 298 K is small in all cases and is essentially zero. Third, the ΔS of transfer is zero for all these compounds at 385 K if the ΔC_p is considered constant *(20)*. (Note that the reference state here is unit mole fraction). These observations allow the transfer of apolar liquids to be approximated by the equation:

$$\Delta G = \Delta C_p \left[(T - 298) - T \ln (T/385)\right] \qquad (6)$$

If the hydrophobic effect is defined as the transfer of apolar compounds from the neat liquid into water, then Eq. (6) can be used to calculate the hydrophobic contribution to protein stability if the hydrophobic contribution to ΔC_p can be determined *(20)*. This approach has been taken by Record and coworkers *(16,48,49)*.

4.2.2. Solid Transfer Studies

Although there seems to be general agreement that the protein interior is more like a crystalline solid than an "oil-drop" *(50,51)*, few data are available on the energetics of dissolution of crystalline amino acid compounds. This is caused in part, by the experimental difficulties arising from the low solubility of such compounds. Recently, we began to address this problem by studying the energetics of dissolution of crystalline cyclic dipeptides *(14,52,53)*, chosen to remove the complication of end-group charge effects.

In considering the compounds containing apolar side chains, as with the apolar liquids, the thermodynamic parameters are proportional to the surface area of the compound *(14)*. Like the proteins, and in contrast to the apolar liquids, it is found that the ΔH of dissolution at 298 K *decreases* with increasing hydrophobicity of the compound, although the ΔH is always positive. Since all these compounds have very similar hydrogen bonding patterns in the crystal, it was concluded that the hydrogen bonding makes a positive contribution to the ΔH, whereas the apolar groups make a negative contribution *(10)*.

The negative contribution of the apolar groups is significant and warrants further elaboration. Transfer of apolar surface from the liquid phase into water has a ΔH near zero. Since enthalpies of fusion are positive, one would expect the ΔH of transferring apolar surface from the solid phase into water to be positive. This is necessarily the case for van der Waals solids, such as solid benzene, but this reasoning ignores the effects of the hydrogen bonds in the amino acid crystals. Hydrogen bonds, unlike van der Waals interactions, require a specific geometry. The negative contribution of the apolar groups in this case indicates that the crystal sacrifices van der Waals interactions between the apolar groups in order to make good, but geometrically constraining, hydrogen bonds *(11,14,54)*. In other words, the apolar surface is less well packed in the amino acid crystals than in a pure hydrocarbon liquid.

5. Interpretations of the Experimental Data

Any attempt at a partitioning of the noncovalent forces that stabilize globular protein structures should take into account the experimental observations presented above. All of the interpretations of the data that will be discussed share several features. In particular, most investigators assume that a group-additivity analysis is valid *(55,56)*. Group additivity states that a functional group, for example, a hydroxyl –OH, makes the same contribution to the thermodynamics independent of its specific environment in the protein. This assumption is implicit in considering energetic terms to be proportional to changes in accessible surface area (ASA) on unfolding of the protein chain *(56)*. The utility of group-additivity has been demonstrated for the dissolution of apolar liquids *(46,47)*, crystalline cyclic dipeptides *(14)*, and gaseous nonelectrolytes *(57)*, and has been applied to the energetics of protein folding by numerous investigators *(10,11,16,49,58–60)*.

Before proceeding, it is important to note that in spite of the common application of ASA correlations in estimating energetic parameters, the results should be used with caution and a clear understanding of their limitations. Correlations of energetics with ASA naturally involve the averaging of contributions over many Å^2 of surface. Consequently, although this approach can give very good estimates of the thermodynamics of large systems, such as the folding/unfolding transitions of proteins that involve thousands of Å^2 of surface area, estimates should not be expected to be valid for small systems, such as changes resulting from point mutations. Although the average contribution of a square angstrom of apolar surface area can be estimated quite well, the contribution of a *specific* methyl group can differ significantly from the average because of its unique van der Waals interactions, conformational strain, and so on.

In order to partition the energetics of protein stability into the various noncovalent interactions, it is necessary to select a reference state, either the gas, liquid, or solid phase. The different approaches that various investigators have taken can be classified according to the reference state utilized.

5.1. Gas Phase Reference

The unfolding process can be described in terms of the following steps:

1. The transfer of the native state from water into the gas phase;
2. The gas phase *(in vacuo)* unfolding of the protein; and
3. The transfer of the unfolded state from the gas phase into water.

This process is illustrated in Fig. 4. The transfer energetics of model compounds from the gas phase into water were tabulated by Cabani et al. *(57)* and by Makhatadze and Privalov *(24,25)*, and can be used to estimate the energetics of steps 1 and 3 in the above scheme. Knowing the energetics of unfolding in solution from experimental studies, one can then calculate the energetics of unfolding *in vacuo* by completing the thermodynamic cycle.

This approach was applied by Oobatake and Ooi *(58,60)*, who showed that the free energy change of hydration is negative (i.e., favors unfolding) for all but aliphatic functional groups. This negative hydration free energy change results from the negative enthalpy of hydration but is opposed by the entropy change of hydration, which is generally negative (i.e., giving a positive contribution to ΔG). This negative hydration ΔG results in a sum of steps 1 and 3 above that is destabilizing. Because the

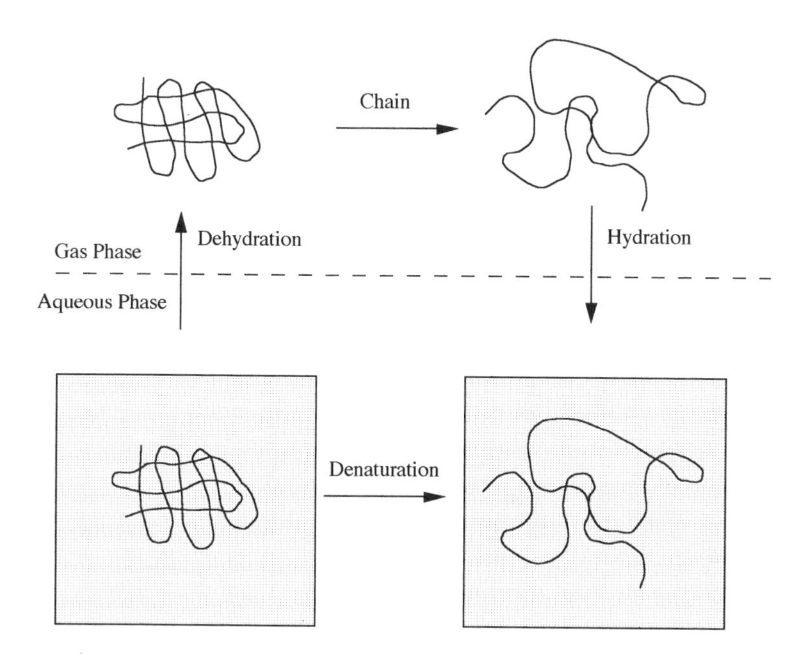

Fig. 4. Use of the gas reference state in forming a thermodynamic cycle for protein unfolding.

configurational entropy change also favors unfolding, the protein must then be stabilized by the enthalpically favorable van der Waals interactions and hydrogen bonding in the native state, which oppose step 2 in the cycle.

The interpretation of the entropy contributions in this scheme is problematic. Oobatake and Ooi defined the hydration entropy change, ΔS_h, as the sum of the ASA of each atom type times the area normalized group contribution of that atom type from the gas-to-water transfer process *(58)*. The hydration contribution to the unfolding ΔS is then the difference between the hydration entropy of the folded and unfolded states. This definition assumes that there are no nonhydration contributions to ΔS of the gas-to-water transfer process. The chain entropy change, ΔS_c, is calculated as the difference between the experimental entropy change and ΔS_h. The nonhydration contributions to ΔS of transfer from gas to water will thus have the greatest effect on ΔS_c of those residues that are most buried in the native state, since buried surface is defined as having $\Delta S_h = 0$. In consequence, the "chain" entropy as defined by this analysis is not the same as the configurational entropy defined earlier. This is apparent in noting that the chain entropy change is given *(58)* as 1.85 cal/K/mol for Gly and 7.55 cal/K/mol for Ala, whereas the configurational entropy change has been shown to be greater for Gly than for Ala in both theoretical *(61)* and experimental *(62,63)* studies.

The difficulties in the interpretation of the entropic terms with the above approach are a result of the choice of standard state. For the transfer of a compound from the gas phase into water, ΔS is composed of two contributions: one that reflects the interactions of water with the solute, and one that reflects the changes in the translational degrees of freedom of the solute in the two phases *(64)*. This second term has been referred to as the liberation or roaming entropy. Since there is a large difference in the translational degrees of freedom available to a solute molecule in the gas and aqueous phases, one would like to exclude this term from consideration. In an attempt to do this, Privalov and Makhatdze *(24,25)* used a set of hydration coefficients (gas-to-water transfer) calculated using Ben Naim's local standard state *(64)*, which is expected to correct for the liberation entropy changes. They refer to this as transfer from the "compact" phase into water.

Configurational entropy calculations again appear to be problematic, however. For example, Privalov and Makhatdze *(25)* found that the configurational entropy change for myoglobin at 298 K is approx 2.4 kcal/K/mol. On a per residue basis, this value lies between the ΔS of sublimation and the ΔS of vaporization for organic molecules. The configurational entropy change can be calculated using the standard relationship $\Delta S = R \ln (\Omega)$, where R is the gas constant and Ω is the ratio of degrees of freedom between the final and initial states. Myoglobin contains 745 effective dihedral angles *(60)*, each of which has up to three degrees of freedom *(61)*. The upper limit of the configurational entropy change can be calculated assuming that all rotational degrees of freedom are lost in the native state and available in the denatured state. This would then give $\Delta S = 745\ R \ln (3) = 1.6$ kcal/K/mol. The actual value in fact should be considerably smaller, since the approximation of zero degrees of freedom in the native state is too stringent even for buried residues, not to mention residues on the protein surface, and since excluded volume effects must also be considered.

Privalov and Makhatdze *(24,25)* found that the hydration of all functional groups, except methyl and methylene, is favorable at 298 K. They also found that enthalpic interactions in the native state, i.e., hydrogen bonding and van der Waals interactions, are stabilizing.

In order to partition the stabilizing enthalpic terms into hydrogen bonding and van der Waals, Privalov and Makhatadze made the assumption that the van der Waals interactions between apolar surfaces are equivalent to the van der Waals interactions in an apolar organic crystal. Therefore, the ΔH from apolar van der Waals is given by the enthalpy of sublimation, ΔH_{sub} of organic crystals. Because of the uncertainty of this assumption *(see below)*, they also looked at the case in which the van der Waals ΔH is taken as 50% of ΔH_{sub}. The ΔH owing to hydrogen bonding is then taken as the difference between the experimental ΔH and the sum of the hydration and van der Waals ΔH terms. The estimation of hydrogen bond strength is therefore dependent on the estimation of the apolar van der Waals interactions.

The ΔC_p terms are analyzed from the changes in accessible surface area of the various functional groups on unfolding *(65)*. Again, the authors corrected the gas-to-water ΔC_p values to the local standard state. This correction essentially results in the use of the constant volume heat capacity change, ΔC_v.

The analysis results in the conclusion that the overall ΔH associated with the transfer of apolar surface from the protein interior into water is positive above 298 K and stabilizes the native state, whereas the overall ΔH for transfer of polar surface is generally negative (or destabilizing) above 298 K. Although they find that the ΔH of polar (i.e., hydrogen bonding) groups is negative, the ΔG is positive because of a negative ΔS of hydration of polar groups.

5.2. Liquid Reference State

A common approach to partitioning the energetics of protein stability is the use of the "liquid hydrocarbon model" *(16,20)*. This approach assumes that the apolar contribution to the energetics can be evaluated from the transfer of model compounds from the pure liquid into water. Thus, the apolar ΔH is zero at 298 K (25°C) and the apolar ΔS is zero at 385 K (112°C). This model accounts for the convergence of $\Delta \overline{S}$ values; the $\Delta \overline{S}$ values converge where the apolar ΔS is zero and only the average configurational entropy change per residue is observed. In contrast, the implicit assumption of a "liquid-like" protein interior does not explain the $\Delta \overline{H}$ convergence *(20)*.

Several investigators recently noted that since the entropy changes converge, and the melting temperatures, T_m, of globular proteins do not vary greatly on the absolute temperature scale, then the enthalpy changes must also converge *(66)*. Although this observation is indeed valid, the result primarily shifts the question from "Why do the $\Delta\overline{H}$ values converge?" to "Why are the T_ms about the same?" Regardless, a physical basis for the observation of the enthalpy convergence is still required.

The liquid hydrocarbon model assumes that the apolar contribution to ΔH is zero at 298 K. Consequently, some other contribution must be found to explain why $\Delta\overline{H}$ decreases with increasing $\Delta\overline{C}_p$. This problem is exacerbated if a ΔH_{sub} is assumed to apply to the apolar surface. Yang et al. *(23)*, as well as Makhatadze and Privalov *(24)*, postulated that the negative contribution to ΔH necessary to compensate for the predicted negligible (or positive) contribution from the apolar groups arises from the hydration of hydrogen bonding groups. This hydration can persist even for buried hydrogen bonds if they are not too distant from the solvent, and is supposed to be extremely destabilizing for fully desolvated hydrogen bonds. However, this point of view does not appear to be consistent with the solid model compound dissolution data *(see below)*.

Another interpretation of the convergence phenomenon was put forth by Lee *(19)*, who pointed out that convergence can occur at the temperature where the apolar and polar contributions are equal. Recently, we suggested that this analysis requires normalizing the protein unfolding data to the total buried surface area rather than to the number of residues *(11,17)*. Comparison of the two normalization approaches seems to be consistent with the interpretation of the convergence temperatures given in the next section.

Another point regarding the energetics of the liquid hydrocarbon model merits mention. Honig and coworkers recently proposed that an additional correction must be made to the ΔS of exposing hydrophobic groups before it can be applied to proteins *(67)*. This correction is based on lattice theory *(68)* and results in a ΔS per unit area at 298 K of about 0.17 cal/K/(mol-Å^2), which is about twice the value generally assumed without the volume correction. If this correction is valid for protein folding/unfolding transitions, it will require the configurational entropy change to be significantly larger than suggested by both theoretical and experimental studies.

5.3. Solid Reference State

Use of either of the above reference states requires ΔH associated with hydrogen bonds to be negative in order to account for the experimentally observed values. Although the hydration of the hydrogen bonding groups would seem to provide a ready explanation of this negative value, it is not consistent with the available model compound data. As discussed above, the dissolution of crystalline cyclic dipeptides shows a decreasing ΔH with increasing apolar surface at 298 K, although the overall ΔH remains positive. These compounds contain the same hydrogen bonding groups as the proteins and should have nearly identical hydration changes. Nevertheless, the polar ΔH remains large and positive. The suggestion that the geometric constraints of the hydrogen bonds result in weakened van der Waals interactions seems to be the best explanation of these results. It is difficult to interpret the model compound results with a negative hydrogen bonding term.

In the solid model compounds, the apolar van der Waals interactions are less than in an apolar liquid. The observation that packing densities and packing volumes in proteins are similar to those observed in amino acid crystals *(40,41)* suggests that the interactions should be similar. To the extent that this is true, the use of ΔH_{sub} from apolar compounds in estimating van der Waals interactions in proteins is inappropriate. The data on the cyclic dipeptides indicate that the solid-like nature of the protein interior may actually result in a decrease of apolar van der Waals interactions relative to hydrocarbon liquids.

The relative contributions of apolar and polar groups to the ΔH and ΔC_p of dissolution of the solid cyclic dipeptides results in an enthalpy convergence in which the convergence enthalpy, ΔH^*, arises solely from polar interactions, i.e., the apolar contribution to the ΔH at the convergence temperature T_H^* is zero *(10,14)*. This means that at 298 K, the apolar ΔH is negative while the polar ΔH is positive. The same reasoning seems to be applicable to the protein data.

Several lines of evidence support this conclusion. It has been shown that ΔH^* for proteins scales linearly with the buried polar surface area *(11,17)*. The addition of low concentrations of alcohol, which is expected to mainly affect apolar contributions, results in a decreased ΔC_p and an increased ΔH of denaturation *(69,70)*, as expected if the apolar ΔH is negative *(11,70)*. Even under solvent conditions in which $\Delta C_p = 0$, and the solvent ordering contribution to the hydrophobic effect appears to be zero, a stable protein structure is observed in ubiquitin *(71)*.

It has often been argued that the tight packing of the protein interior indicates good van der Waals interactions of the apolar groups *(23,25,51,72)*, which has led to the use of ΔH_{sub} in estimating these terms. How can the apolar van der Waals interactions be less in the protein interior than in a liquid hydrocarbon if the interior is packed as well as an organic crystal? The answer to this question apparently is that it is the *average* packing density that is the same in the protein interior as in organic crystals, but the packing along the backbone is considerably higher than the average *(73)*, so that the packing of the side chains, which is the major contribution to apolar interactions, is *less* than that of an organic crystal.

In this empirical breakdown of the energetics, the contribution of polar and apolar surface is considered from an operational point of view. Thus the polar energetic terms include the disruption of hydrogen bonds and van der Waals interactions of polar groups within the protein interior and the subsequent hydration of those groups, including the formation of van der Waals interactions and hydrogen bonds, in solvent water. Electrostatic contributions are also averaged into these terms. Likewise, the apolar contribution includes both the disruption of apolar van der Waals interactions and the subsequent hydration of those groups, including van der Waals interactions and any restructuring of the solvent molecules. To emphasize the operational definitions being applied, the energetics are referred to as apolar and polar contributions rather than the hydrophobic effect and hydrogen bonding.

Within the context of the solid reference state, it is convenient to write the expression for the enthalpy change with reference to the temperature at which the apolar contribution is zero:

$$\Delta H = \Delta H^* + \Delta C_p (T - T_H^*) \tag{7}$$

where T_H^* is that reference temperature and ΔH^* is then the polar contribution to the enthalpy change evaluated at T_H^*. The heat capacity change is given as the sum of the polar and apolar contributions as:

$$\Delta C_p = \Delta C_{p,pol} + \Delta C_{p,ap} \tag{8}$$

Likewise, the entropy change can also be written with respect to the temperature at which the apolar contribution is zero:

$$\Delta S = \Delta S^* + \Delta C_p \ln (T/T_S^*) \quad (9)$$

Here, ΔS^* is the nonapolar contribution to the entropy change at T_S^* and is predominantly owing to configurational entropy changes. The free energy change is then given as:

$$\Delta G = \Delta H^* - T\,\Delta S^* + \Delta C_p[(T - T_H^*) - T \ln T/T_S^*)] \quad (10)$$

All the parameters in Eqs. (7–10) can be related to structural features, namely the buried apolar and polar surface areas and the number and types of residues, as will be discussed later. The application of this formalism results in a picture of protein stability in which the favorable ΔH associated with the burial of polar surface area is the primary stabilizing contribution. The apolar contribution to the stability is positive (although small) above ~20°C (293 K), but becomes negative below this temperature, although the ΔS associated with apolar transfer is always stabilizing. As noted above, this is consistent with the effects of methanol on protein stability *(70,71)*. The configurational entropy change is, of course, destabilizing at all temperatures. The breakdown of the energetic contributions is illustrated for myoglobin in Fig. 5.

The energetic parametrization described here has several implications for protein folding. The "driving force" for protein folding may depend on the temperature at which folding occurs *(11)* because the polar and apolar contributions to the energetics have opposite temperature dependencies. This might be an important consideration in studying the kinetics of folding, because the pathway(s) might be different at different temperatures. Additionally, since the apolar contribution to protein stability seems to be decreased by the geometric constraints of hydrogen bonding, it is conceivable that the energetics are different at various stages of folding. If, for example, the first step in folding is a nonspecific hydrophobic collapse, in which specific hydrogen bonds are not present, then the energetics of these hydrophobic interactions would be stronger (more like the liquid hydrocarbon energetics) than after the hydrogen bonds are formed and the van der Waals interactions between the apolar groups are concomitantly diminished.

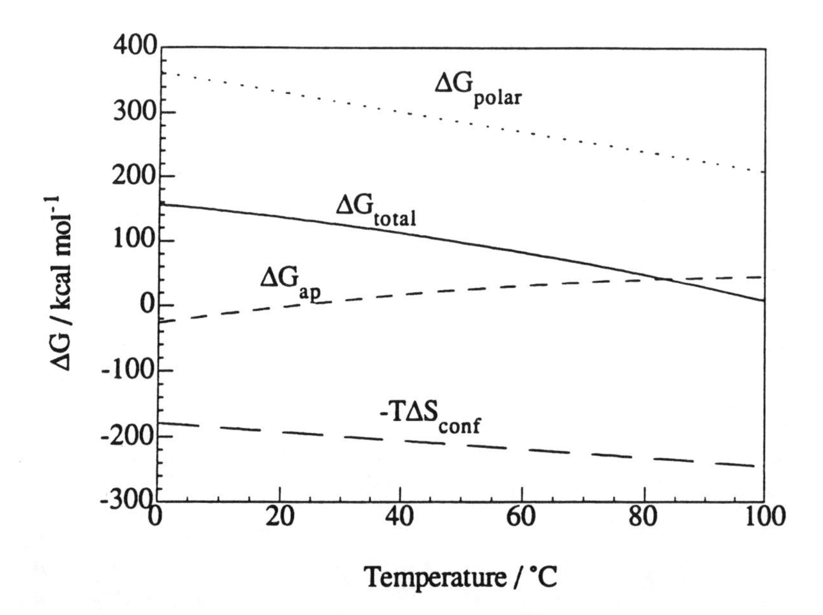

Fig. 5. Breakdown of the various interactions to the overall stabilization of myoglobin according to the parameters listed in Table 1.

5.4. Comparison of the Various Viewpoints

It is difficult to directly compare the various approaches just discussed. Several of the approaches use an intermediate phase (the gas phase or the "compact" phase) in describing the energetics as opposed to directly considering the overall energetic consequence of transferring different functional groups from the protein interior into solvent.

The differences in the various viewpoints seem fairly large when one considers that the contribution of burying apolar surface is considered as nearly the sole stabilizing contribution in the case of Yang et al. *(23)*, but considered as making only a minor contribution in our own work *(11)*.

Nevertheless, there are many areas of agreement between most of the approaches. It is clear that the ΔH of hydration (i.e., the transfer from either the gas or compact phases into water) of all functional groups is destabilizing. Also, the entropy change of exposing apolar groups to water favors folding. The magnitude of this effect is generally agreed to be $\Delta C_{p,ap} \ln (T/T_S^*)$ assuming a temperature-independent ΔC_p *(16,18,20)*, although some investigators argue that it is larger *(67)*. It is also now generally accepted that exposure of both apolar and polar surface area contribute to ΔC_p, but with opposite signs *(10,14,16,59)*.

The primary point of disagreement appears to be the partitioning of the stabilizing ΔH between hydrogen bonding and van der Waals interactions. Use of sublimation enthalpies in estimating the van der Waals contribution leads to a smaller enthalpy contribution of hydrogen bonding groups *(24,25)*, whereas analysis of the convergence phenomenon, in accord with the cyclic dipeptide dissolution data, suggests a more substantial hydrogen bond contribution *(10,11)*. Since all the approaches are essentially empirical, resolution of these differences must await new data.

6. Calculating Energetics from Crystal Structures

6.1. Basic Description

The parameterization of the energetics based on the solid reference state recently has been applied successfully to calculating the energetics of protein folding/unfolding transitions *(11,17)* and a protein–hormone interaction *(74)*. As noted above, the thermodynamics are related to the buried polar and apolar surface areas and the number and types of residues through fundamental thermodynamic parameters evaluated from model compound and protein unfolding data. The buried surface area is taken as the difference in the accessible surface area (ASA) of the unfolded state of the protein and the ASA of the native state as determined from the crystal structure. Accessible surface areas are evaluated using the Lee and Richards algorithm *(75)* with a 1.4 Å probe radius and a slice width of 0.25 Å using a program written by Scott Presnell (UCSF). The ASA of the unfolded state is taken as the sum of the residue areas in an extended tripeptide Ala-Xaa-Ala.

The ΔC_p is given as:

$$\Delta C_p = A_{ap}\, \Delta C_p{}^{\circ}{}_{ap} + A_{pol}\, \Delta C_p{}^{\circ}{}_{pol} \qquad (11)$$

where A_{ap} and A_{pol} are the buried apolar and polar surface areas, and $\Delta C_p{}^{\circ}{}_{ap}$ and $\Delta C_p{}^{\circ}{}_{pol}$ are the fundamental heat capacity contributions normalized to the surface area. The ΔH at the reference temperature (*see* Eq. [7]) is given as:

$$\Delta H^* = A_{pol}\, \Delta H^{\bullet} \qquad (12)$$

Table 1
Fundamental Thermodynamic Parameters
for the Calculation of Folding/Unfolding Energetics
from High-Resolution Structures

Parameter	Value
$\Delta C_p{}^{\circ}{}_{ap}$	0.45 cal K/(mol-Å^2) apolar surface
$\Delta C_p{}^{\circ}{}_{pol}$	–0.26 cal K/(mol-Å^2) polar surface
$\Delta H^{\bullet}$	35 cal/(mol-Å^2) polar surface
$\Delta S^{\bullet}$	4.3 cal/K/(mol-res)
$T_H{}^*$	373 K
$T_S{}^*$	385 K

where $\Delta H^{\bullet}$ is the area-normalized polar contribution to the ΔH at $T_H{}^*$. Finally, the configurational ΔS is given as:

$$\Delta S^* = \sum_i N_i \, \Delta S^{\bullet}{}_i \tag{13}$$

where N_i is the number of residues of type i, and $\Delta S^{\bullet}{}_i$ is the configurational entropy change for that residue type. Currently there are no good $\Delta S^{\bullet}$ values for different residue types and an average value is used. The best current values for the fundamental parameters are given in Table 1.

6.2. Evaluation of Fundamental Parameters

With the exception of the configurational entropy change, $\Delta S^{\bullet}$, and the reference temperatures, the parameters given in Table 1 are dependent on the algorithm used to determine ASA values. Since several algorithms are widely used in ASA calculations it is important to be able to derive a set of parameters for any algorithm.

The parameters for $\Delta C_p{}^{\circ}{}_{ap}$ and $\Delta C_p{}^{\circ}{}_{pol}$ were determined from the cyclic dipeptide dissolution data. The $\Delta C_p{}^{\circ}{}_{ap}$ value was originally determined per apolar hydrogen (aH), that is, per hydrogen atom bound to carbon, as 6.7 cal/K/(mol-aH) *(14)*. Using the Lee and Richards algorithm as described above, a value of 14.7 Å^2/aH was determined from the slope of ASA vs aH for linear alkanes. Values between 14.2 and 16.8 Å^2/aH can be expected from various algorithms using different atomic radii *(11)*. The value in Table 1 is then 6.7 cal/K/(mol-aH)/14.7 Å^2/aH.

Likewise, the value of $\Delta C_p{}^{\circ}{}_{pol}$ was originally given per amide bond as –14.3 cal/K/mol. In the cyclic dipeptides, each amide makes a hydrogen bond, so that this value can be considered as the heat capacity per hydrogen bond. In order to normalize this parameter to the polar ASA, the ASA per hydrogen bond in globular proteins was determined *(11)*. The ASA per hydrogen bond was taken as the slope of the polar ASA vs the number of hydrogen bonds, and this value (55 $Å^2$ per hydrogen bond) was used to obtain the value in Table 1.

The final, normalized parameter is $\Delta H^{\bullet}$. This parameter is obtained by dividing the convergence enthalpy change, $\Delta \overline{H}^{*}$, of 1.35 kcal/(mol-res) by the average buried polar surface area per residue. Using our ASA calculations, the average buried polar surface area per residue is 38.6 $Å^2$.

These same steps can be taken to renormalize the fundamental thermodynamic parameters for any ASA algorithm. This renormalization is required before this approach can be applied.

6.3. Applications of Structural Thermodynamic Calculations

The values in Table 1 have been applied to the thermal denaturation of globular proteins for which both good thermodynamic and structural data are available. The fractional deviation between the calculated and experimental values is graphically illustrated in Fig. 6. For both ΔH and ΔS, the comparison is made at 333 K (60°C), the median melting temperature, in order to minimize extrapolation errors. The calculated and experimental values agree nearly within experimental error, as indicated by the horizontal lines.

Even though the thermodynamic parameters can be predicted nearly within experimental error, the calculation is not precise enough to predict melting temperatures. The difficulty in predicting T_ms is largely the result of insufficient precision in calculating ΔS, since the free energy contribution of ΔS is multiplied by the absolute temperature. Determining more precise values for ΔS, which take into account residue type, degree of burial, and so on, is a major area in which the method described here can be improved.

The structural energetic calculation recently was applied to the binding of angiotensin II to a monoclonal antibody that serves as a receptor analog *(74)*. Very good agreement was found between the predicted and

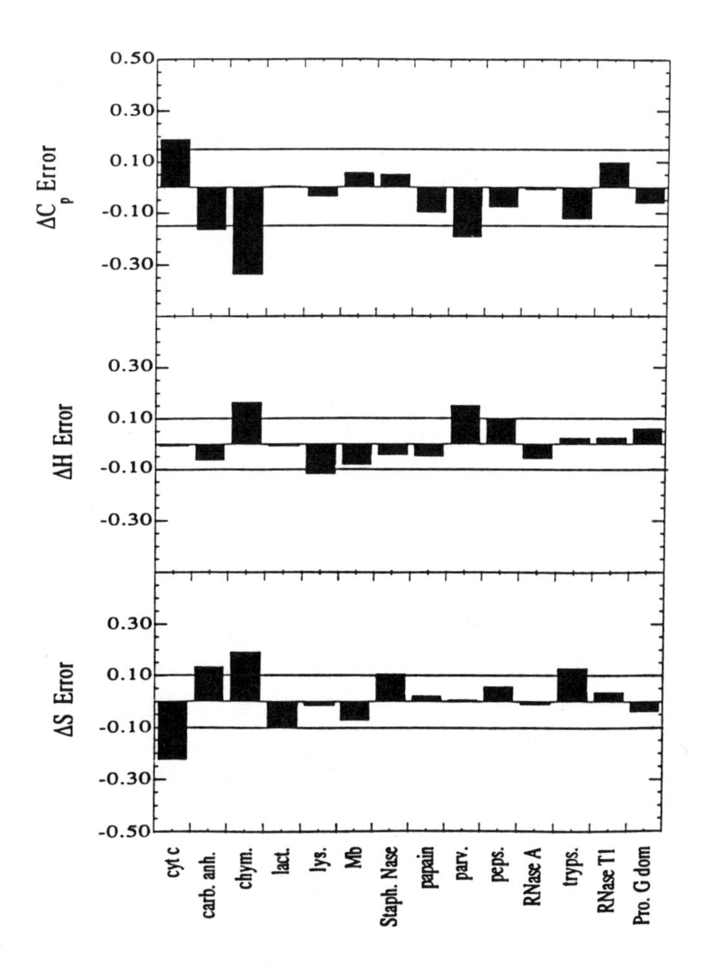

Fig. 6. Comparison of calculated and experimental thermodynamic parameters for several globular proteins. The dashed lines indicate the experimental error. The ΔH and ΔS comparisons are at 333 K (60°C), the median melting temperature, in order to minimize extrapolation errors. The experimental values are from the following: cytochrome c *(22)*; carbonic anhydrase *(86)*; chymotrypsin *(87)*; apo-α-lactalbumin *(76)*; lysozyme *(88)*; myoglobin *(89)*; *Staph.* nuclease *(90)*; papain *(91)*; parvalbumin *(92)*; pepsinogen *(93)*; RNase A *(94,95)*; trypsin *(96)*; protein G B2 domain *(97)*.

experimentally determined values of ΔC_p and ΔH. The agreement in ΔS was also good, but several assumptions in the calculation, particularly regarding the configurational entropy effects of side chains, remain to be verified. These results illustrate the utility of the structural energetic calculations for reactions other than folding/unfolding transitions.*

The close agreement between the calculated and experimental values is important in several respects. First, it suggests that the breakdown of the energetics on which the calculation is based is likely to be valid. Second, regardless of the validity of the energetic breakdown, the ability to accurately predict overall energetic values provides a potentially useful tool in modeling protein folding intermediates *(17,76)*, protein–ligand complexes, and novel protein structures. Third, it can serve as a diagnostic approach for assessing the degree of unfolding of a given protein, thus suggesting models for unfolded states.

The latter idea recently was applied in studying a β-sheet protein, III^{Glc} *(77)*. The experimentally determined energetic parameters for III^{Glc} were significantly smaller than predicted from structural energetic calculations that assume complete unfolding. This observation suggested that the thermally denatured form of the protein contains significant residual structure. This has since been confirmed by CD measurements and by hydrogen exchange kinetics as monitored by NMR (Robertson, unpublished results).

The structural energetic parameters described here have also been used in understanding the origins of cooperative unfolding of globular proteins *(11,17,78,79)*. For example, the parameters describing the apolar contribution to protein stability can explain the observation that the high temperature melting of phosphoglycerate kinase (PGK) shows a single cooperative transition, whereas the cold denaturation is characterized by two transitions indicating independent unfolding of the two domains *(79)*.

7. Implications for Contributions of Fundamental Forces

As seen above, this empirical approach appears to be reasonably successful at correlating overall energetics with structure. From a practical point of view, it should prove useful for protein design and for structure-based drug design. In using this kind of empirical method, however, the

*We have since applied similar calculations to the dimerization of interleukin 8. The predicted values of $\Delta G°$, ΔH, ΔS, and ΔC_p are in excellent agreement with values determined by calorimetry and analytical ultracentrifugation *(98)*.

observation that the parameters work is not the same as understanding why they work. The reasons for the success of this approach, and the relationship between the parameters listed in Table 1 and the atomic interactions occurring within a protein or a complex, remain to be explored.

What does the approach just described tell us about the fundamental energetics of protein stability? The parameters in Table 1 describe the overall energetics associated with apolar transfer, polar transfer, and configurational entropy, rather than directly describing hydrogen bonding and the hydrophobic effect. These terms, nevertheless, do provide qualitative information regarding the basic forces described in Section 2.

7.1. van der Waals Interactions

The overall ΔH associated with the transfer of apolar surface from the protein interior into water includes a ΔH from the difference in van der Waals interaction of the apolar surface between the protein interior and water, and a ΔH associated with the restructuring of water around the apolar surface. The observation that the apolar ΔH is negative at 298 K for proteins indicates that the van der Waals interactions of the apolar surface in the protein interior are less favorable than the sum of the water restructuring and solute–solvent van der Waals interactions.

It has been suggested that the difference in van der Waals interactions between a hydrocarbon in water and a hydrocarbon in the liquid phase is negligible. If this is correct, then the small ΔH of dissolution of liquid hydrocarbons at 298 K would indicate that the water restructuring term is also small at 298 K *(80)*. Given these assumptions, the negative ΔH at 298 K entirely would result from the improved van der Waals interactions of the apolar surface with water.

It is also important to note that the apolar ΔH for protein denaturation is somewhat more negative than that observed for the cyclic dipeptide dissolution. This is apparent by noting that T_H* for the proteins is about 373 K, whereas it is about 343 K for the dipeptide dissolution *(14)*. This would suggest that the apolar van der Waals interactions are weaker in the protein interior than in the amino acid crystals. This difference might reflect the additional geometric constraints of the protein being a polymer. It is also likely that the discrepancy reflects the inclusion of both aromatic and aliphatic surface in the protein (thus providing a weighted average), whereas it is solely from aliphatic surface in the dipeptides.

7.2. Hydrogen Bonds

The polar contribution to protein stability reflects the difference in ΔH of the protein hydrogen bonding groups solvated in water and buried in the protein. Kauzmann *(3)* suggested early on that intramolecular hydrogen bonds were unlikely to contribute to protein stability since they were competing with hydrogen bonds to 55*M* water. It is clear, however, that hydrogen bonds provide a significant stabilizing ΔH to dipeptide crystals relative to water, as well as to amide hydrogen bonded dimers in water *(81–83)*.

The polar ΔH appears to be the major contributor to the stability of globular proteins, highlighting the importance of hydrogen bonds. This is consistent with the observed ΔH of unfolding of a synthetic α-helix *(84)*. The large ΔH of amide hydrogen bond formation in water indicates that the peptide hydrogen bond is more energetically favorable than the hydroxyl hydrogen bond that can be formed with water. This might reflect a difference in the quantum mechanics of the two interactions. It might also result from cooperativity in hydrogen bond formation *(85)*.[†]

It is also important to realize that hydrogen bonds, in the context discussed here, are considered as purely enthalpic. It is not possible currently to separate any entropy change specifically owing to hydrogen bonds from the overall configurational ΔS. It is therefore difficult to compare directly the polar ΔH contribution to estimates of hydrogen bond free energies.

7.3. Configurational Entropy

The average configurational entropy change per residue from Table 1 is 4.3 cal/K/mol. According to data given by Oobatake and Ooi *(60)*, there are an average of 4.6 ± 0.2 effective rotors per residue in globular proteins. This would indicate that each rotor in the unfolded state has about 1.9 times the degrees of freedom that it has in the folded state. As mentioned above, each rotor can be assumed to have a maximum of three degrees of freedom. The value of 1.9 suggests that the rotors do not go from a native state in which there are no degrees of freedom (i.e., zero

[†]This idea is supported by recent *ab initio* calculations by Guo and Karplus on peptide hydrogen bonds *(99)*.

entropy) to an unfolded state in which there are three per rotor. Rather, the smaller value is consistent with having moderate flexibility in the native state and having some rotational restrictions in the unfolded state.

7.4. The Role of Water

Within the context of the structural energetic calculations, the role of water in protein stability can only be explicitly considered in regard to entropy effects. The solvent contribution to ΔS is given as ΔC_p ln $(T/T_S{}^*)$. This results in a solvent contribution of about 25 cal (mol-Å^2), often considered as the "hydrophobic effect." Indeed, the difference between the apolar contribution to the stability, as used here, and the common conception of the hydrophobic effect, is the assumption in the latter of a zero ΔH for hydrophobic groups. If the hydrophobic effect is defined as the ΔG arising from the *entropic* contribution of exposing apolar surface, then the hydrophobic effect can be considered as the major driving force for protein folding. With regard to the overall effects on protein stability, it is important to keep in mind that there are enthalpic consequences from apolar groups.

8. Conclusions

Significant advances have been made over the last several years in our understanding of the contributions of noncovalent interactions to the stability of globular proteins. It is now clear that the "textbook" view of protein stability as being solely determined by the hydrophobic effect is inadequate, since increasing the "hydrophobicity" of proteins does not lead to increased stability. Rather, protein stability is a complex balance of often compensating interactions. Although some disagreement remains as to the magnitude of certain contributions, much of this disagreement is the result of how the various interactions are operationally defined, particularly the choice of reference states. Nevertheless, many investigators now agree on significant points. The configurational entropy is destabilizing and it appears that it can be evaluated as the ΔS extrapolated to 385 K. The unfavorable configurational entropy is compensated by the stabilizing hydrophobic entropy, associated with apolar exposure to water. The hydration of all groups in proteins is enthalpically destabilizing, and the experimentally observed positive ΔH is the result of favorable hydrogen bonds and van der Waals interactions in the protein interior. Perhaps the most significant problem in protein energetics at this point is to find new experimental approaches with which we can dissect the relative contributions of hydrogen bonding and van der Waals interactions.

In spite of remaining questions regarding the partitioning of various interactions, it is now clear that practical, empirical approaches can provide accurate estimates of thermodynamic parameters from high resolution structures. These approaches promise to be of great utility, not only in attempts at protein design and redesign, but also in studying protein–protein interactions and in structure-based drug design.

Acknowledgments

The author would like to thank the many colleagues whose careful reading of this manuscript led to many improvements. This work was partially supported by NIH grant RR-04328 and by the Roy J. Carver Charitable Trust.

References

1. Baker, E. N. and Hubbard, R. E. (1984) *Prog. Biophys. Molec. Biol.* **44,** 97–179.
2. Stickle, D. F., Presta, L. G., Dill, K. A., and Rose, G. D. (1992) *J. Mol. Biol.* **226,** 1143–1159.
3. Kauzmann, W. (1959) *Adv. Protein Chem.* **14,** 1–63.
4. Damaschun, G., Damaschun, H., Gast, K., Zirwer, D., and Bychkova, V. E. (1991) *Int. J. Biol. Macromol.* **13,** 217–221.
5. Stigter, D., Alonso, D. O. V., and Dill, K. A. (1991) *Proc. Natl. Acad. Sci. USA* **88,** 4176–4180.
6. Goto, Y. and Nishikiori, S. (1991) *J. Mol. Biol.* **222,** 679–686.
7. Dill, K. A. (1990) *Science* **250,** 297.
8. Privalov, P. L., Gill, S. J., and Murphy, K. P. (1990) *Science* **250,** 297–298.
9. Herzfeld, J. (1991) *Science* **253,** 88.
10. Murphy, K. P. and Gill, S. J. (1991) *J. Mol. Biol.* **222,** 699–709.
11. Murphy, K. P. and Freire, E. (1992) *Adv. Protein Chem.* **43,** 313–361.
12. Privalov, P. L. (1979) *Adv. Protein Chem.* **33,** 167–239.
13. Privalov, P. L. and Gill, S. J. (1988) *Adv. Protein Chem.* **39,** 191–234.
14. Murphy, K. P. and Gill, S. J. (1990) *Thermochim. Acta* **172,** 11–20.
15. Privalov, P. L. and Makhatadze, G. I. (1990) *J. Mol. Biol.* **213,** 385–391.
16. Spolar, R. S., Livingstone, J. R., and Record, M. T., Jr. (1992) *Biochemistry* **31,** 3947–3955.
17. Murphy, K. P., Bhakuni, V., Xie, D., and Freire, E. (1992) *J. Mol. Biol.* **227,** 293–306.
18. Murphy, K. P., Privalov, P. L., and Gill, S. J. (1990) *Science* **247,** 559–561.
19. Lee, B. K. (1991) *Proc. Natl. Acad. Sci. USA* **88,** 5154–5158.

20. Baldwin, R. L. (1986) *Proc. Natl. Acad. Sci. USA* **83,** 8069–8072.
21. Frank, H. S. and Evans, M. W. (1945) *J. Chem. Phys.* **13,** 507.
22. Privalov, P. L. and Khechinashvili, N. N. (1974) *J. Mol. Biol.* **86,** 665–684.
23. Yang, A.-S., Sharp, K. A., and Honig, B. (1992) *J. Mol. Biol.* **227,** 889–900.
24. Makhatadze, G. I. and Privalov, P. L. (1993) *J. Mol. Biol.,* **232,** 639–659.
25. Privalov, P. L. and Makhatadze, G. I. (1993) *J. Mol. Biol.,* **232,** 660–679.
26. Matthews, B. W. (1987) *Biochemistry* **26,** 6885–6888.
27. Shirley, B. A., Stanssens, P., Hahn, U., and Pace, C. N. (1992) *Biochemistry* **31,** 725–732.
28. Kellis, J. T., Jr., Nyberg, K., Sali, D., and Fersht, A. R. (1988) *Nature* **333,** 784–786.
29. Kellis, J. T., Jr., Nyberg, K., and Fersht, A. R. (1989) *Biochemistry* **28,** 4914–4922.
30. Karpusas, M., Baase, W. A., Matsumura, M., and Matthews, B. W. (1989) *Proc. Natl. Acad. Sci. USA* **86,** 8237–8241.
31. Shortle, D., Stites, W. E., and Meeker, A. K. (1990) *Biochemistry* **29,** 8033–8041.
32. Eriksson, A. E., Baase, W. A., Zhang, X., Heinz, D. W., Blaber, M., Baldwin, E. P., and Matthews, B. W. (1992) *Science* **255,** 178–183.
33. Lee, B. (1993) *Protein Science* **2,** 733–738.
34. Lumry, R. and Rajender, S. (1970) *Biopolymers* **9,** 1125–1227.
35. Eftink, M. R., Anusiem, A. C., and Biltonen, R. L. (1983) *Biochemistry* **22,** 3884–3896.
36. Connelly, P. R., Ghosaini, L., Cui-Qing, H., Kitamura, S., Tanaka, A., and Sturtevant, J. M. (1991) *Biochemistry* **30,** 1887–1891.
37. Shortle, D. and Meeker, A. K. (1986) *Proteins* **1,** 81–89.
38. Alber, T., Dao-pin, S., Wilson, K., Wozniak, J. A., Cook, S. P., and Matthews, B. W. (1987) *Nature* **330,** 41–46.
39. Dill, K. A. and Shortle, D. (1991) *Annu. Rev. Biochem.* **60**, 795–825.
40. Richards, F. M. (1974) *J. Mol. Biol.* **82,** 1.
41. Richards, F. M. (1977) *Annu. Rev. Biophys. Bioeng.* **6,** 151–176.
42. Nozaki, Y. and Tanford, C. (1971) *J. Biol. Chem.* **246,** 2211.
43. Fauchere, J. L. and Pliska, V. (1983) *Eur. J. Chem.-Chim. Ther.* **18,** 369–375.
44. Gill, S. J., Nichols, N. F., and Wadsö, I. (1975) *J. Chem. Thermodynamics* **7,** 175–183.
45. Gill, S. J., Nichols, N. F., and Wadsö, I. (1976) *J. Chem. Thermodynamics* **8,** 445–452.
46. Gill, S. J. and Wadsö, I. (1976) *Proc. Natl. Acad. Sci. USA* **73,** 2955–2958.
47. Hermann, R. B. (1972) *J. Phys. Chem.* **76,** 2754–2759.
48. Spolar, R. S., Ha, I.-H., and Record, M. T., Jr. (1989) *Proc. Natl. Acad. Sci. USA* **86,** 8382–8385.
49. Livingstone, J. R., Spolar, R. S., and Record, M. T., Jr. (1991) *Biochemistry* **30,** 4237–4244.
50. Chothia, C. (1984) *Annu. Rev. Biochem.* **53,** 537–572.
51. Bello, J. (1977) *J. Theor. Biol.* **68,** 139–142.
52. Murphy, K. P. and Gill, S. J. (1989) *Thermochim. Acta* **139,** 279–290.
53. Murphy, K. P. and Gill, S. J. (1989) *J. Chem. Thermodynamics* **21,** 903–913.
54. Creighton, T. E. (1991) *Curr. Opinion Struct. Biol.* **1,** 5–16.
55. Eisenberg, D. and McLachlan, A. D. (1986) *Nature* **319,** 199–203.

56. Ooi, T., Oobatake, M., Némethy, G., and Scheraga, H. A. (1987) *Proc. Natl. Acad. Sci. USA* **84,** 3086–3090.
57. Cabani, S., Gianni, P., Mollica, V., and Lepori, L. (1981) *J. Solution Chem.* **10,** 563–595.
58. Ooi, T. and Oobatake, M. (1988) *J. Biochem.* **103,** 114–120.
59. Makhatadze, G. I. and Privalov, P. L. (1990) *J. Mol. Biol.* **213,** 375–384.
60. Oobatake, M. and Ooi, T. (1992) *Prog. Biophys. Mol. Biol.* **59,** 237–284.
61. Némethy, G., Leach, S. J., and Scheraga, H. A. (1966) *J. Phys. Chem.* **70,** 998–1004.
62. Merutka, G., Lipton, W., Shalongo, W., Park, S.-H., and Stellwagen, E. (1990) *Biochemistry* **29,** 7511–7515.
63. O'Neil, K. and DeGrado, W. (1990) *Science* **250,** 646–651.
64. Ben-Naim, A. (1978) *J. Phys. Chem.* **82,** 792–803.
65. Privalov, P. L. and Makhatadze, G. I. (1992) *J. Mol. Biol.* **224,** 715–723.
66. Baldwin, R. L. and Muller, N. (1992) *Proc. Natl. Acad. Sci. USA* **89,** 7110–7113.
67. Sharp, K. A., Nicholls, A., Fine, R. F., and Honig, B. (1991) *Biochemistry* **30,** 9686–9697.
68. DeYoung, L. R. and Dill, K. A. (1990) *J. Phys. Chem.* **94,** 801–809.
69. Velicelebi, G. and Sturtevant, J. M. (1979) *Biochemistry* **18,** 1180–1186.
70. Fu, L. and Freire, E. (1992) *Proc. Natl. Acad. Sci. USA* **89,** 9335–9338.
71. Woolfson, D. N., Cooper, A., Harding, M. M., Williams, D. H., and Evans, P. A. (1993) *J. Mol. Biol.* **229,** 502–511.
72. Bello, J. (1978) *Int. J. Peptide Protein Res.* **12,** 38–41.
73. Richards, F. M. (1992) in *Protein Folding* (Creighton, T. E., ed.), Freeman, New York, pp. 1–58.
74. Murphy, K. P., Xie, D., Garcia, K. C., Amzel, L. M., and Freire, E. (1993) *Proteins* **15,** 113–120.
75. Lee, B. and Richards, F. M. (1971) *J. Mol. Biol.* **55,** 379–400.
76. Xie, D., Bhakuni, V., and Freire, E. (1991) *Biochemistry* **30,** 10,673–10,678.
77. Murphy, K. P., Meadow, N. D., Roseman, S., and Freire, E. (1993) *Biophys. J.* **64,** A172.
78. Freire, E. and Murphy, K. P. (1991) *J. Mol. Biol.* **222,** 687–698.
79. Freire, E., Murphy, K. P., Sanchez-Ruiz, J. M., Galisteo, M. L., and Privalov, P. L. (1992) *Biochemistry* **31,** 250–256.
80. Lee, B. (1991) *Biopolymers* **31,** 993–1008.
81. Schellman, J. A. (1955) *C. R. Trav. Lab. Carlsberg Ser. Chim.* **29,** 223–229.
82. Susi, H., Timasheff, S. N., and Ard, J. S. (1964) *J. Biol. Chem.* **239,** 3051–3054.
83. Gill, S. I. and Noll, L. (1972) *J. Phys. Chem.* **76,** 3065–3068.
84. Scholtz, J. M., Marqusee, S., Baldwin, R. L., York, E. J., Stewart, J. M., Santoro, M., and Bolen, D. W. (1991) *Proc. Natl. Acad. Sci. USA* **88,** 2854–2858.
85. Maes, G. and Smets, J. (1993) *J. Phys. Chem.* **97,** 1818–1825.
86. Tatunashvili, L. V. and Privalov, P. L. (1986) *Biofizika (USSR)* **31,** 578–581.
87. Tischenko, V. M., Tiktopulo, E. I., and Privalov, P. L. (1974) *Biofizika (USSR)* **19,** 400–404.
88. Khechinashvili, N. N., Privalov, P. L., and Tiktopulo, E. I. (1973) *FEBS Lett.* **30,** 57–60.

89. Privalov, P. L., Griko, Y. V., Venyaminov, S. Y., and Kutyshenko, V. P. (1986) *J. Mol. Biol.* **190,** 487–498.
90. Calderon, R. O., Stolowich, N. J., Gerlt, J. A., and Sturtevant, J. M. (1985) *Biochemistry* **24,** 6044–6049.
91. Tiktopulo, E. I. and Privalov, P. L. (1978) *FEBS Lett.* **91,** 57–58.
92. Filimonov, V. V., Pfeil, W., Tsalkova, T. N., and Privalov, P. L. (1978) *Biophys. Chem.* **8,** 117–122.
93. Mateo, P. L. and Privalov, P. L. (1981) *FEBS Lett.* **123,** 189–192.
94. Privalov, P. L., Tiktopulo, E. I., and Khechinashvili, N. N. (1973) *Int. J. Pept. Protein Res.* **5,** 229–237.
95. Straume, M. and Freire, E. (1992) *Anal. Biochem.* **203,** 259–268.
96. Tischenko, V. M. and Gorodnov, B. G. (1979) *Biofizika (USSR)* **24,** 334–335.
97. Alexander, P., Fahnestock, S., Lee, T., Orban, J., and Bryan, P. (1992) *Biochemistry* **31,** 3597–3603.
98. Burrows, S. D., Doyle, M. D., Murphy, K. P., Franklin, S. G., White, J. R., Brooks, I., et al. (1994) *Biochemistry,* **33,** 12,741–12,745.
99. Guo, H. and Karplus, M. (1994) *J. Phys. Chem.* **98,** 7104,7105.

CHAPTER 2

Degradative Covalent Reactions Important to Protein Stability

David B. Volkin, Henryk Mach, and C. Russell Middaugh

1. Introduction

The covalent modification of proteins in vivo has been proposed as a natural mechanism to designate enzymes for turnover *(1)*. Both enzymatic and nonenzymatic pathways of posttranslational modification of proteins have been identified. Spontaneous, nonenzymatic reactions include the deamidation of asparaginyl residues, racemization of aspartyl residues, isomerization of prolyl residues, and glycation of amino groups, as well as site-specific metal catalyzed oxidations (interaction of H_2O_2 and Fe(II) at metal binding sites on proteins) *(1)*. Enzymes have been identified in vivo that specifically interact with covalently modified proteins, including carboxymethyl transferase (which methylates isoaspartyl [iso-Asp] residues) and alkaline protease (which degrades oxidized proteins). It has been proposed that covalent changes caused by in vivo protein oxidation are primarily responsible for the accumulation of catalytically compromised and structurally altered enzymes during aging *(2)*. In addition, protein oxidation may play a role in several pathological states, including inflammatory disease, atherosclerosis, neurological disorders, and cataractogenesis *(3)*.

This review will emphasize the general characteristics of commonly observed chemical modifications in proteins occurring during their in vitro purification, storage, and handling. Although some of these in vitro degradative covalent reactions are similar or identical to ones observed

From: *Methods in Molecular Biology, Vol. 40: Protein Stability and Folding: Theory and Practice*
Edited by: B. A. Shirley

in biological systems, the identity of the most labile amino acids within a protein and their relative reaction rates often can be significantly different. These covalent modifications are potential problems during any protein unfolding and refolding experiment, since the examination of protein stability is often carried out under rather nonphysiological conditions (elevated temperatures, acidic and alkaline pH, the presence of denaturants, exposure to light, and so on).

Advances in our understanding of the weak chemical links of a protein molecule in vitro originate primarily from three areas in which scientists and engineers are attempting to utilize biomolecules for technological purposes:

1. Food scientists are examining changes in food proteins during processing and heating, especially at extremes of temperature and pH *(4)*.
2. Chemists are increasingly exploring the use of enzymes as specific catalysts for organic synthesis *(5)*, while chemical engineers are utilizing enzymes as biocatalysts to manufacture a variety of chemicals, sweeteners, and detergents *(6,7)*. These studies examine mechanisms of protein inactivation at elevated temperatures and in nonaqueous solvent systems.
3. Recently, the biotechnology and pharmaceutical industries are actively developing naturally occurring peptides and proteins as therapeutic agents. The purification and formulation of these protein drugs requires an understanding of the causes and mechanisms of inactivation in order to develop rational strategies for their stabilization *(8–11)*.

In each of these applications, protein molecules are exposed to nonphysiological conditions resulting in stresses on their structural and chemical integrity that may lead to both their covalent and noncovalent alteration.

The relationship between the conformational stability and chemical integrity of a protein is of particular importance to the understanding of the mechanisms of protein inactivation. On exposure to changes in environmental conditions (elevated temperature, acidic/basic conditions, or the presence of structure perturbing solutes), protein molecules may undergo either conformational changes (local changes in secondary and tertiary structure), reversible unfolding (cooperative loss of higher ordered structure), or inactivation (irreversible changes in the structural or chemical integrity of the molecule). Perturbation of protein structure often leads to the exposure of previously buried amino acid residues, facilitating their chemical reactivity. In fact, partial unfolding of a protein is often observed prior to the onset of irreversible chemical or conformational processes *(8,9)*. Moreover, protein conformation generally may control the rate and extent of deleterious chemical reactions. Conversely, chemical changes to the polypeptide backbone or amino acid side chains of a protein may lead to loss of conformational stability. For example, the reduction of disulfides or the oxidation of cysteine residues can induce protein unfolding and aggregation. During the upcoming discussion of the chemical lability of proteins, the interplay between these reactions and protein conformation will be emphasized. Obviously, this coupled interaction between these two phenomena has the potential to complicate studies of protein folding and unfolding significantly.

This chapter is not intended to be a comprehensive survey of the literature. Rather, the orientation of this discussion is toward the types of problems that can be encountered during the typical handling of protein solutions by the laboratory scientist, with particular emphasis on the conditions used to purify proteins and study their stability and protein folding pathways. Illustrative examples from the literature have been selected to emphasize the strategies used to identify chemical degradation in proteins and minimize its occurrence.

2. Deamidation and Isoaspartate Formation

The spontaneous, nonenzymatic deamidation of asparagine residues is one of the most commonly encountered chemical modifications of proteins. Numerous papers and reviews recently have appeared describing examples of deamidation events during either the isolation or storage of proteins *(12–14)*. Deamidation can occur under acidic, neutral, or alkaline conditions, although the chemical mechanism of hydrolysis is strongly dependent on pH *(see below)*. The biological purpose of deamidation in vivo may involve the regulation of protein degradation and clearance, thus serving as a type of "biological clock" *(15)*. In fact, naturally occurring protein methyl transferases have been identified that specifically modify deamidated byproducts, perhaps tagging damaged protein for either repair or clearance *(16)*.

By examining the rate of amide loss for a large series of synthetic pentapeptides of sequence (Gly-X-Asn-X-Gly and Gly-X-Gln-X-Gly) under physiological conditions, Robinson and coworkers clearly demonstrated the enhanced lability of peptide amides compared to simple aliphatic amides *(14,15)*. The asparagine containing peptides were

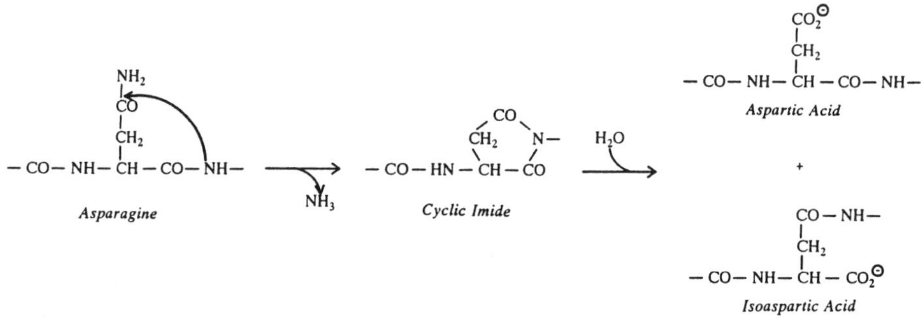

Fig. 1. Deamidation and isoaspartate formation in proteins proceeds from asparaginyl residues through a cyclic imide intermediate. *See* ref. *12* for mechanistic details.

observed to deamidate about five- to tenfold faster than their glutamine counterparts (mean half-life of deamidation of 70 vs 410 d). The deamidation rates for both the aspartyl and glutamyl containing pentapeptides were also shown to be sequence specific (half-life of deamidation ranging from 6–507 and 96–3409 d, respectively) depending on the identity of residue "X" *(16,17)*.

Although some cases of protein deamidation may be caused by direct hydrolysis of amide linkages *(14)*, this reaction is too slow to account for the deamidation of particularly labile peptide/protein bound asparagine residues. This suggests an intramolecular mechanism *(12)* in which, under neutral to basic solution conditions, the peptide bond nitrogen attacks asparaginyl carbonyl groups, causing ring closure with the concomitant release of ammonia. The resulting five-membered succinimide ring is unstable and susceptible to subsequent hydrolysis. This, in turn, leads to a mixture of α- and β-aspartyl residues (*see* Fig. 1). Under acidic conditions, deamidation is thought to proceed by direct hydrolysis, resulting in the formation of α-aspartyl residues alone *(18)*.

By examining both the temperature and sequence dependence of the deamidation of a series of Asn containing hexapeptides at pH 7.4, Asn-Gly and Asn-Ser sequences were found to be particularly labile owing to decreased steric hindrance of succinimide formation by the C-terminal residue *(19)*. For example, the Asn-Gly sequence manifested a half-life of only 1.4 d under physiological conditions. The authors also isolated and characterized the two main byproducts of the deamidation reaction, a succinimide intermediate (cyclic imide) and an isoaspartic acid residue. It was demonstrated that the succinimide product may racemize prior to hydrolysis (leading to the formation of D-amino acids) and that the overall ratio of iso-Asp to Asp residues after hydrolysis of the succinimide intermediate is typically 3:1. The corresponding Asp-Gly sequence can also form the succinimide product, although about 30-fold more slowly than Asn-Gly. A comprehensive examination of Asn and Asp hexapeptides showed a 232-fold sequence dependent range in the rate of succinimide formation under physiological conditions *(20)*. The relative infrequency of detectable glutaminyl deamidations can be understood as a consequence of the instability of the corresponding six-membered glutarimide intermediate *(12)*.

Are these sequence-dependent rates of Asn deamidation in peptides applicable to proteins with higher ordered structure? By comparing the rates of Asn deamidation in pentapeptides to the identical sequences in cytochrome C, Robinson and coworkers *(15)* concluded that secondary and tertiary structure does play an important role in dictating deamidation rates. More recently, comparisons of the deamidation rates of the abovementioned Asn-Gly peptides to two Asn-Gly sequences in triosephosphate isomerase found about a tenfold slower rate in the enzyme *(21)*. Neutron diffraction studies showed that the sites of deamidation in protein crystals of trypsin do not occur at the sites predicted from peptide studies *(22)*. Thus, protein tertiary structure, perhaps by dictating the flexibility of the polypeptide chain in the region of susceptible asparagine side chains, can either enhance or inhibit succinimide formation *(23)*. Nevertheless, many of the published examples of protein deamidation do, in fact, occur at Asn-Gly or Asn-Ser sequences *(12)*. Thus, although conformationally rigid regions of a protein molecule may inhibit deamidation at labile Asn residues, the presence of a sensitive Asn sequence in a particularly flexible region may enhance the susceptibility of this site to deamidation.

The detection of deamidation in proteins can be accomplished in a variety of ways based on either charge or molecular weight differences or by directly monitoring the formation of succinimide or isoaspartic acid residues. To illustrate the different strategies available to identify and characterize deamidation in proteins, examples using ribonuclease (RNase) as a model protein will be summarized briefly. First, deamidation has been shown to contribute to the irreversible thermal

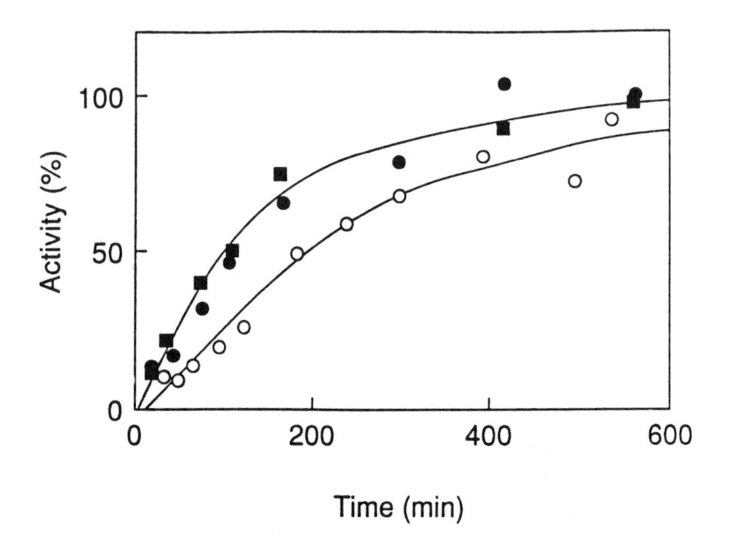

Fig. 2. Effect of deamidation on the renaturation of unfolded and reduced ribonuclease A (RNase): native RNase (Asn7) ■, deamidated RNase (Asp-67) ●, and (iso-Asp 67) ○. Experimental details in *(27)* courtesy of American Society of Biochemistry and Molecular Biology.

inactivation of RNase as detected by purification of deamidated species employing preparative isoelectric focusing. The rate of formation of multiple deamidated RNase species was monitored by IEF and correlated well with the evolution of ammonia measured enzymatically *(24)*. Second, incubation of RNase under acidic conditions (0.1*N* HCl, 30°C) led to protein deamidation with a resulting change in the enzyme's susceptibility to proteolytic cleavage *(25)*. Third, a specific deamidation event in RNase at Asn 67-Gly, in both native and unfolded enzyme, was monitored at a variety of temperatures and pH values by ion exchange chromatography. It was found that this residue deamidates 30-fold more slowly in the folded form of the enzyme *(26)*. Finally, deamidated N67D-RNase recently was further purified into Asp and iso-Asp fractions using hydrophobic interaction chromatography; each species was then examined separately in terms of catalytic activity and renaturation kinetics *(27)*. Deamidation at Asn 67 to either Asp 67 or iso-Asp 67 lowered the catalytic activity of the enzyme toward a variety of substrates. Furthermore, as illustrated in Fig. 2, the Asp 67 derivative refolded at the same rate as the native enzyme, whereas the iso-Asp 67 form refolded 50% more slowly.

Two other well-characterized examples of protein deamidation involve recombinant proteins under investigation as therapeutic agents, human growth hormone (rhGH), and the leech anticoagulant protein hirudin. Incubation of rhGH at pH 7.4, 37°C for up to 14 d resulted in the specific deamidation of Asn 149-Ser and isomerization of Asp 130-Gly *(28)*. Protein methyl transferase was used to label the isoaspartic acid residues in the altered rhGH with a radioactive methyl group. A combination of peptide mapping and subsequent amino acid analysis and mass spectroscopy was utilized to identify the specific deamidation sites in the purified peptides *(28)*. Capillary zone electrophoresis also was used to analyze deamidation in rhGH *(29)*. Interestingly, succinimide formation also was demonstrated in a lyophilized form of rhGH during storage at 45°C by using fast atom bombardment-mass spectroscopy to analyze both the intact protein and purified peptides from tryptic digestion (after resolubilization). An 18 U atomic weight difference corresponding to the loss of a water molecule was observed at the Asn 130 site in freeze-dried rhGH *(30)*.

Another example of the power of recent advances in bioanalytical technology to probe deamidation events is the use of anion exchange and reverse-phase HPLC to isolate two deamidated forms of recombinant hirudin (pH 9, 37°C). Acid-catalyzed carboxyl methylation was used to introduce a +15 U mass shift in each deamidated residue with subsequent detection by liquid secondary mass spectroscopy *(31)*, resulting in the identification of two labile Asn-Gly sequences. During the purification of a mutant form of recombinant hirudin from yeast, a 1% impurity was identified in the final product by RP-HPLC. Further work separated this impurity into two peaks that were identified as succinimide-containing forms of hirudin corresponding to alterations of the two Asn-Gly sites described above *(32)*. Identification was confirmed by both mass spectroscopy and by the presence of isoaspartic acid residues detected by sequence analysis (iso-Asp residues are resistant to Edman degradation).

3. Cleavage of Peptide Bonds

Three major mechanisms of peptide bond cleavage have been identified:

1. Preferential hydrolysis of peptide bonds at aspartic acid residues under acidic conditions;
2. At more physiological pH, C-terminal succinimide formation at Asn residues; and
3. Enzymatic proteolysis including autolysis.

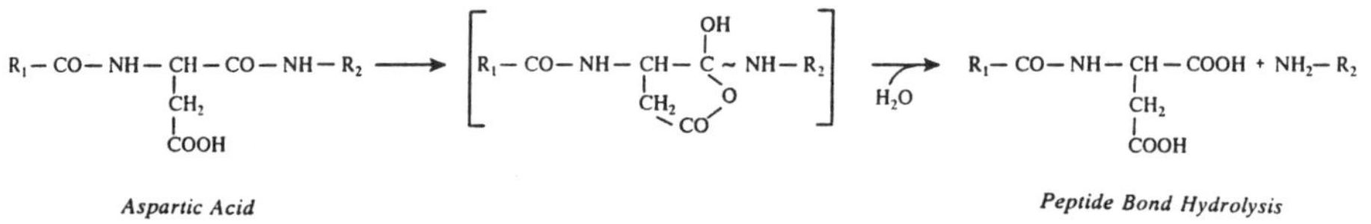

$$R_1-CO-NH-\underset{\underset{COOH}{|}}{\underset{CH_2}{|}}{CH}-CO-NH-R_2 \longrightarrow \left[R_1-CO-NH-CH(CH_2)-\overset{OH}{\overset{|}{C}}\sim NH-R_2 \;\; (CH_2-CO-O-C\text{ ring})\right] \xrightarrow[H_2O]{} R_1-CO-NH-\underset{\underset{COOH}{|}}{\underset{CH_2}{|}}{CH}-COOH + NH_2-R_2$$

Aspartic Acid *Peptide Bond Hydrolysis*

Fig. 3. Hydrolysis of peptide bonds at the C-terminal side of protein bound aspartic acid residues. *See* ref. *35* for mechanistic details.

The cleavage of a peptide bond obviously disrupts the linear sequence of amino acid residues within a protein chain. This covalent modification, however, may or may not affect the higher ordered structure of a protein and its biological activity. For example, there are numerous examples of both nonspecific hydrolysis or proteolysis leading to extensive protein degradation as well as specific proteolytic clips activating precursor forms of enzymes. Conversely, since the intramolecular interactions responsible for tertiary structure formation are sufficiently strong (cooperative), the introduction of a single intrachain clip in the polypeptide backbone may have little or no effect on a protein's structure or function.

Partridge and Davis *(33)* first reported the preferential release of aspartic acid residues after boiling a number of proteins in weak acid. This initial observation led to a series of papers examining hydrolysis of Asp-X sequences to determine whether this type of cleavage was sufficiently specific for sequence determination. When a series of proteins was heated in 0.03*N* HCl (pH 2.0) at 105°C for up to 24 h and the amount of free Asp measured by amino acid analysis, it was found that Asp is preferentially cleaved and released at a rate at least 100 times greater than any other amino acid (for review, *see* ref. *34*). The $t_{1/2}$ for this reaction (time required for the release of 50% of the total Asp and Asn residues) was 4 h for glucagon, 6 h for tobacco mosaic virus protein, 7.5 h for ribonuclease, and 24 h for insulin.

The preferential hydrolysis of a peptide bond at Asp residues is generally believed to occur at the C-terminal side of this residue in polypeptide chains, as shown in Fig. 3. The carboxyl group side chain of Asp catalyzes the cleavage reaction by acting as a proton donor at pH values below the pKa of the carboxyl group *(35)*. The Asp-Pro bond is known to be particularly labile. Studies with model dipeptides (110°C in 0.015*N* HCl) have shown that the Asp-Pro bond is 8–20 times more labile than other Asp-X or X-Asp sequences and over 100-fold less stable than the backbone of peptides lacking Asp *(36)*. The greater acid lability of Asp-Pro peptide bonds is thought to be owing to either an enhanced α/β isomerization of aspartyl residues or the more basic nature of the proline nitrogen *(37)*.

The cleavage of the C-terminal Asp peptide bond at acidic to neutral pH has been demonstrated to contribute to the irreversible thermal inactivation of enzymes, such as lysozyme and ribonuclease *(38)*. For example, mechanistic studies have shown that Asp-X peptide bond hydrolysis accounts for 77% of the total loss of RNase enzymatic activity when heated at pH 4, 90°C with a half-life of about 7 h *(24)*. The hydrolysis at Asp residues recently was reported to contribute to the thermal degradation of more complex proteins, such as endoglucanase I and glyceraldehyde-3-phosphate dehydrogenase *(40)*. The long-term storage of recombinant human epidermal growth factor under moderately accelerated conditions (pH 3, 45°C) results in the partial cleavage of an Asp-Ser peptide bond *(41)*. Although the cleavages described above generally require moderately accelerated conditions of low pH and high temperature, these are, in fact, solution conditions under which protein folding and unfolding studies may often be conducted. Attempts to actually utilize the acid lability of Asp sequences have included construction of *E. coli* expression vectors encoding bovine growth hormone fusion proteins in which an Asp-Pro sequence was inserted between the two proteins. This permitted recovery of bGH from the fusion protein under acidic conditions *(42)*.

Cleavage of polypeptide chains can also occur under physiological conditions. Analogous to the deamidation reaction discussed in the previous section, succinimide formation at asparagine residues can potentially led to the spontaneous cleavage of polypeptide chains. In this case, the side chain amide nitrogen attacks the peptide bond to form a C-terminal succinimide residue and a newly formed amino terminus *(12)*. This type of cleavage has been reported to occur in both model peptides *(18)* and in proteins (for review, *see* ref. *12*). More commonly, contaminating proteases are often found to cleave recombinant proteins during both fermentation and purification *(43)*. Strategies to limit proteolysis include the addition of protease inhibitors, careful selection of cell host including protease negative mutants, sequence modification of susceptible sites in target proteins, and optimization of fermentation and purification conditions *(44)*. Storage of purified proteases under certain conditions may also lead to peptide bond cleavage (autolysis). The stabilization of subtilisin BPN against thermally induced autolysis by removal of a susceptible cleavage site through site-directed mutagenesis has been attempted with partial success *(45)*. A related approach was employed with tissue plasminogen activator (t-PA), a serine protease that contains an activation cleavage site at arginine 275. The product of the t-PA reaction, plasmin, clips t-PA resulting in conversion of the enzyme from a one-chain to a two-chain form. Site-directed mutagenesis was performed by replacing this Arg residue with the 19 other amino acids, thereby preventing the proteolytic clipping *(46)*. The "one-chain" mutant forms of t-PA had equivalent plasminogen activating activity in the presence of a fibrin(ogen) cofactor, yet diminished activity in the absence of the cofactor.

The detection and quantitation of peptide bond cleavage in proteins usually employ SDS-PAGE with gel scanning densitometry or measurements of the appearance of new amino and carboxy termini with subsequent sequence identification *(38)*. Particular caution is required during SDS-PAGE analysis. Sample preparation for electrophoresis often involves boiling proteins in an SDS solution, which, in turn, can cause artifactual peptide bond cleavage at Asp-X residues. For example, the preparation of human fibronectin *(47)*, β-galactosidase *(48)*, and fructose 1,6 biphosphatases *(49)* for SDS-PAGE has been reported to cause the appearance of multiple fragmentation byproducts. Recently, advances in both electrospray and laser desorption time of flight mass spectroscopy has permitted the direct detection of peptide bond hydrolysis products in both peptides and proteins. These techniques are emerging as the most accurate and sensitive methods to monitor such covalent changes *(50)*, since the resultant precise determination of the molecular weight of the peptide cleavage products allows their unambiguous identification.

4. Cystine Destruction and Thiol–Disulfide Interchange

Cystine residues (disulfides) are naturally occurring crosslinks that covalently connect polypeptide chains either intra- or intermolecularly. Disulfides are formed by the oxidation of thiol groups of cysteine residues by either thiol disulfide interchange or direct oxidation (*see* next section). The probability of formation of a disulfide bond will depend on both the intrinsic stability of potential cystine residues compared to free cysteines and the conformation of the protein molecule *(51,52)*. Intracellular proteins usually lack such crosslinks and their atypical presence commonly reflects a role in an enzyme's catalytic mechanism or involvement in the regulation of its activity *(53)*. In contrast, extracellular proteins frequently contain disulfide bonds, probably reflecting the need for the increased stability of such proteins. The thermodynamic basis for this stabilization remains controversial, but cystine residues are believed to both directly stabilize the folded form of a protein and reduce the conformational entropy of the unfolded state *(53)*. Nevertheless, cystine residues are also "weak links" in protein molecules, since they are quite labile under certain conditions, especially when exposed to reducing potentials, heat, and alkaline pH.

The destruction of cystine residues in proteins have been shown to proceed by a base catalyzed β-elimination reaction in alkaline media (pH 12–13) *(54–56)*. The proton on polypeptide α-carbon atoms is relatively labile at high pH, since it is attached to two electron withdrawing groups (-CONH- and -NHCO-). This β-elimination results in the formation of two unstable intermediates, dehydroalanine and thiocysteine, as shown in Fig. 4A. The persulfide species breaks down into a variety of sulfur containing species, including hydrosulfide. Dehydroalanine is susceptible to nucleophilic attack, especially from the ε-amino group of lysine residues leading to the formation of the nonnatural crosslink, lysinoalanine. This same reaction can occur at neutral pH and elevated temperatures and has been shown to contribute to the irreversible thermoinactivation of ribonuclease and lysozyme at pH 6–8 and 90–100°C *(38)*. Another interesting example occurs in recombinant hirudin, a remarkably stable 65 amino acid protein with three disulfide linkages resistant to temperature, denaturant, and low pH induced unfolding. The combination of mildly elevated pH and temperature (pH 9, 50°C) leads to inactivation and loss of activity through β-elimination of the cystine residues and consequent formation of a mixture of atypical, unnatural crosslinks *(57)*.

Cystine destruction has been shown to be a quite general phenomena in a series of 12 different proteins heated a 100°C. Average half-lives of about 1 h at pH 8, 12 h at pH 6, and 6 d at pH 4 were encountered for these proteins *(58)*. In these studies it was also found that β-elimination leads to the generation of free thiol groups that can in turn catalyze a second degradative reaction, disulfide exchange *(24,58)*. For example,

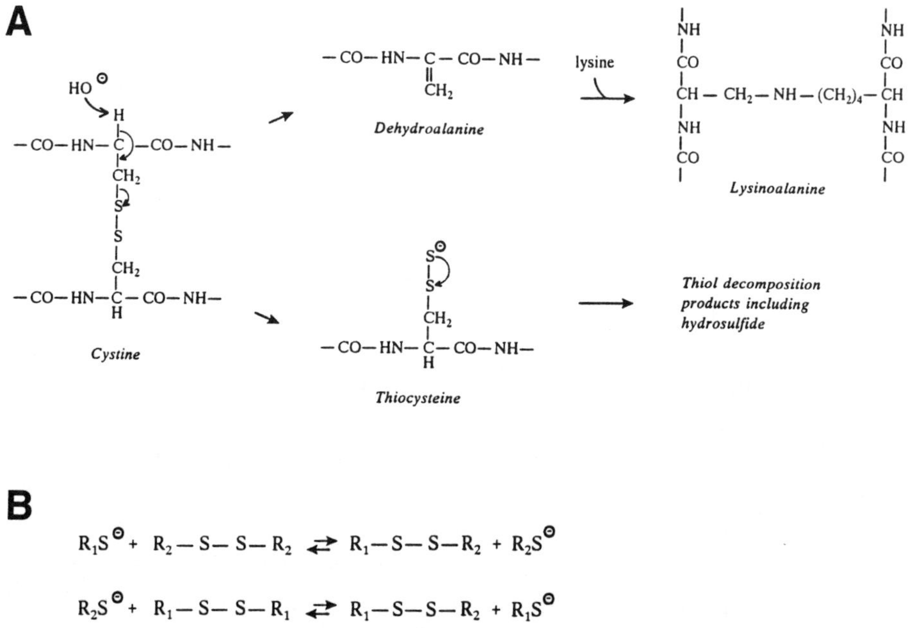

A

Cystine

Dehydroalanine

lysine

Lysinoalanine

Thiocysteine

Thiol decomposition products including hydrosulfide

B

$$R_1S^{\ominus} + R_2-S-S-R_2 \rightleftarrows R_1-S-S-R_2 + R_2S^{\ominus}$$

$$R_2S^{\ominus} + R_1-S-S-R_1 \rightleftarrows R_1-S-S-R_2 + R_1S^{\ominus}$$

Fig. 4. Degradative reactions involving protein bound cystine residues. (**A**) base catalyzed β-elimination, and (**B**) thiol catalyzed disulfide interchange reaction. *See* refs. *55* and *61*, respectively, for mechanistic details.

heat-induced thiol formation accelerated the release of glutathione from a synthetically prepared protein-glutathione mixed disulfide *(58)*. A similar inactivation pathway has been elucidated for the inactivation of insulin analogs during storage at pH 7.4 between 30 and 50°C *(59)*. This reaction was shown to be dependent on the conformational stability of insulin with the native state apparently protecting disulfides from chemical degradation (β-elimination and disulfide interchange).

Ryle and Sanger *(60)* first reported that model peptides such as cystine and oxidized glutathione undergo thiol catalyzed disulfide interchange at neutral to alkaline pH at 35°C. The reaction was accelerated by exogenous thiols and inhibited by reagents that block thiol groups. Thiols act as nucleophiles attacking the sulfur atom of the disulfide bond leading to mixed disulfide formation and subsequent scrambling of cystines as shown in Fig. 4B *(61)*. Since the reaction proceeds from the thiolate form, thiol–disulfide interchange is pH dependent with acceleration under alkaline conditions. Mixtures of reduced and oxidized thiols, such as glu-

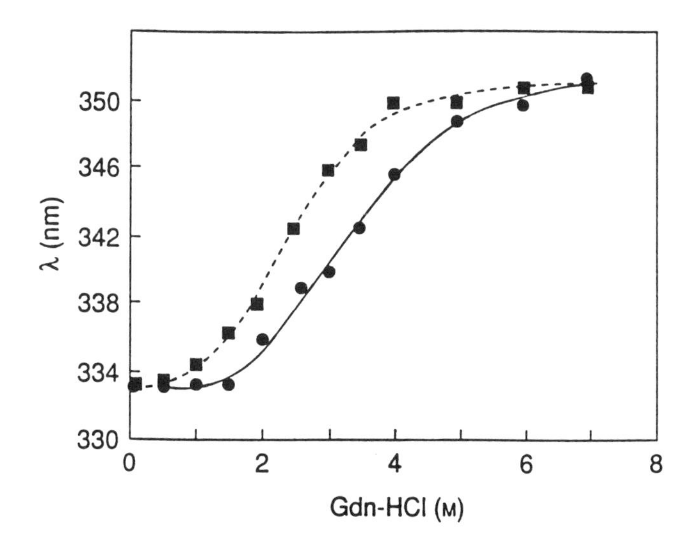

Fig. 5. Conformational destabilization of γ-crystallin owing to mixed disulfide formation with glutathione as detected by unfolding of the protein in guanidine hydrochloride: native γ-II crystallin (●) and GSSG-treated protein (■). Experimental details in ref. *65* courtesy of Academic Press.

tathione, are commonly used to establish redox potentials that catalyze the reformation of protein disulfides during protein refolding experiments via thiol–disulfide interchange reactions.

In vivo, the intracellular concentration of thiols is quite high *(62)*, present primarily in the form of reduced glutathione. This reducing environment is assumed to inhibit intracellular oxidation of protein cysteines to either protein disulfides or mixed disulfides formation with endogenous low-mol-wt thiols. Mixed disulfide formation between protein cysteine residues and reduced glutathione in vitro can lead to enzyme inactivation *(62)*. For example, the enzymatic activity of glutathione S-transferase is inhibited by biological disulfides, such as cystine and cystamine, which form active site mixed-disulfides that presumably sterically occlude the substrate binding region *(63)*. A simple fluorescence-based assay has been used to monitor the concentration of mixed disulfides in α_1-protease inhibitor under a variety of redox conditions *(64)*.

Both mixed disulfide formation and disulfide interchange can lead to altered structure and/or reduced conformational stability of proteins. For example, as shown in Fig. 5, mixed disulfide formation between lens crystallin proteins and glutathione (appearing in human cataracts) leads to the formation of modified proteins with reduced conformational stability as detected by guanidine hydrochloride unfolding experiments *(65)*. Porcine ribonuclease inhibitor, a protein containing 30 cysteine residues, can undergo thiol–disulfide interchange leading to the formation of 15 nonnative disulfide bonds. This "all or nothing" reaction is proposed to be the result of a large conformational change induced by an initial thiol-disulfide interchange event that consequently results in the exposure of the other cysteine residues. The resultant protein has a much more open conformation than the native state as measured by gel filtration and susceptibility to proteolysis *(66)*. The storage of proteins in an anhydrous state would be expected to slow down or eliminate these types of thiol disulfide interchange reactions. Despite this, freeze-dried proteins containing both disulfide bonds and reduced cysteine residues, such as bovine serum albumin, have been reported to undergo moisture-induced aggregation in the solid state owing to intermolecular disulfide formation *(67)*. This, in turn, prevents resolubilization after rehydration.

5. Oxidation of Cysteine Residues

The relative stability of a reduced cysteine residue and its oxidized disulfide counterpart depends on the redox potential of a protein's environment *(51,52)*. In vivo the electron donors and acceptors that interact with protein thiols and disulfides are primarily other thiols and disulfides (e.g., as reduced and oxidized glutathione). These compounds catalyze disulfide exchange reactions, as described in the previous section, resulting in the most thermodynamically favorable redox status of the protein's cysteine residues (free thiols vs disulfides). Similarly, in vitro protein refolding experiments require either redox buffers containing a mixture of oxidized and reduced thiol compounds to catalyze the oxidation of cysteine residues with the resultant reshuffling of disulfide bonds leading to the formation of native protein; or reducing agents, such as dithiothreitol to maintain cysteine residues in their active, reduced form.

Free thiol groups also spontaneously react with dissolved oxygen to form inter- and intramolecular disulfide bonds and monomolecular byproducts such as sulfenic acid as shown in Fig. 6A *(61)*. This autooxidation reaction is catalyzed by divalent cations, such as copper and iron ions, proceeds at enhanced rates at elevated pH, and is potentially reversible on reduction. In fact, both thiol exchange and thiol oxidation reactions proceed via the thiolate ion that has a pKa of 8–9 for

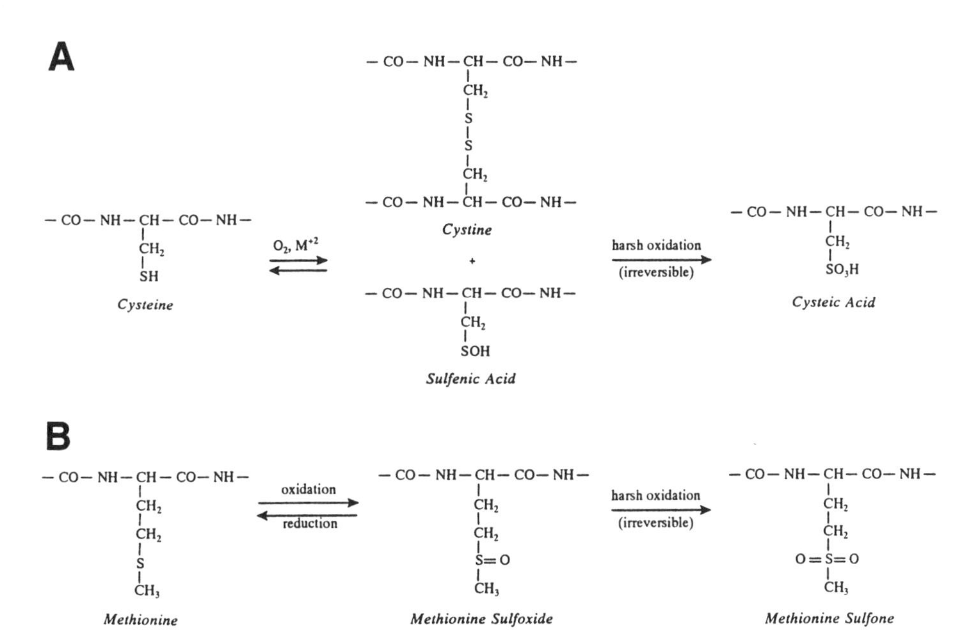

Fig. 6. Commonly encountered in vitro protein oxidation reactions: (**A**) conversion of cysteine residues to cystine, sulfenic acid, or cysteic acid, and (**B**) conversion of a methionine residue to its sulfoxide and sulfone counterparts. *See* refs. *61* and *80*, respectively, for mechanistic details.

most biological thiols. It should be noted, however, that some protein thiols can exhibit pKa values that differ significantly from this range owing to local environmental effects. Each pH unit change toward the acidic decreases the oxidative reactivity of thiols by approximately an order of magnitude *(52)*. Cysteine oxidation can also occur when protein solutions are exposed to other environmental conditions. Increased hydrostatic pressure has been shown to enhance oxidation of sulfhydryl groups in lactate dehydrogenase *(68)*, whereas freezing of protein solutions concentrates solutes that may result in elevated levels of dissolved oxygen leading to an increase in oxidative reactions *(69)*. Harsher oxidation conditions (e.g., performic acid) lead to the irreversible formation of cysteic acid (Fig. 6A).

The formation of disulfide bonds during the purification or storage of proteins containing naturally reduced cysteine residues often produces inactivation. For example, acidic fibroblast growth factor (aFGF), a potent mitogen for a variety of cells, normally contains three reduced

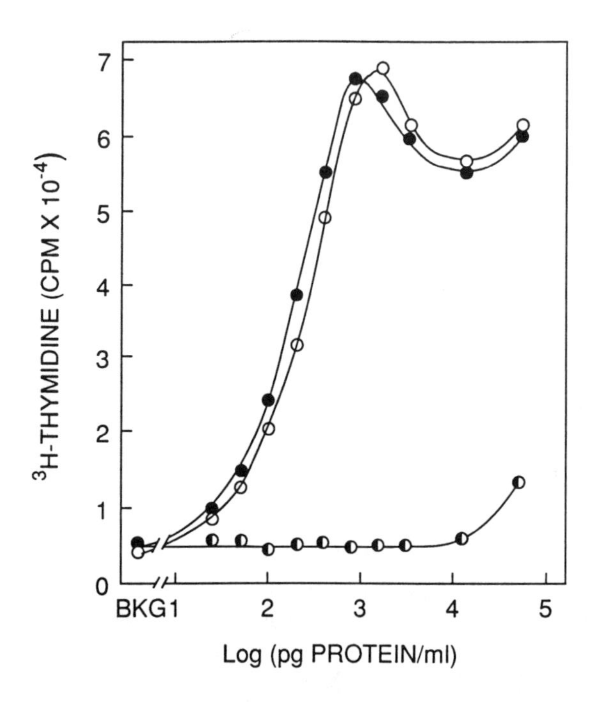

Fig. 7. Reversible inactivation of acidic fibroblast growth factor (aFGF) by oxidation of cysteine residues to cystine as measured by its ability to stimulate the growth of BALB/c3T3 cells monitored by incorporation of tritiated thymidine. (○) Native protein, (◑) oxidized protein, and (●) oxidized protein treated with the reducing agent dithiothreitol. Experimental details in ref. *70* courtesy of Harwood Academic Publishers.

cysteine residues and no disulfide bonds. As shown in Fig. 7, the copper catalyzed formation of an intramolecular disulfide bond inhibits the mitogenic activity of the protein, an effect reversed on reduction with dithiothreitol *(70)*. Similar results have been demonstrated with intermolecular (dimer) formation in aFGF *(71)*. Site-directed mutagenesis has been used to replace the cysteine residues with Ser and these aFGF mutants have increased stability and enhanced mitogenic activity *(72)*. Interestingly, binding of various polyanions to aFGF can also substantially reduce this susceptibility to oxidation, apparently by sterically occluding the most sensitive of the thiol groups *(73)*.

The irreversible thermal inactivation of both bacterial α-amylases and T4 lysozyme further illustrate the potential contribution of cysteine reactivity to protein inactivation. The irreversible thermal inactivation of microbial α-amylases is partially owing to autooxidation of the enzyme's single cysteine residue (pH 8, 90°C). Biochemical analysis of this oxidized cysteine residue revealed that 30% was present as sulfenic acid, whereas 70% had formed intramolecular disulfides *(74)*. Wild-type T4 lysozyme contains two cysteine residues at positions 54 and 97 and no disulfide bonds. Inactivation of the wild-type enzyme has been shown by SDS-PAGE to be caused by cysteine oxidation to disulfide-linked oligomers *(75–78)*. On the introduction of a third cysteine residue at position 3 by site-directed mutagenesis, a disulfide bond is formed between Cys 3 and Cys 97, resulting in a protein with increased stability. In contrast, this disulfide containing mutant is now susceptible to thiol–disulfide interchange reactions with Cys 54. This protein can therefore be further stabilized by mutation of Cys 54 to Thr or Val. Despite the greatly improved stability of this double mutant, the enzyme still undergoes significant inactivation from a combination of noncovalent aggregation and chemical degradation *(78)*.

6. Oxidation of Methionine Residues

The oxidation of methionine residues has been associated with the loss of biological activity in a number of peptides and proteins *(79)*. As seen in Fig. 6B, the oxidation of methionine results in the conversion of this thioether to its sulfoxide counterpart *(80,81)*. This is a reversible reaction in which the methionine residue can be regenerated either by reducing agents or enzymatically. Harsher oxidative conditions cause irreversible formation of methionine sulfone. In vitro, proteins are commonly treated with dilute hydrogen peroxide (H_2O_2) solution or stronger oxidizers, such as chloramine T, to achieve methionine oxidation. In vivo, oxygen containing radicals, such as superoxide, hydroxyl, and H_2O_2, are generated in a variety of cells (e.g., neutrophils), leading to the oxidation of several amino acids, including methionine, with potential implications for various aging or disease-related processes *(79)*.

Oxygen radicals can also be generated in vitro by compounds commonly used in protein folding/unfolding studies. For example, small amounts of copper in the presence of glucose oxidizes a particular methionine residue in α_1-proteinase inhibitor *(82)*, whereas the autooxidation of reducing sugars can inactivate the enzyme rhodanese with a concomitant loss in sulfhydryl titer *(83)*. In addition, air oxidation of DTT can lead to H_2O_2 generation and subsequent protein oxidation *(84)*. Methionine oxidation can also occur during the purification and storage of proteins. For example, the fermentation and purification of recombinant antistasin from yeast *(85)* and the storage of lyophilized recombinant human growth hormone *(86)* result in the oxidation of specific methionine residues.

The oxidation of methionine residues in proteins does not necessarily cause either structural changes or loss of biological activity. There are numerous examples of sulfoxide formation of specific methionine residues in proteins, some leading to complete inactivation and others having little or no effect *(79)*. Perhaps the best studied example of the effects of methionine oxidation on the structure and activity of a protein is the enzyme subtilisin. Treatment of subtilisin with H_2O_2 leads to the formation of a methionine sulfoxide residue at position 222 near the catalytic site of the enzyme (Ser 221). This oxidation directly correlates with loss of enzymatic activity *(87)*. Site-directed mutagenesis has been used to replace this Met residue with nonoxidizable amino acids, resulting in dramatically improved resistance to oxidative degradation *(88)*. The three-dimensional structure of both native and peroxide inactivated subtilisin from *Bacillus amyloliquefaciens* has been determined by X-ray crystallography to examine the structural effects of peroxide oxidation on the enzyme *(89)*. In addition to Met 222, two of the remaining four Met residues were also observed to be partially oxidized as well as the hydroxyl groups of two of the enzyme's tyrosine residues. The oxidation of these Met and Tyr residues in subtilisin did not result in any global structural changes and, surprisingly, the reactivity of these sidechains with oxygen did not appear to correlate with their solvent accessibility.

7. Photodegradation of Proteins

Both ionizing and nonionizing radiation can cause protein inactivation. The effects of different types of ionizing radiations (γ-rays, X-rays, electrons, α-particles) on a protein molecule (in both solid and solution states) have been examined in detail because of interest in the use of radiation as a potential sterilization technique in the food industry *(90)*. Both direct effects (covalent changes, such as amino acid destruction or crosslinking) and indirect effects (radiolysis of water or buffer salts and subsequent protein alterations) have been extensively documented and recently reviewed *(91)*. Nonionizing radiation, such as UV light, also

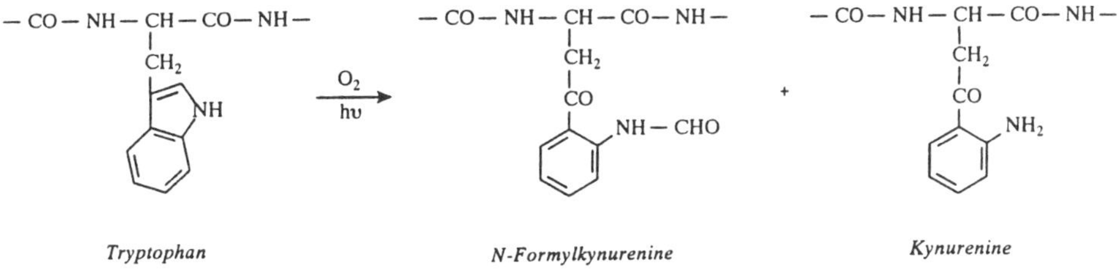

Tryptophan *N-Formylkynurenine* *Kynurenine*

Fig. 8. Photooxidation of a tryptophan residue in proteins resulting in the formation of kynurenine (Kyn) and N-formylkynurenine (NFK). *See* refs. *92–94* for further discussion.

may cause irreversible damage to protein molecules. These effects are of particular concern biologically in understanding the mechanisms of cataract formation and sunburn damage. In addition, protein unfolding/refolding studies frequently utilize UV/visible and fluorescence spectroscopy as methods of detection in which the potential adverse effects of incident light on proteins must be controlled and minimized.

The amino acids tryptophan, tyrosine, and cysteine are particularly susceptible to UV-A (320–400 nm) and UV-B (250–320 nm) photolysis *(92–94)*. The absorption of photons leads to photoionization and the formation of photodegradation products through either direct interaction with an amino acid or indirectly via various sensitizing agents (such as dyes, riboflavin, or oxygen). Commonly observed photodegradation products in an aerated, neutral pH, aqueous protein solution include S–S bond fission, conversion of tyrosine to DOPA, 3-(4-hydroxyphenyl)lactic acid, and dityrosine as well as fragmentation byproducts, and the conversion of tryptophan residues to kynurenine (Kyn) and *N*-formylkynurenine (NFK) *(92–94)*. The latter reaction is the most commonly observed and has been studied in the greatest detail (*see* Fig. 8). For example, model peptide studies have demonstrated that the rate of Gly-Trp photolysis is ten times greater than the corresponding reaction with Trp-Gly *(95)*. The location of a Trp residue within the three-dimensional structure of a protein also influences its reactivity toward photolysis with increasing solvent exposure generally correlating with elevated reactivity *(96)*.

The effect of potential photochemical modifications of tryptophan residues on the structure, stability, and activity of a protein must always be considered during protein unfolding/refolding studies. For example, the

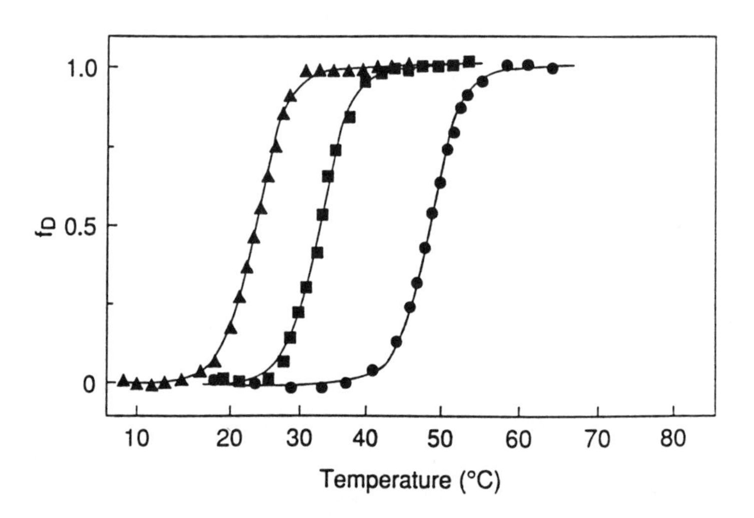

Fig. 9. Thermal unfolding of ribonuclease T1 and its derivatives as measured by circular dichroism: (●) Native protein (Trp 59), (■) Kyn-RNase T1, and (▲) NFK-RNase T1. Experimental details in ref. *99* courtesy of the American Chemical Society.

photooxidation of hydrophobic tryptophan residues causes significant crosslinking and conformational changes in lens crystallins *(97)*. The UV irradiation of the *lac* repressor of *E. coli* leads to the photodegradation of Trp residues, resulting in both decreased fluorescence of the protein and loss of inducer activity *(98)*. The chemical conversion of specific tryptophan residues (to Kyn and NFK) in lysozyme, ribonuclease T1, and a λ type immunoglobulin light chain have recently been examined to determine effects on the conformation and thermodynamic stability of these proteins *(99)*. These modified proteins were found to have nearly identical overall structures as judged by CD and fluorescence spectroscopy, yet they displayed significantly reduced stability toward thermal and GuHCl induced unfolding (*see* Fig. 9). Moreover, the more hydrophobic the microenvironment of the tryptophan residue, the greater the effect of photolysis of this residue on decreasing the protein's stability. We have found that certain monoclonal immunoglobulins display remarkably large decreases in their intrinsic fluorescence emission intensities on exposure to UV light. This has been found to be a result of photolysis of single Trp residues in immunoglobulin hypervariable regions, with the large spectroscopic changes the result of the formation of NFK and Kyn residues that quench the overall fluorescence by a nonradiative energy transfer mechanism. Any protein molecule is potentially susceptible to photodamage during CD, UV, or fluorescent measurements. Thus, comparisons of consecutively collected spectra to assure stability of the CD, UV, or fluorescence signal (and hence aromatic residues) should be carried out routinely during such studies. In many cases, minimization of the intensity of the incident beam by reduction of slit widths and rapid scan rates combined with thorough degassing of solutions can prevent damage from photooxidation during optical measurements.

8. Glycation and Carbamylation of Protein Amino Groups

Sugars are frequently used as stabilizers of proteins during storage in solution or as lyophilized powders *(100,101)*. Reducing sugars, however, can covalently react with protein amino groups (e.g., the ε-amino groups of lysine residues or the amino group at the *N*-terminus of polypeptide chains), which may lead to irreversible changes in the conformation and stability of proteins. When a reducing sugar, such as glucose, is incubated with proteins over long periods, the spontaneous formation of a Schiff's base between protein amino groups and glucose is often observed. Through a series of subsequent reactions known as the Amadori rearrangement, covalent adducts are then formed *(102)*. This process is frequently referred to as the Maillard reaction or nonenzymatic browning (Fig. 10A). These Maillard adducts can further degrade to form so-called "advanced glycosylation end products" (AGEs), resulting in both protein crosslinking and the appearance of fluorescent byproducts. These glycation reactions are believed to be involved in degenerative processes in vivo. For example, the nonenzymatic browning of lens crystallins may play a role in cataract formation *(103)*, and accumulation of AGEs is believed to correlate with the development of diabetic complications *(104)*. The appearance of glucosylated hemoglobin (HbA1c) in diabetics is particularly well known *(105)*. The kinetics of these glycation reactions as well as the analysis of various endproducts have been examined in detail with several proteins. For example, the glucosylation of ribonuclease A leads to the formation of dimers and trimers *(106)* and the frucation of bovine serum albumin to the formation of protein bound fluorescence *(107)*. In the case of α-crystallins,

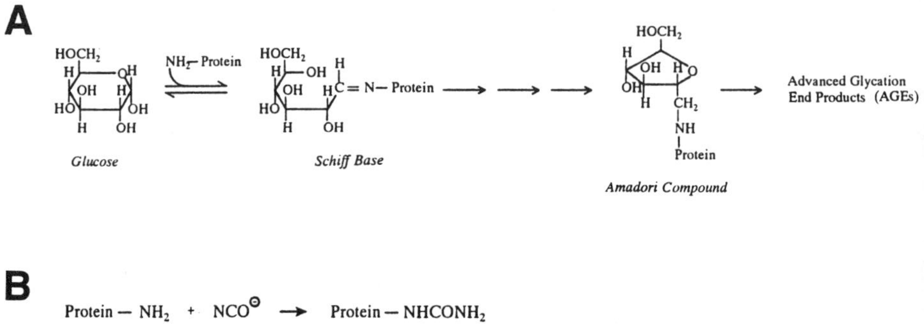

A

Glucose ⇌ (NH_2–Protein) Schiff Base → → → Amadori Compound → Advanced Glycation End Products (AGEs)

B

$$\text{Protein}-NH_2 + NCO^{\ominus} \rightarrow \text{Protein}-NHCONH_2$$

Fig. 10. Addition reactions at protein amino groups such as the ε-amino groups of lysine residues or the α-amino terminus of polypeptide chains. (**A**) Nonenzymatic browning reactions with reducing sugars such as glucose, and (**B**) carbamylation reaction with isocynate ions that are found in equilibrium with urea in aqueous solution. *See* refs. *102* and *109*, respectively, for further elaboration.

glycation causes both conformational changes and destabilization of the protein against urea induced unfolding, as shown in Fig. 11 *(108)*.

Protein amino groups are also reactive with isocynate ions leading to the carbamylation of proteins *(109,110)* as shown in Fig. 10B. Urea is in equilibrium with isocynate ions. Therefore, protein unfolding experiments that employ high concentrations of this denaturant are always susceptible to this modification if proper precautions are not taken. These include minimization of the period of contact between urea and protein and the use of freshly prepared urea solutions. Carbamylation is believed to play a role in cataract formation in certain disease states in vivo in which elevated levels of urea (and therefore isocynate) are present *(103,111)*. Under conditions of neutral pH and moderate temperature, isocynate levels approach about 1% the amount of urea present *(111)*. The kinetics of carbamylation of specific lysine residues in α-crystallin have recently been analyzed by a combination of IEF (isoelectric focusing) and fast atom bombardment mass spectroscopy *(111)*. It was found that the protein's seven lysyl residues each had unique rate constants for carbamylation in which enhanced rates of individual lysine residues correlated with their greater solvent accessibility and, presumably, surface exposure. The carbamylation of several proteins, such as bovine growth hormone *(112)*, insulin *(113)*, and hemoglobin *(114)* has been

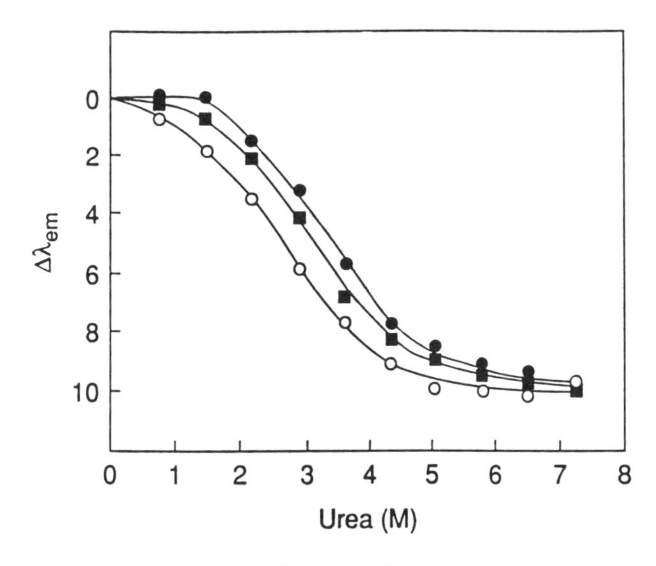

Fig. 11. Conformational destabilization of α-crystallins as a consequence of glycation of protein amino groups as measured by urea-induced unfolding: Native protein (●) and glucose 6-phosphate treated protein (■) and ribose treated protein (○). Experimental details are described in ref. *108* courtesy of Academic Press.

shown to decrease their biological activities. The generation of specific antibodies against both glycated and carbamylated proteins has also been reported using modified proteins as immunogens *(115,116)*. These antibodies may serve as the basis for alternative methods of detection for these protein amino group addition reactions to supplement the usual chemical approaches.

9. Conclusions

A variety of environmental conditions can lead to irreversible, covalent changes in protein structure. These alterations often adversely affect the conformation, stability, and bioactivity of a protein molecule. The purpose of this review is to heighten the reader's awareness of these potential reactions since many of the experimental conditions frequently used either during protein unfolding/refolding studies or during the storage of protein solutions can themselves cause irreversible, covalent changes to protein structure. For example, the near-UV light employed in spectroscopic measurements (e.g., UV absorption, CD, fluorescence) may degrade aromatic residues, denaturants such as urea may cause carbamylation at lysine residues or at the *N*-terminus, and trace metal ions can autooxidize cysteine residues. Protein unfolding studies also are carried out frequently under nonphysiological conditions, such as high or low pH, in order to increase solubility and enhance reversibility of the folding process. Unfortunately, acidic pH can lead to hydrolysis of polypeptide chains (especially at Asp-Pro sequences), whereas alkaline pH can catalyze cystine destruction and asparagine deamidation. Compounds often used as protein stabilizers, such as sugars or reducing agents, can also react with proteins if the proper precautions are not taken. Thus, studies of what is assumed to be native protein may, in fact, actually be a mixture of modified and unmodified protein. Clearly, attempts to draw conclusions about fundamental processes in the folding/unfolding of proteins in such situations are fraught with significant hazards.

The conformation of a protein molecule can be a key factor affecting both the rate and extent of these covalent reactions. Thus, certain stages of a protein unfolding/refolding experiment are more "high risk" for covalent degradation of proteins. In the case of deamidation of asparagine residues and isoaspartate formation, increased flexibility of the polypeptide chain backbone greatly enhances the formation of the cyclic imide intermediate. Similarly, greater solvent accessibility (either through intrinsic surface exposure or partial unfolding) of cysteine, lysine, or tryptophan residues increases their lability toward oxidation, addition reactions, and photolysis, respectively.

Technological advances, such as mass spectroscopy and capillary zone electrophoresis, have greatly enhanced our ability to detect and quantitate the accumulation of covalently altered protein molecules. Recently, a monoclonal antibody against a specific isoaspartyl residue containing sequence in bovine growth hormone was generated *(117)*. Clearly, the potential of developing monoclonal antibodies against a variety of chemical modifications in proteins now exists, both specific to individual proteins or particular modified residues. These antibodies could be used for the next generation of sensitive methods to detect covalent modifications of proteins. At the very least, we believe that any protein that has been subjected to significant stress in a protein folding/unfolding experiment should be examined by IEF. Although there are a few modifications that could go undetected by this method, most commonly observed covalent alterations do produce a change in charge that should produce a shift in the position of an IEF band. Importantly, lack of changes in biological activity and mobility on SDS-PAGE are not adequate for this purpose. Increasing use of capillary electrophoresis of native proteins should eventually enable this technique to be routinely used in place of IEF. Finally, mass spectroscopy, especially in the laser desorption time-of-flight format, with its 0.1–0.01% M_r accuracy, promises to become the method of choice to test for the presence of modified protein.

By understanding the causes and mechanisms of these protein modification reactions, strategies for minimizing their occurrence can also be implemented. We have found that it is extremely useful to carefully examine the amino acid sequence of a protein (if available) and identify amino acid "hot spots" before initiating protein unfolding/refolding studies. What strategies can be offered to minimize degradative, covalent reactions during the handling and storage of proteins? If appropriate, site-directed mutagenesis can be used to replace labile amino acid residues. For example, site-directed mutagenesis has been used to replace a labile Asn residue with Ser in recombinant interleukin-1α *(118)* and isoaspartate formation in recombinant human epidermal growth factor has been eliminated by the substitution of a particularly labile Asp residue with Glu *(41)*.

Such time- and labor-intensive strategies, however, may not be necessary. One straightforward method is careful control of the temperature and pH of the protein solution. For example, the rate of deamidation and isoaspartate formation as well as cysteine residue reactivity are enhanced under basic conditions. Elevated temperatures not only increase reaction rates but may increase protein flexibility and thus solvent exposure of labile amino acid residues. Since oxygen plays a role in both photolysis and autooxidation processes, these problems often can be minimized by avoidance of aerated solutions or removal of oxygen. Thus, the careful selection of experimental and storage solution conditions (including temperature, pH, and additives), based on the particular properties of the specific protein molecule under consideration, often can minimize successfully or even eliminate the occurrence of these deleterious processes. This suggests direct advantages in conducting folding/unfolding reactions at neutral pH and lower temperature using agents, such as guanidine hydrochloride and chaotropic salts, to induce structural perturbation.

Acknowledgments

The authors wish to thank Laurie Rittle for her assistance in typing the manuscript and preparing many of the figures.

References

1. Stadtman, E. R. (1990) *Biochemistry* **29,** 6323–6331.
2. Stadtman, E. R. (1992) *Science* **257,** 1220–1224.
3. Stadtman, E. R. and Oliver, C. N. (1991) *J. Biol. Chem.* **266,** 2005–2008.
4. Feeney, R. E. (1980) in *Chemical Deterioration of Proteins* (Whitaker, J. R. and Fujimaki, M., eds.), American Chemical Society, Washington, DC Symposium Series 123, pp. 1–47.
5. Klibanov, A. M. (1989) *TIBS* **14,** 141–144.
6. Bucke, C. (1981) in *Enzymes and Food Processing* (Birch, G. G., Blakebrough, N., and Parker, K. J., eds.), Applied Science, London, pp. 51–72.
7. Ng, T. K. and Kenealy, W. R. (1986) in *General, Molecular and Applied Microbiology of Thermophiles* (Brock, T. D., ed.), Wiley, New York, pp. 197–215.
8. Volkin, D. B. and Klibanov, A. M. (1989) in *Protein Function: A Practical Approach* (Creighton, T. E., ed.), IRL, Oxford, pp. 1–24.
9. Volkin, D. B. and Middaugh, C. R. (1992) in *Stability of Protein Pharmaceuticals Part A—Chemical and Physical Pathways of Protein Degradation* (Ahern, T. J. and Manning, M. C., eds.), Plenum, New York, pp. 215–247.
10. Pearlman, R. and Nguyen, T. (1992) *J. Pharm. Pharmacol.* **44,** 178–185.
11. Manning, M. C., Patel, K., and Borchardt, R. T. (1989) *Pharm. Res.* **6,** 903–918.
12. Clarke, S., Stephenson, R. C., and Lowenson, J. D. (1992) in *Stability of Protein Pharmaceuticals Part A—Chemical and Physical Pathways of Protein Degradation* (Ahern, T. J. and Manning, M. C., eds.), Plenum, New York, pp. 1–29.
13. Lui, D. T. Y. (1992) *TIBTECH* **10,** 364–369.
14. Wright, T. H. (1991) *Crit. Rev. Biochem. Mol. Biol.* **26,** 1–52.
15. Robinson, A. B. and Rudd, C. J. (1974) in *Current Topics in Cellular Regulation* (Horecker, B. L. and Stadtman, E. R., eds.), Academic, New York, pp. 247–295.
16. Aswad, D. W. and Johnson, B. A. (1987) *TIBS* **12,** 155–158.
17. Wright, H. T. and Robinson, A. B. (1982) in *From Cyclotrons to Cytochromes* (Kaplan, N. O. and Robinson, A. B., eds.), Academic, New York, pp. 727–743.
18. Tyler-Cross, R. and Schirch, V. (1991) *J. Biol. Chem.* **266,** 22,549–22,556.
19. Geiger, T. and Clarke, S. (1987) *J. Biol. Chem.* **262,** 785–794.
20. Stephenson, R. C. and Clarke, S. (1989) *J. Biol. Chem.* **264,** 6164–6170.
21. Yüksel, K. U. and Gracy, R. W. (1986) *Arch. Biochem. Biophys.* **248,** 452–459.
22. Kossiakoff, A. A. (1988) *Science* **240,** 191–194.
23. Clarke, S. (1987) *Int. J. Peptide Protein Res.* **30,** 808–821.
24. Zale, S. E. and Klibanov, A. M. (1986) *Biochemistry* **25,** 5432–5444.
25. Manjula, B. N., Seetharama, A., and Vithayathil, P. J. (1977) *Biochem. J.* **165,** 337–345.
26. Wearne, S. J. and Creighton, T. E. (1989) *Proteins: Struc. Func. Genetics* **5,** 8–12.
27. DiDonato, A., Ciardiello, M. A., Nigris, M. D., Piccoli, R., Mazzarella, L., and D'Alessio, G. (1993) *J. Biol. Chem.* **268,** 4745–4751.

28. Johnson, B. A., Shirokawa, J. M., Hancock, W. S., Spellman, M. W., Basa, L. J., and Aswad, D. W. (1989) *J. Biol. Chem.* **264,** 14,262–14,271.
29. Frenz, J., Wu, S. L., and Hancock, W. S. (1989) *J. Chrom.* **480,** 379–391.
30. Teshima, G., Stults, J. T., Ling, V., and Canova-Davis, E. (1991) *J. Biol. Chem.* **266,** 13,544–13,547.
31. Tuong, A., Maftough, M., Ponthus, C., Whitechurch, O., Roitsch, C., and Picard, C. (1992) *Biochemistry* **31,** 8291–8299.
32. Bischoff, R., Lepage, P., Jaquinod, M., Cauet, G., Acker-Klein, M., Clesse, D., Laporte, M., Bayol, A., Van Dorsselaer, A., and Roitsch, C. (1993) *Biochemistry* **32,** 725–734.
33. Partridge, S. M. and David, H. F. (1950) *Nature* **165,** 62–63.
34. Schultz, J. (1967) *Meth. Enzymol.* **11,** 255–263.
35. Inglis, A. S. (1983) *Meth. Enzymol.* **91,** 324–332.
36. Marcus, F. (1985) *Int. J. Peptide Protein Res.* **25,** 542–546.
37. Piszkiewicz, D., Landon, M., and Smith, E. L. (1970) *Biochem. Biophys. Res. Commun.* **40,** 1173–1178.
38. Ahern, T. J. and Klibanov, A. M. (1988) *Meth. Biochem. Anal.* **33,** 91–127.
39. Dominguez, J. M., Acebal, C., Jimenez, J., Mata, I., Macarron, R., and Castillon, M. P. (1992) *Biochem. J.* **287,** 583–588.
40. Hensel, R., Jakob, I., Scheer, H., and Lottspeich, F. (1992) *Biochem. Soc. Symp.* **58,** 127–133.
41. George-Nascimento, C., Lowenson, J., Borissenko, M., Calderón, M., Medina-Selby, A., Kuo, J., Clarke, S., and Randolph, A. (1990) *Biochemistry* **29,** 9584–9591.
42. Szoka, P. R., Schreiber, A. B., Chan, H., and Murthy, J. (1986) *DNA* **5,** 11–20.
43. Baneyx, F. and Georgiou, G. (1992) in *Stability of Protein Pharmaceuticals Part A—Chemical and Physical Pathways of Protein Degradation* (Ahern, T. J. and Manning, M. C., eds.), Plenum, New York, pp. 69–108.
44. Enfors, S. O. (1992) *TIBTECH* **10,** 310–315.
45. Braxton, S. and Wells, J. A. (1992) *Biochemistry* **31,** 7796–7801.
46. Higgins, D. L., Lamb, M. C., Young, S. L., Powers, D. B., and Anderson, S. (1990) *Thromb. Res.* **57,** 527–539.
47. Miekka, S. I. (1983) *Biochim. Biophys. Acta* **748,** 374–380.
48. Kowit, J. D. and Maloney, J. (1982) *Anal. Biochem.* **123,** 86–93.
49. Rittenhouse, J. and Marcus, F. (1984) *Anal. Biochem.* **138,** 442–448.
50. Geisow, M. J. (1992) *TIBTECH* **10,** 432–441.
51. Creighton, T. E. (1988) *BioEssays* **8,** 57–63.
52. Creighton, T. E. (1989) in *Protein Sructure: A Practical Approach* (Creighton, T. E., ed.), IRL, Oxford, pp. 155–167.
53. Kosen, P. A. (1992) in *Stability of Protein Pharmaceuticals Part A—Chemical and Physical Pathways of Protein Degradation* (Ahern, T. J. and Manning, M. C., eds.), Plenum, New York, pp. 31–67.
54. Whitaker, J. R. and Feeney, R. E. (1983) *CRC Crit. Rev. Food Sci. Nutr.* **19,** 173–212.
55. Feeney, R. E. (1980) in *Chemical Deterioration of Proteins* (Whitaker, J. R. and Fujimaki, M., eds.), American Chemical Society, Washington, DC, pp. 1–47.
56. Florence, T. M. (1980) *Biochem. J.* **189,** 507–520.

57. Chang, J. Y. (1991) *J. Biol. Chem.* **266,** 10,839–10,843.
58. Volkin, D. B. and Klibanov, A. M. (1987) *J. Biol. Chem.* **262,** 2945–2950.
59. Brems, D. N., Brown, P. L., Bryant, C., Chance, R. E., Green, L. K., Long, H. B., Miller, A. A., Millican, R., Shields, J. E., and Frank, B. H. (1992) *Protein Eng.* **5,** 519–525.
60. Ryle, A. P. and Sanger, F. (1955) *Biochem. J.* **60,** 535–540.
61. Torchinsky, Y. M. (1981) *Sulfur in Proteins*, Pergamon, Oxford.
62. Gilbert, H. F. (1984) *Meth. Enzymol.* **107,** 330–351.
63. Nishihara, T., Maeda, H., Okamoto, I., Oshida, T., Mizoguchi, T., and Terada, T. (1991) *Biochem. Biophys. Res. Commun.* **174,** 580–585.
64. Tyagi, S. C. and Simon, S. R. (1992) *Biochemistry* **31,** 10,584–10,590.
65. Liang, J. N. and Pelletier, M. R. (1988) *Exp. Eye Res.* **47,** 17–25.
66. Fominaya, J. M. and Hofsteenge, J. (1992) *J. Biol. Chem.* **267,** 24,655–24,660.
67. Liu, W. R., Langer, R., and Klibanov, A. M. (1991) *Biotechnol. Bioeng.* **37,** 177–184.
68. Schmid, G., Lüdemann, H. D., and Jaenicke, R. (1978) *Eur. J. Biochem.* **86,** 219–224.
69. Murase, N. and Franks, F. (1989) *Biophys. Chem.* **34,** 293–300.
70. Linemeyer, D. L., Menke, J. G., Kelly, L. J., Disalvo, J., Soderman, D., Schaeffer, M. T., Ortega, S., Gimenez-Gallego, G., and Thomas, K. A. (1990) *Growth Factors* **3,** 287–298.
71. Englea, K. A. and Maciag, T. (1992) *J. Biol. Chem.* **267,** 11,307–11,315.
72. Ortega, S., Schaeffer, M. T., Soderman, D., Disalvo, J., Linemeyer, D. L., Gimenez-Gallego, G., and Thomas, K. A. (1991) *J. Biol. Chem.* **266,** 5842–5846.
73. Volkin, D. B., Tsai, P. K., Dabora, J. M., Gress, J. O., Burke, C. J., Linhardt, R. J., and Middaugh, C. R. (1993) *Arch. Biochem. Biophys.* **300,** 30–41.
74. Tomazic, S. J. and Klibanov, A. M. (1988) *J. Biol. Chem.* **263,** 3086–3091.
75. Perry, L. J. and Wetzel, R. (1984) *Science* **266,** 555–557.
76. Perry, L. J. and Wetzel, R. (1986) *Biochemistry* **25,** 733–739.
77. Perry, L. J. and Wetzel, R. (1987) *Protein Eng.* **1,** 101–105.
78. Wetzel, R., Perry, L. J., Mulkerrin, M. G., and Randall, L. M. (1990) in *Protein Design and Development of New Therapeutics and Vaccines* (Hook, J. B. and Poste, G., eds.), Plenum, New York, pp. 79–115.
79. Swaim, M. W. and Pizzo, S. V. (1988) *J. Leuk. Biol.* **43,** 365–379.
80. Brot, N. and Weissbach, H. (1983) *Arch. Biochem. Biophys.* **223,** 271–281.
81. Brot, N., Fliss, H., Coleman, T., and Weissbach, H. (1984) *Meth. Enzymol.* **107,** 352–360.
82. Hall, P. K. and Roberts, R. C. (1992) *Biochim. Biophys. Acta* **1121,** 325–330.
83. Horowitz, P. M., Butler, M., and McClure, G. D. (1992) *J. Biol. Chem.* **267,** 23,596–23,600.
84. Costa, M., Pecci, L., Pensa, B., and Cannella, C. (1977) *Biochem. Biophys. Res. Commun.* **78,** 596–603.
85. Przysiecki, C. T., Joyce, J. G., Keller, P. M., Markus, H. Z., Carty, C. E., Hagopian, A., Sardana, M. K., Dunwiddie, C. T., Ellis, R. W., Miller, W. J., and Lehman, E. D. (1992) *Protein Expr. Purif.* **3,** 185–195.
86. Pikal, M. J., Dellerman, K. M., Roy, M. L., and Riggin, R. M. (1991) *Pharm. Res.* **8,** 427–436.

87. Stauffer, C. E. and Etson, D. (1969) *J. Biol. Chem.* **244,** 5333–5338.
88. Estell, D. A., Graycar, T. P., and Wells, J. A. (1985) *J. Biol. Chem.* **260,** 6518–6521.
89. Bott, R., Ultsch, M., Kossiakoff, A., Graycar, T., Katz, B., and Power, S. (1988) *J. Biol. Chem.* **263,** 7895–7906.
90. Schwimmer, S. (1981) *Source Book of Food Enzymology,* Avi, Westport.
91. Yamamoto, O. (1992) in *Stability of Protein Pharmaceuticals Part A—Chemical and Physical Pathways of Protein Degradation* (Ahern, T. J. and Manning, M. C., eds.), Plenum, New York, pp. 361–421.
92. Creed, D. (1984) *Photochem. Photobiol.* **39,** 537–562.
93. Creed, D. (1984) *Photochem. Photobiol.* **39,** 563–575.
94. Creed, D. (1984) *Photochem. Photobiol.* **39,** 577–583.
95. Tassin, J. D. and Borkman, R. F. (1980) *Photochem. Photobiol.* **32,** 577–585.
96. Pigault, C. and Gerard, D. (1984) *Photochem. Photobiol.* **40,** 291–296.
97. Andley, U. P. and Clark, B. A. (1988) *Curr. Eye Res.* **7,** 571–579.
98. Spodheim-Maurizot, M., Charlier, M., and Helene, C. (1985) *Photochem. Photobiol.* **42,** 353–359.
99. Okajima, T., Kawata, Y., and Hamaguchi, K. (1990) *Biochemistry* **29,** 9168–9175.
100. Timasheff, S. N. and Arakawa, T. (1992) in *Protein Structure: A Practical Approach* (Creighton, T. E., ed.), IRL, Oxford, pp. 331–345.
101. Carpenter, J. F. and Crowe, J. H. (1988) *Cryobiol.* **25,** 244–255.
102. Gottschalk, A. (1972) in *Glycoproteins: Their Composition, Structure and Function* (Gottschalk, A., ed.), Elsevier, Amsterdam, pp. 141–157.
103. Harding J. J. (1991) *Lens Eye Tox. Res.* **8,** 245–250.
104. Vlassara, H., Fuh, H., Makita, Z., Krungkrai, S., Cerami, A., and Bucala, R. (1992) *Proc. Natl. Acad. Sci. USA* **89,** 12,043–12,047.
105. Bunn, H. F., Gabbay, K. H., and Gallop, P. M. (1978) *Science* **200,** 21–27.
106. Elbe, A. S., Thorpe, S. R., and Baynes, J. W. (1983) *J. Biol. Chem.* **258,** 9406–9412.
107. Suárez, G., Rajaram, R., Oronsky, A. L., and Gawinowicz, M. A. (1989) *J. Biol. Chem.* **264,** 3674–3679.
108. Liang, J. N. and Rossi, M. T. (1990) *Exp. Eye Res.* **50,** 367–371.
109. Stark, G. R., Stein, W. H., and Moore, S. (1960) *J. Biol. Chem.* **235,** 3177–3181.
110. Stark, G. R. (1965) *Biochemistry* **4,** 1030–1036.
111. Qin, W., Smith, J. B., and Smith, D. L. (1992) *J. Biol. Chem.* **267,** 26,128–26,133.
112. Nowicki, C. and Santomé, J. A. (1981) *Int. J. Peptide Protein Res.* **18,** 52–60.
113. Oimomi, M., Hatanaka, H., Yoshimura, Y., Yokono, K., Baba, S., and Taketomi, Y. (1987) *Nephron* **46,** 63–66.
114. Lee, T. C. K. and Gibson, Q. H. (1981) *J. Biol. Chem.* **256,** 4570–4577.
115. Steinbrecher, U. P., Fischer, M., Witztum, J. L., and Curtiss, L. K. (1984) *J. Lipid Res.* **25,** 1109–1116.
116. Horiuchi, S., Araki, N., and Morino, Y. (1991) *J. Biol. Chem.* **266,** 7329–7332.
117. Lehrman, S. R., Hamlin, D. M., Lund, M. E., and Walker, G. A. (1992) *J. Protein Chem.* **11,** 657–663.
118. Wingfield, P. T., Mattaliano, R. J., MacDonald, H. R., Craig, S., Clore, G. M., Gronenborn, A. M., and Schmeissner, U. (1987) *Protein Eng.* **1,** 413–417.

CHAPTER 3

Fluorescence Spectroscopy

Catherine A. Royer

1. Introduction

The intrinsic fluorescence of aromatic amino acids in proteins has long been used as a means of monitoring unfolding/refolding transitions induced by chemical denaturants, temperature, pH changes, and pressure. The fluorescence properties of tryptophan residues in particular are exquisitely sensitive to perturbations of protein structure, whereas the low quantum yields of phenylalanine and tyrosine render these probes somewhat less useful for such studies. Accordingly, the following discussion of the application of fluorescence techniques to the study of protein folding will be limited to the intrinsic fluorescence of tryptophan in proteins and a few examples of the use of extrinsic fluorescent dyes.

Although changes in fluorescence properties have been widely used to construct equilibrium and kinetic protein folding and unfolding profiles, rarely is there a detailed physical interpretation of these changes. Generally, protein folding is well described by a two-state process, and the profiles obtained monitoring fluorescence are coincident with those obtained by circular dichroism. In these cases, the fluorescence profiles are taken as indicators of general loss of protein structure, even though they arise from changes that are local to the tryptophan residue(s). In a number of studies, however, the fluorescence unfolding/refolding profiles, because they arise from structural changes local to particular tryptophan residues, have revealed non-two-state behavior *(1–3)*. Thus, in some cases, fluorescence profiles can yield information about transient and stable intermediates in the folding/unfolding transition.

From: *Methods in Molecular Biology, Vol. 40: Protein Stability and Folding: Theory and Practice*
Edited by: B. A. Shirley

The extent to which changes in these fluorescence observables on unfolding can be given a physical interpretation depends on the extent to which the structural and dynamic factors contributing to the fluorescence properties of the folded/unfolded protein are resolved. In recent years, a great deal of progress has been made in characterizing the intrinsic fluorescence of the native states of single and two-tryptophan containing proteins *(4–12)*. The fluorescence properties of proteins are highly individual, a sort of fingerprint of the protein's structure, and their behavior on unfolding is not necessarily predictable. Relative quantum yield ratios between the native and denatured states of proteins vary widely *(13,14)*, owing to differences among proteins in both the native as well as the denatured state. For example, the tryptophan emission in some proteins is highly quenched in the native state (e.g., heme proteins; *[15]*), such that unfolding results in a large increase in the fluorescence lifetime and intensity. On the other hand, in other proteins, such as staphylococcal nuclease *(16,17)*, the tryptophan has a relatively high quantum yield in the native state, whereas its fluorescence is efficiently quenched in the unfolded state. The studies referenced above also have demonstrated that the fluorescence decay of single tryptophan-containing proteins can only rarely be described by a single exponential decay, and that such heterogeneity of decay rates reflects a heterogeneity of conformational states in the folded structure.

Studies of protein folding using fluorescence as the observable have generally relied on the simple, yet relatively uninformative measurement of changes in the steady-state fluorescence intensity at a single wavelength of emission. A more thorough characterization of all of the fluorescence properties can lead to increased insight into the structural and dynamic characteristics of folding intermediates and pathways. Three major fluorescence parameters can and should be monitored as indicators of protein structure. These three parameters are the intensity of the emission, the average energy or average wavelength of the emission, and the polarization. In order to obtain the most detailed and well-resolved information, these parameters should be investigated in time-resolved as well as steady-state mode. In addition to these measurements of the tryptophan fluorescence properties, studies involving selective quenching by external chemical agents, energy transfer from the tryptophan to covalently bound extrinsic probes and the emission properties of the extrinsic probes themselves can provide even greater characterization of folding pathways and intermediates.

1.1. Intensity

As noted above, there is no general response of intrinsic tryptophan fluorescence intensity and lifetime of proteins to unfolding of the polypeptide chain. Measurement of both can aid in determining the relative contributions of changes in dynamic and static quenching processes to the observed changes in total fluorescence intensity for an unfolding or refolding transition. Dynamic quenching of fluorescence results from the excited-state encounter of the tryptophan with a quenching moiety, such as a disulfide, an amine, or other electron accepting group on a neighboring amino acid *(7)*. The result of dynamic quenching is to lead to a decrease in the fluorescence lifetime owing to the competing excited state process. Static quenching results from the formation of ground-state nonfluorescent complexes between the tryptophan and such quenching groups. Although a decrease in the intensity is observed with an increase in static quenching, the fluorescence lifetime remains unchanged. These two processes may report on distinct steps in the unfolding of the native structure or refolding of the denatured protein, and thus differentiating between the two can lead to the identification of folding intermediates.

1.2. Average Energy

The average energy of emission, ν_g or the intensity (F_i) weighted average of the inverse wavelengths (ν_i) scanned in a spectrum, is indicative of the solvent exposure of the tryptophan.

$$\bar{\nu}_g = \sum_{i=\nu_1}^{\nu_N} F_i\nu_i \Big/ \sum_{i=\nu_1}^{\nu_N} F_i \qquad (1)$$

Solvent relaxation around the excited state dipole of the tryptophan leads to a lowering of the energy of the excited state and thus to a less energetic, or red-shifted emission.

Regardless of its intensity response to unfolding, tryptophan emission invariably shifts red (to longer wavelength) on protein denaturation to an emission spectrum similar to that of *N*-acetyl-tryptophamide (NATA) in water, with a maximum near 355 nm. Because of the aromatic nature of tryptophan, it is nearly always at least partially buried and thus at least partially inaccessible to solvent when the protein is in its native, folded form. The magnitude of the shift to the red observed on denaturation depends on the extent to which the tryptophan residue is buried in the native protein and exposed on denaturation. Tryptophan residues buried in the hydrophobic interior of proteins can have emission maxima as low as 308 nm (e.g., azurin *[4]*). The shifts in the emission maximum on unfolding can significantly contribute to changes in the steady-state intensity measurements obtained at a single wavelength. Observation on the red edge of the emission spectrum (above 360 nm) can lead to an observed increase in intensity on unfolding simply because of a shift to the red in the emission spectrum, whereas observations on the blue edge can yield an apparent loss in intensity, even in situations where the quantum yield is actually higher in the unfolded state.

1.3. Fluorescence Polarization

The polarization of fluorescence of intrinsic tryptophan emission in proteins is another measure of structural integrity. Generally speaking, the folded structure of the protein constrains the tryptophan to small amplitude motions that result in relatively little depolarization of the fluorescence emission with respect to polarized excitation. Much of the depolarization of tryptophan in native proteins arises from the Brownian tumbling of the macromolecule in solution. However, on denaturation and disruption of the protein's structure, the tryptophan's indole moiety can rotate freely around the C_α and C_β bonds and depolarization owing to local mobility becomes quite efficient. For this reason, polarization is a good measure of the loss of structural integrity. Because there are changes in the shape of the molecule on unfolding as well as the size for oligomeric unfolding profiles, it is wise to carry out both steady-state and time-resolved anisotropy measurements. In a time-resolved experiment the local and global motions of the tryptophan residue can be resolved since global tumbling occurs on a timescale of 5–100 ns, whereas the local motions are fast, on the order of 10–100 ps.

Anisotropy, *A,* is defined as the difference between the parallel and perpendicular emission intensity with respect to the total intensity when parallel polarized excitation is used.

$$A = (I_{\parallel} - I_{\perp}) / (I_{\parallel} + 2I_{\perp}) \quad (2)$$

The anisotropy, A, is related to the polarization, p, as shown below.

$$A = 2/3(1/p - 1/3)^{-1} \tag{3}$$

Because changes in the fluorescence lifetime as well as changes in the probe's rotational properties can lead to changes in the observed fluorescence anisotropy, lifetime values should also be measured as function of denaturation or refolding. The relationship between the anisotropy, A, and the lifetime and rotational correlation time is given by the Perrin equation *(18)*:

$$(A_o/A) - 1 = \tau / \tau_c \tag{4}$$

where A_o is the limiting anisotropy, τ is the lifetime, and τ_c is the rotational correlation time. Thus, one should measure the anisotropy and lifetime values and use these to calculate the rotational correlation time. Since the probe will most probably exhibit multiple rotational modes, such as local rotational depolarization and global macromolecular tumbling, the measured steady-state anisotropy is an intensity weighted average of depolarization arising from these different modes of depolarization. Likewise, if the probe exhibits multiple lifetime components, then the fractional intensity (not amplitude) weighted average lifetime should be used to calculate an average rotational correlation time.

Thus, there are a number of fluorescence parameters that can be used to monitor and characterize the denaturation or renaturation of proteins. Changes in fluorescence intensity and lifetime will invariably occur, although whether an increase or a decrease will be observed is often difficult to predict. The spectrum will shift to the red on exposure of the tryptophan residue to solvent and the polarization of the emission will decrease because of an increase in the local rotational mobility of the tryptophan residue in the denatured state. Since the physical basis for changes in these various parameters is different, comparison of equilibrium or kinetic profiles of all of these parameters can lead to identification of folding intermediates and can inform on their physical characteristics. Often tryptophan residues in different protein domains will exhibit different responses to unfolding and independent behavior of these domains can sometimes be resolved *(2)*. Given this background in fluorescence-based studies of protein folding, the following will treat the more practical aspects of equilibrium and kinetic measurements of steady-state and time-resolved fluorescence properties, including intensity, emission wavelength, and polarization.

2. Materials

All fluorescence experiments, and UV fluorescence experiments in particular, necessitate the use of ultrapure reagents for buffers and other solutions. In particular, the water used in such studies should not only be deionized but should also be void of organics. This is usually accomplished by the inclusion of a charcoal filter in the water purification system. In addition, all buffers should be prepared immediately before use, since bacterial growth can introduce significant fluorescence contamination. It is the experience of this investigator that reagent grade salts (KCl and NaCl) and phosphate buffer components are sufficiently void of fluorescent contaminants. However, Tris, Bis-Tris, Mops, urea, and guanidine hydrochloride in ultrapure form must be used. These can be purchased from ICN (Cleveland, OH) or Aldrich (Milwaukee, WI). It is essential that the purity of the buffer be verified by the measurement of its background fluorescence relative to the fluorescence intensity of the protein at the lowest concentration one intends to use. Despite the use of ultrapure chemical denaturants, at high concentrations (up to 9*M* urea, for example) one observes significant contaminating fluorescence in the ultraviolet region. Thus, background measurements should be carried out for each concentration of denaturant used. Protein, of course, must be very pure, preferably purified to a single silver stained band in a denaturing polyacrylamide gel. Quartz cuvets, not glass, must be used for ultraviolet fluorescence measurements. Cuvets can be purchased from Uvonic Instruments, Inc. (Plainview, NY). If high pressure unfolding is carried out, then the pressure generating fluid, generally ethanol, must also be the highest purity, distilled ethanol. For temperature measurements, the sample compartment should be flushed with high purity nitrogen such that at temperatures below ambient, vapor condensation does not interfere with fluorescence.

3. Methods

3.1. Solution Preparation

Equilibrium unfolding measurements using chemical denaturant can be carried out either by sequential addition of denaturant to the protein and background buffer solutions or by preparation of a series of solutions containing increasing amounts of chemical denaturant. The former approach necessitates less protein than the latter, but mixing inside the cuvet is not always efficient and the intensity values must be normalized for the dilution factor. In addition, such an approach cannot be used when studying the chemical denaturation of oligomeric proteins, since dilution will perturb the subunit equilibria. When using a series of solutions, reduced volume cuvets are used for both the solution containing the protein and the buffer and urea containing background solution to minimize the amount of protein required per denaturation curve. In an SLM (Urbana, IL) or ISS (Champaign, IL) commercial fluorometer, a minimum volume of 500 µL (with spacers under the cuvet in the cuvet holder) is required for accurate detection using a 1 cm pathlength 0.4 cm width reduced volume cuvet. Generally, in preparing the solutions it is preferable to add the buffer to all tubes (Eppendorf) followed by the urea containing buffer solution, and finally, the protein. Pipeting must be carried out with extreme care and all solutions should be mixed with pipetmen and allowed to equilibrate 1 h before measurement. A Pasteur pipet is used to extract the solution from each cuvet (sample and background) after each measurement, taking care not to scratch the cuvet walls. Pipeting should be done slowly so as to minimize the volume of solution remaining from the previous measurement.

For denaturation by temperature or pressure, only one protein solution is necessary. In temperature measurements, background subtraction can be performed, although generally there is very little contamination from buffers that do not contain chemical denaturants. The necessity will depend on the concentration of the protein solution used. Above 1 µ*M* in protein, background subtraction is probably not necessary in a temperature study, and depending on the instrumentation available, can significantly lengthen the time of the experiment. The major time-limiting factor in a temperature denaturation experiment, however, is the amount of time required for the protein solution to equilibrate at each temperature. It is useful to monitor the temperature of the solution by maintaining a cuvet with buffer in the background cuvet holder and to immerse in it a microthermocouple wire connected to a digital thermometer on the exterior of the sample compartment. It may be necessary to drill a hole in the sample compartment of your fluorometer in order to feed in the thermocouple wire without breaking it.

Unfortunately, it is not feasible to carry out background subtraction in a pressure experiment, but, as in the case of temperature measurements, it is not usually necessary. The pressure denaturation behavior of concentrations of 1–10 µ*M* protein is easily studied. Equilibration from adiabatic heating or cooling on compression or decompression requires <10 min/Kbar. Thus, for steps of 100 bar, 1 min at each pressure is sufficient.

3.2. Steady-State Measurements

3.2.1. Equilibrium

Equilibrium unfolding profiles using chemical denaturants such as pH, guanidinium hydrochloride, and urea are the most common type of fluorescence unfolding experiments. Typically, the emission intensity is monitored at a single wavelength as a function of increasing denaturant concentration. Such an experiment will generally yield a quite reasonable denaturation profile. However, as mentioned in Section 1., fluorescence intensity measurements at a single wavelength of observation can be misleading. As shown in Fig. 1A, the total intensity of the entire fluorescence spectrum for a single tryptophan-containing mutant of the tryptophan repressor protein, TRW99F, shows quite complex behavior as a function of urea *(19)*. However, comparing the intensity observed at 320 nm (Fig. 1B) and that observed at 380 nm (Fig. 1D), one sees an increase in intensity on the red edge of the spectrum and a decrease in intensity on the blue edge. The intensity at 350 nm (Fig. 1C), in the middle of the spectrum follows the complex behavior observed for the entire spectrum. The reasons for these differences can be understood from the evolution of the emission spectrum with increasing urea concentration, as shown in Fig. 2. Not only is there a change in intensity at the peak, but there is a large red shift and a broadening of the spectrum as the protein unfolds. The loss of intensity on the blue edge arises because of the shift of the spectrum toward longer wavelength. The increase in intensity on the red edge is also caused, in part, by this shift to the red with urea. In addition, above 6*M* urea, there is a direct effect of the urea on the tryptophan fluorescence, contributing to the increase in both intensity and lifetime *(20)*. This phenomenon is observed for free NATA, indicating that there is a direct effect of urea on the emission of the fluorophore. The effect is small, however, about 3% increase in intensity/1*M* urea.

Simple measurements of intensity as a function of increasing denaturant concentration can be obtained by monitoring emission through a monochromator set at a particular wavelength, through a bandpass filter centered around a particular wavelength or through a highpass filter, which allows light of wavelength higher than the cutoff to be monitored. However, since these simple intensity measurements can be misleading,

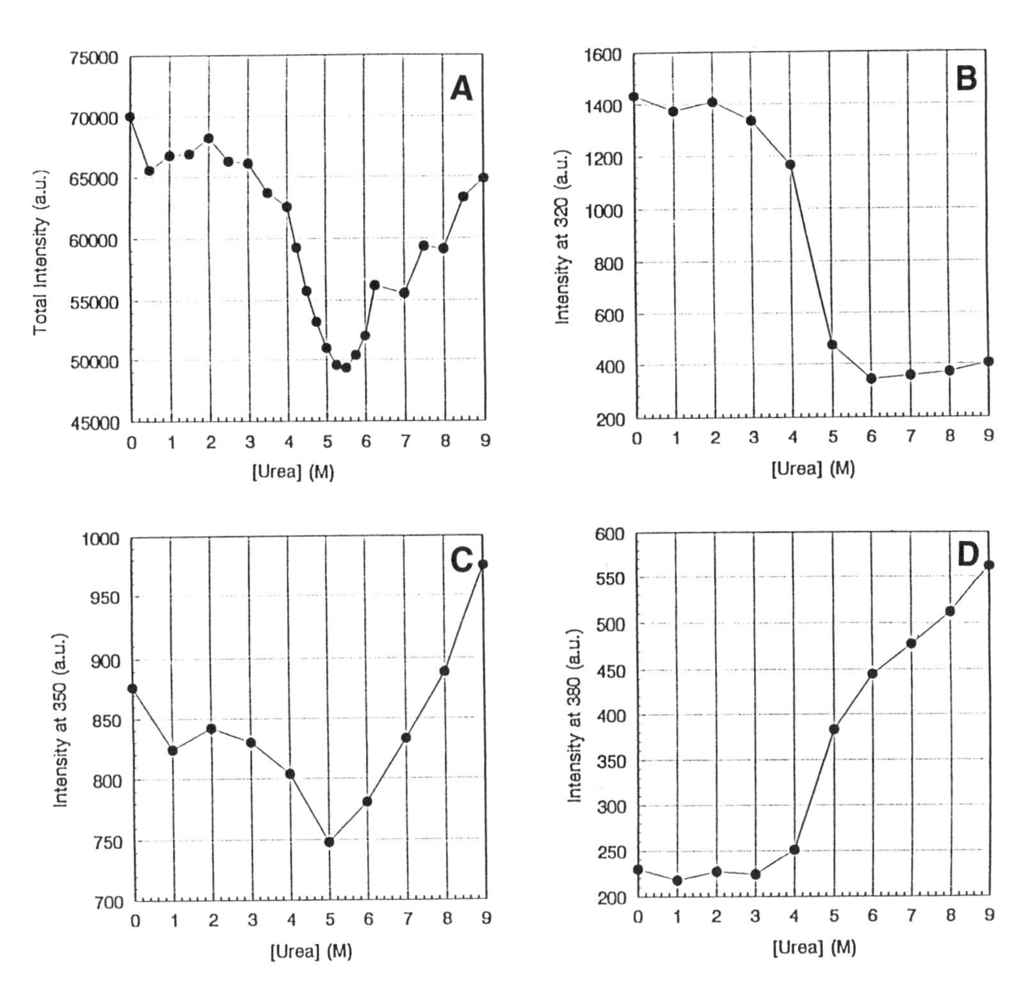

Fig. 1. Evolution of the fluorescence intensity of W99F as a function of urea concentration. (**A**) total intensity of the spectra in 2A; (**B**) intensity observed at 320 nm emission; (**C**) intensity observed at 350 nm emission; and (**D**) intensity observed at 380 nm emission. Excitation was at 295 nm (taken from ref. *19*).

it is more instructive to acquire the entire emission spectrum at each denaturant concentration examined. This experiment takes considerably longer than a single point acquisition, but allows one to obtain the average emission wavelength as well as the total intensity and the intensity at each individual wavelength scanned. If one has access to an optical multichannel analyzer (OMA, Princeton Instruments, Trenton, NJ) the time limitation is avoided, since all of the emission wavelengths are collected

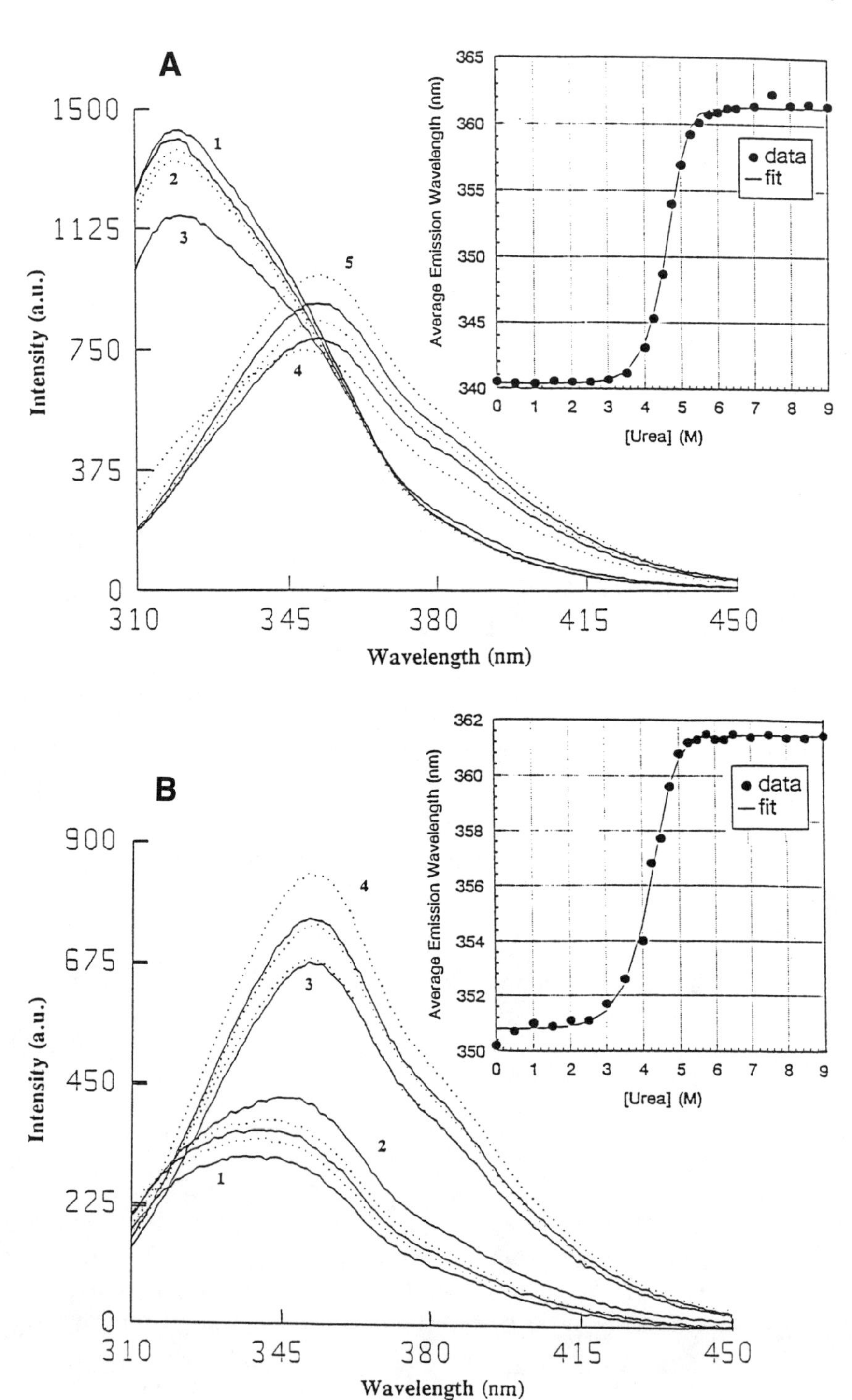
A
1500
1125
750
375
0
310
345
380
415
450
Intensity (a.u.)
Wavelength (nm)
1
2
3
4
5
Average Emission Wavelength (nm)
[Urea] (M)
data
fit
B
900
675
450
225
0
Intensity (a.u.)
Wavelength (nm)
1
2
3
4
Average Emission Wavelength (nm)
[Urea] (M)
data
fit

simultaneously, rather than one at a time, as with a monochromator. The average emission wavelength, which is simply the intensity weighted average of all of the wavelengths scanned, is a much better measure of any shift or change in shape of the emission spectrum than is the maximum. For example, a component can appear on the red edge of the spectrum without changing the emission maximum. However, since a shoulder is present, the average emission wavelength will be larger. In addition, since the average emission wavelength is an integral measurement it is much less susceptible to noise in the data. The inset in Fig. 2 shows the evolution of the average emission wavelength as a function of urea concentration for the TRW99F protein. The red shift (to longer wavelength) is quite evident in this plot and the data are of very high quality.

The profiles obtained by the red shift in fluorescence, besides providing noise-free denaturation profiles, probably are most indicative of the global unfolding transition. However, this is not always the case. The profiles obtained from fluorescence observables should always be compared to those obtained from circular dichroism or other independent methods. In many cases these do not overlap. When comparing any fluorescence parameter other than intensity with unfolding profiles obtained by other methods, the data must be quantum yield weighed for any intensity differences observed. In particular, it has been shown for TRW99F and another single tryptophan mutant of the *trp* repressor, TRW19F, that the CD and fluorescence red-shift profiles did not overlap, indicating the existence of a stable intermediate in unfolding (Fig. 3A,B) *(3)*. Since the midpoint of the fluorescence red shift profiles occurred at lower concentrations of urea than the change in CD, it was concluded that the intermediates exhibited a tryptophan residue exposed to solvent (W99 or W19), yet retained a significant degree of secondary structure. Since both tryptophan 99 and 19 are located at the subunit interface, it was proposed that the intermediates were partially folded monomers.

Another probe of tryptophan exposure to solvent is the accessibility of the tryptophan residue to external quenching reagents. Potassium iodide,

---

Fig. 2. *(previous page)* Evolution of the intrinsic tryptophan emission spectrum as a function of urea concentration between 0 and 9*M* urea of the **(A)** TRW99F and **(B)** TRW19F single tryptophan mutants. Inset represents the average emission wavelength calculated from the spectra and fit to a two-state denaturation transition. Excitation was at 295 nm (taken from ref. *19*).

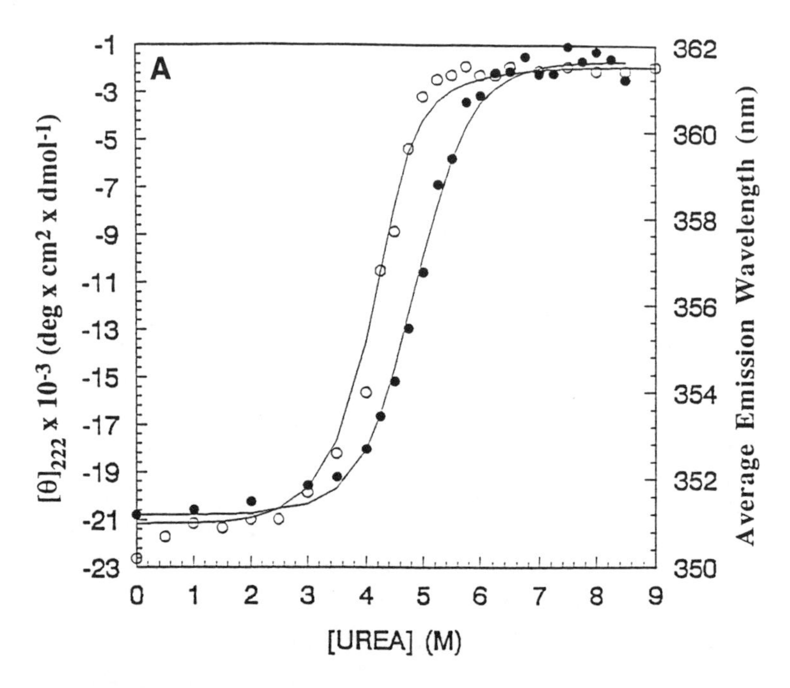

Fig. 3. (**A**) Fluorescence (○) and CD (●) profiles for the equilibrium urea induced unfolding of TRW19F.

in particular, is quite useful since its large negative charge precludes migration into the protein interior. Based on KI quenching, emission from a buried tryptophan can often be resolved from that of an exposed tryptophan, since the quenching on the blue edge of the emission spectrum is much less efficient than quenching on the red edge. Care must be taken to compare the fluorescence in presence of varying amounts of potassium iodide to that in presence of equal amounts of sodium chloride to eliminate ionic strength as the basis for observed changes in fluorescence intensity. Such an approach is most useful in the study of multitryptophan containing proteins, and has been employed to partially resolve the response of the two tryptophan residues in the wild-type *trp* repressor to unfolding by urea *(21)*. A more in-depth treatment of fluorescence quenching can be found in ref. *8*.

Steady-state polarization (or anisotropy) measurements can also be useful in determining the unfolding profile of proteins. Either the intrinsic tryptophan fluorescence or that of an extrinsic probe can be used to characterize protein folding. This can be particularly useful in discrimi-

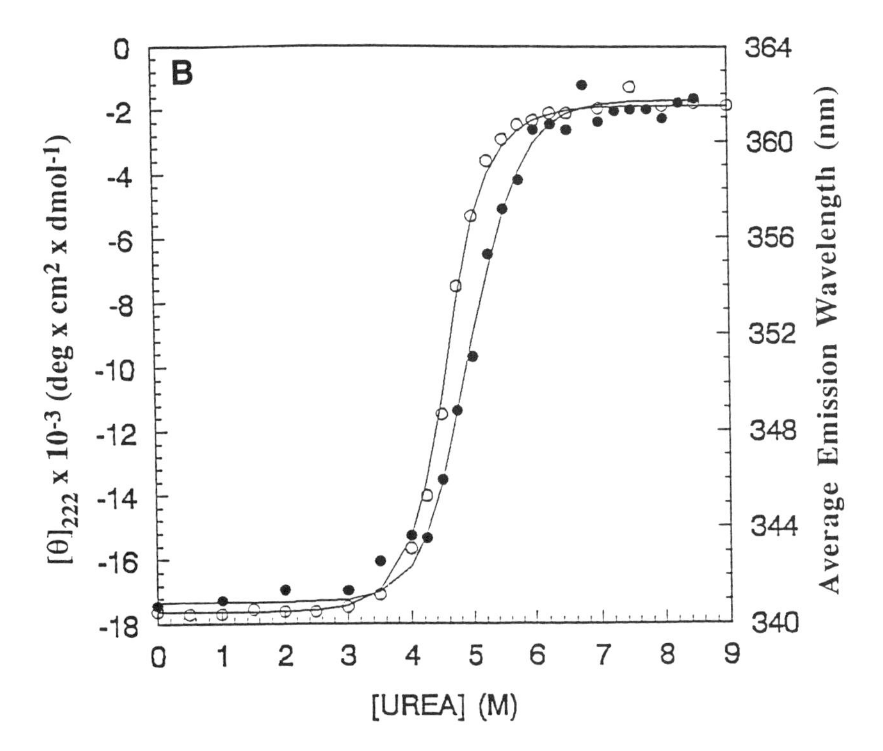

(**B**) Fluorescence and CD profiles for the equilibrium urea induced unfolding of TRW99F (taken from ref. *3*).

nating between dissociation and denaturation in multimeric proteins. An example of such an application is shown below. Fernando and Royer *(20)* used DNS (5-dimethylaminonaphthalene-1-sulfonate) covalently bound to the *trp* repressor to reveal dissociation of the protein tetramer at lower urea concentrations than the concerted denaturation/dissociation transition of the dimer. The effect of tryptophan on this transition was also studied. The steady-state anisotropy for protein in absence and in presence of tryptophan was monitored as a function of urea concentration. As can be seen in Fig. 4A both in absence and in presence of corepressor there are two phases to the urea-induced decrease in the average rotational correlation time for DNS labeled *trp* repressor. It was further demonstrated that these two phases were both protein concentration dependent, leading to the conclusion that the first phase corresponded to dissociation of the protein tetramer, whereas the second phase corresponded to the concerted dissociation/denaturation transition of the dimer. It can be seen (Fig. 4A)

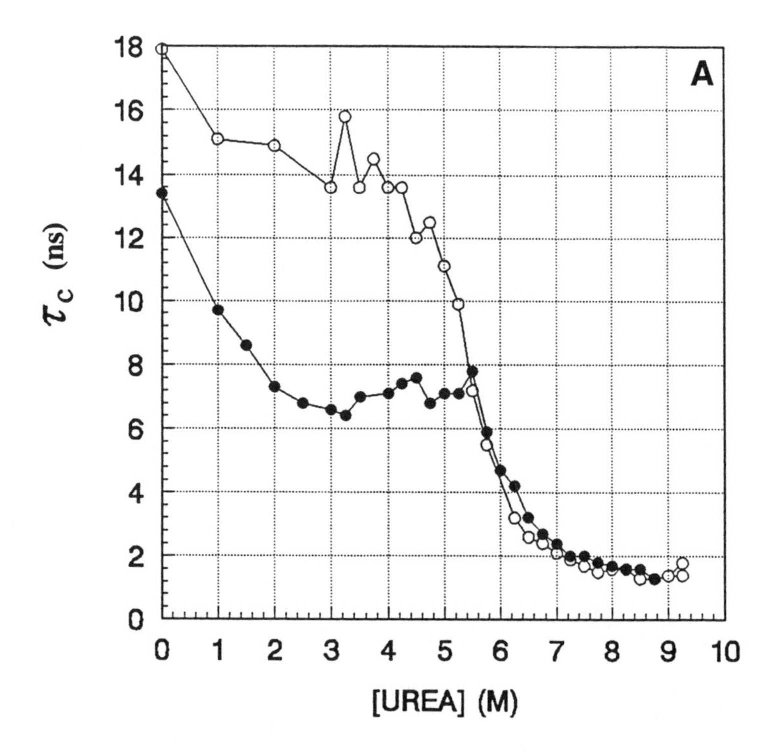

Fig. 4. The dependence of the average correlation time of the DNS labeled repressor as a function of urea. (**A**) Plots of average correlation time vs [urea] in absence (○) and in presence (●) of 0.4 m*M* L-tryptophan.

that tryptophan binding shifts the first phase of the rotational correlation time profiles to lower urea concentration and this was interpreted as a tryptophan induced destabilization of the repressor tetramer. However, the corepressor has the opposite effect on the second phase of urea denaturation (Fig. 4B), shifting it to higher urea concentration, indicating a corepressor-based stabilization of the repressor dimer.

3.2.2. Kinetics

The equilibrium denaturation profiles generated using fluorescence techniques can sometimes bring to light the existence of stable intermediates in protein folding or denaturation. However, protein folding is such a cooperative process that there are often no detectable intermediates by any physical methods under equilibrium unfolding conditions. On the other hand, transient intermediates can often be identified in

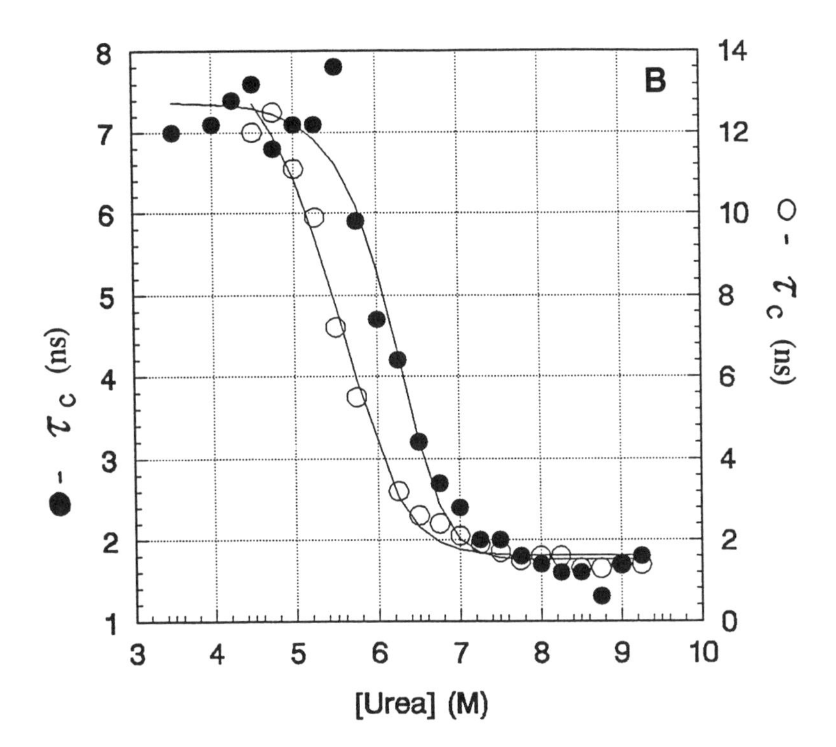

(**B**) To highlight the effect of tryptophan on the main unfolding transition the average rotational correlation time of the DNS labeled repressor in absence (○) and in presence (●) of tryptophan are plotted to scale. Data are from the main transition in Fig. 4A, whereas the lines correspond to the fit calculated from the nonlinear least squares analysis of the data. The χ^2 values for the two fits were 1.86 and 1.80, respectively, in absence and in presence of tryptophan (taken from ref. *20*).

kinetics experiments. All of the steady-state fluorescence parameters that have been discussed in detail for the equilibrium unfolding experiments can also be monitored in kinetic experiments. A number of companies (i.e., Molecular Kinetics, Pullman, WA) offer a stopped-flow apparatus for use with optical detection electronics on the millisecond timescale. The dead time of the Bio-Logique SFM-3 stopped-flow module is approx 6 ms. Steady-state emission intensity is generally acquired using a long-pass filter to eliminate scattered exciting light. However, a stopped-flow apparatus such as this can be coupled via fiberoptics to a polychromator and OMA such that the entire steady-state emission spectrum can be collected every few milliseconds *(22)*. Such measurements allow one to determine if changes in intensity coincide with changes in the average emission wavelength. Since different physical processes can give rise to changes in these two observables, intermediates can be identified. A number of investigators *(23–25)* have carried out stopped-flow CD measurements of protein folding as well. As in the case of equilibrium studies, comparison of the fluorescence and CD kinetic profiles is highly useful in identifying and characterizing folding intermediates. Steady-state anisotropy measurements can also be carried out on a millisecond timescale by collecting data either first parallel and then perpendicular in a two-shot stopped-flow experiment, or alternately, using T-format, in a one shot experiment *(26,27)*. In these polarization experiments, the emission is collected through a long-pass filter that eliminates scattered exciting light. Examples of kinetic acquisition of total intensity and steady-state anisotropy carried out by Beechem and coworkers *(27)* on a slow-folding mutant of staphylococcal nuclease can be found in Fig. 5. It is clear from the figure that the two observables are not coincident, indicating the existence of intermediates in the refolding reaction, although quantum yield weighting of the native and denatured states accounts necessarily for some of the difference.

Another useful probe of kinetic intermediates in refolding experiments is based on the binding of the hydrophobic dye, ANS (anilino-naphthalene sulfonate) to intermediates in protein folding. Since ANS exhibits very little fluorescence in water, one can monitor an increase in the fluorescence intensity of the probe, as well as a shift to bluer emission (520–480 nm maximum) if it binds transiently to hydrophobic patches on the protein during the refolding reaction *(28)*.

3.3. Time-Resolved Measurements

It is clear from the above discussion that reasonably detailed interpretations of steady-state fluorescence-based protein folding profiles are possible if a number of different fluorescence experiments are carried out and compared. However, a much more thorough picture of the folding or unfolding process can be obtained if the observed fluorescence parameters are time-resolved. The sophistication and level of interpretation of time-resolved fluorescence unfolding experiments has greatly increased recently with advances in the quality and speed of data collection and analysis methodologies. In addition, the availability of three-

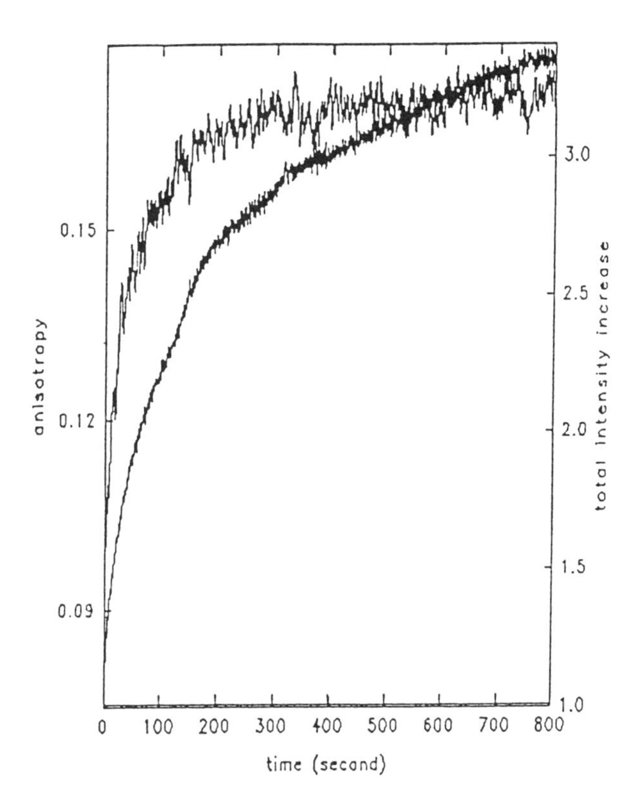

Fig. 5. Steady-state anisotropy and total intensity changes on refolding of staphylococcal nuclease (taken from ref. *27*).

dimensional structures for a number of proteins whose fluorescence has been studied has also been invaluable in our ability to give structural interpretations to the time-resolved fluorescence results. It should be noted that single tryptophan containing proteins allow for the most detailed interpretations of the fluorescence decay parameters.

3.3.1. Equilibrium

To make the most complete analysis and interpretation of fluorescence unfolding studies, it has become the rule to create multidimensional time-resolved fluorescence data surfaces. These multidimensional data surfaces can be analyzed simultaneously using global analysis software *(28)* for a consistent set of decay parameters that describe the entire surface. For example, measurement of the emission wavelength dependence of the fluorescence decay as a function of unfolding (e.g., urea concentration, pressure, temperature) yields the evolution of the Decay Associated Spectra (DAS), which represent the emission spectra associated with each decay component of the single and/or multiple tryptophan residues in the protein. As pointed out in Section 1., there is no particular expected behavior of the fluorescence lifetimes of intrinsic tryptophan residues as a function of denaturation. Interestingly, it has been possible in at least two instances (*trp* repressor *[19]* and phosphoglycerate kinase [Maas and Beechem, submitted]) to analyze the entire surface of decay vs wavelength vs denaturant concentration in terms of linked lifetimes across all data sets, allowing only the preexponential factors to vary as a function of wavelength and urea. Successful application of the lifetime linked analysis of equilibrium unfolding experiments was first carried out on the *trp* repressor single tryptophan mutants TRW19F and TRW99F *(19)*. Both were analyzed with two decay components. In the native state, TRW19F exhibits a component with a lifetime of 0.5 ns and a relative concentration of 90% and a minor component with a 2–3 ns lifetime. The urea concentration dependence of the decay of this single tryptophan over the entire emission spectrum could be very reasonably analyzed in terms of these two lifetime components present in the native state *(11)* and a very red, long-lived component that contributed little in the native state, but that became prominent in the denatured state. The DAS recovered from this analysis can be found in Fig. 6B. This linked global analysis scheme reveals that all three DAS shift to the red with increasing urea concentration, and the 0.5 ns component decreases in its relative concentration and intensity contribution, whereas the longer components increase. An analogous treatment of TRW99F lifetime denaturation profiles yields the DAS plotted in Fig. 6A. The relative success of such an analysis scheme implies that even in the folded protein a small percentage of the molecules are in a conformation that may not be completely unfolded, but that exhibits enough disorder for the tryptophan decay to be similar to the fully denatured state. Further studies on a number of different proteins will be necessary to determine whether this is a general phenomenon.

The evolution of the individual DAS for a single tryptophan as a function of denaturation is not always superimposable (Beechem, personal communication). This noncoincidence of transitions for different lifetime components in single tryptophan-containing proteins is indicative of differential responses of the various conformational substrates to the

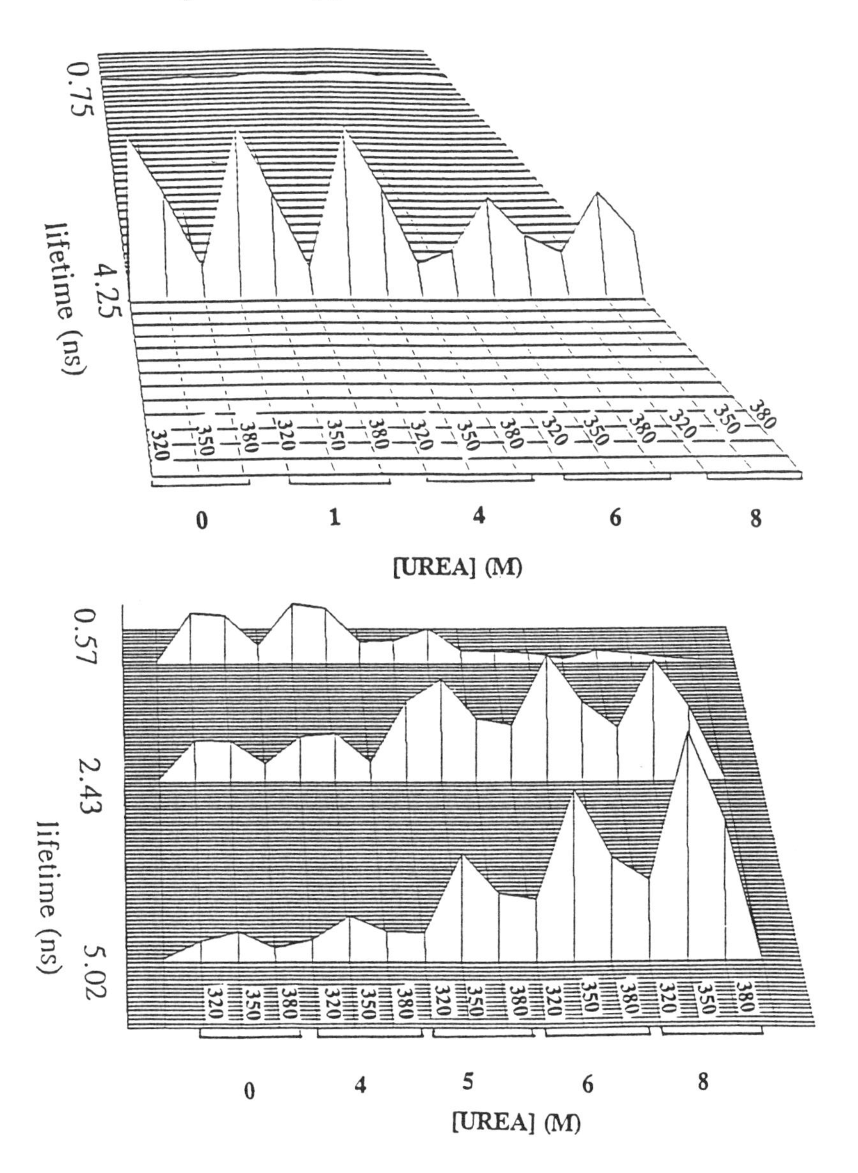

Fig. 6. Decay Associated Spectra (DAS) as a function of urea concentration for (**A**) TRW99F at 0, 1, 4, 6, and 8*M* urea; and (**B**) TRW19F at 0, 4, 5, 6, 7, and 8*M* urea. Emission wavelengths are 320, 350, and 380 nm at each urea concentration. These DAS result from a global analysis of 27 data sets in which lifetimes were linked across urea concentration and across emission wavelength, whereas the preexponential factors were allowed to vary between data sets. Global χ^2 was 3.9. The longest lifetime for the W99F mutant is not pictured so that the evolution of the dominant components may be better visualized (taken from ref. *19*).

unfolding perturbation. In multiple tryptophan containing proteins, such noncoincidence can be indicative of differential unfolding of discrete structural domains. One useful approach to determine whether domains fold independently in such two-tryptophan containing proteins has been to use site-directed mutagenesis to replace each of the tryptophan residues with a tyrosine or phenylalanine *(19)*. This allows for the resolution of the wild-type fluorescence unfolding profiles in terms of each of the two contributing tryptophan residues by studying the profiles for the single tryptophan mutants *(1,3,19)*. Where mutagenesis is not possible, time-resolved studies of the urea dependence of the potassium iodide quenching of the DAS (decay vs wavelength vs urea vs [KI]) can also be used to differentiate between the emission from two intrinsic tryptophan residues. Finally, one should always take care to compare the denaturation profiles obtained observing the total steady-state fluorescence intensity and those obtained from the time-resolved experiments. It has been noted in the case of staphylococcal nuclease that both static and dynamic quenching mechanisms contribute to the observed decrease in total intensity on denaturation *(27,30)*.

Extrinsic fluorescence probes can also be used to gather low-level structural information about the native and denatured states as well as intermediates in folding. Energy transfer from tryptophan to an extrinsic fluorophore bound covalently to a known site on the protein, or between two extrinsic fluorophores, has been used to measure distances as a function of denaturant concentration *(31–34)*. These are rather complex experiments in which the lifetime of both probes must be measured and analyzed simultaneously for energy transfer efficiencies that are converted into distances or distance distributions. The orientation and mobility factors must also be confirmed to approximate $\kappa^2 = 2/3$ (approximating freely rotating probe) through measurements of the time-resolved anisotropy decay of both probes. Amir and Haas *(31–33)* and Brand and coworkers *(34)* observed distance distributions consistent with a compact denatured state in bovine pancreatic trypsin inhibitor and staphylococcal nuclease, respectively.

3.3.2. Kinetics

Although equilibrium nanosecond time-resolved fluorescence denaturation profiles can yield useful information for the characterization of intermediates in folding or unfolding, as was noted in the discussion of the steady-state experiments, this is only true if these intermediates are

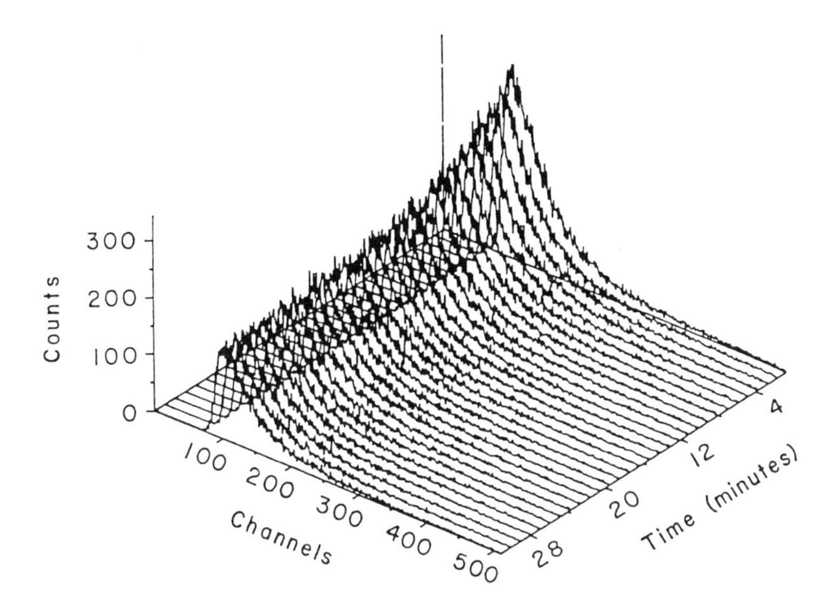

Fig. 7. The consecutive fluorescence decay curves obtained during the first half-hour of the acid unfolding of HLADH. Each decay curve represents a 1-min accumulation. The curves are plotted at the midpoints of their collection times. Each point along the channel axis represents a unit of time equal to 0.102 ns. The heights as well as the areas can be seen to decrease with time. The change in the areas corresponds to the change in relative intensity (taken from ref. *35*).

stable enough to be populated to any significant degree. The highly cooperative nature of the folding transition often precludes the observation of equilibrium intermediates. Thus, one would like to be able to carry out these high resolution nanosecond time-resolved experiments on a millisecond time-scale. These so called double kinetics experiments were first carried out by Brand and coworkers *(35)* in the time-domain on a time-scale of several minutes to study the acid denaturation of liver alcohol dehydrogenase. An example of the evolution of the fluorescence intensity decay profiles as a function of time for horse liver alcohol dehydrogenase acid unfolding obtained by these investigators can be found in Fig. 7. Since that time Beechem *(22,26,27)* has developed time-domain hardware and software for such acquisition on the millisecond timescale. A frequency domain version of the millisecond/nanosecond double kinetics experiment was developed by Gratton and coworkers *(36)*. Beechem and coworkers *(26,27)* have monitored the ms evolution of the

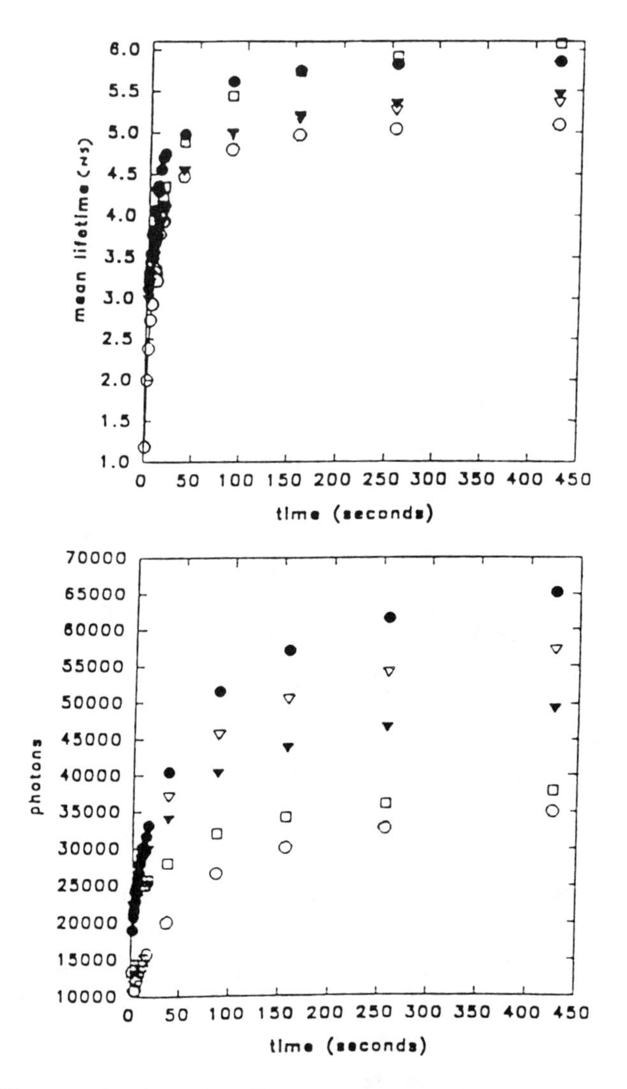

Fig. 8. Changes in the mean fluorescence lifetime (upper) and steady-state intensity (lower) on staphylococcal nuclease unfolding (taken from ref. *27*).

fluorescence intensity and anisotropy decay for the acid unfolding of the same slow-folding mutant of staphylococcal nuclease, for which the intensity and anisotropy refolding profiles were compared in Fig. 5. From comparison of the kinetics of the refolding profiles of the average lifetimes and total steady-state intensities obtained by Beechem and coworkers *(27)* and shown in Fig. 8, these transitions do not appear to overlap, indicating the population of a folding intermediate. Using multianode microchannel plate detectors has also allowed these investigators to monitor emission at multiple wavelengths simultaneously. In addition, Seethaler and Beechem *(26)* have observed the millisecond evolution of the nanosecond time-resolved anisotropy of wild-type staphylococcal nuclease on unfolding and refolding. Such experiments allow for a very intimate view of the intrinsic tryptophan's response to the unfolding/folding transitions of proteins and can also be used with extrinsic probes.

4. Summary

Although providing extensive detail, the time-resolved experiments described here can be quite complex and the instrumentation is not always readily available. However, a number of national fluorescence user facilities funded by the National Institutes of Health or the National Science Foundation are currently in operation and are dedicated to the use of fluorescence spectroscopy in the biomedical and biophysical sciences. These centers include the Laboratory for Fluorescence Dynamics at the University of Illinois and the Center for Fluorescence Spectroscopy at the University of Maryland. Even in the absence of the sophisticated equipment necessary for carrying out the time-resolved experiments, a great deal of information can be obtained from steady-state fluorescence profiles if one is careful to monitor all of the available fluorescence observables, namely intensity, wavelength or color, and polarization. Steady-state measurements of ANS binding are also quite informative. The combination of kinetic as well as equilibrium approaches, as with folding studies using any technique, will provide further insight into the pathways and stable and transient intermediates in the folding and unfolding reactions.

Fluorescence spectroscopy offers a very sensitive window into the structural and dynamic characteristics of macromolecules. Recent advances in data acquisition and analysis combined with available structure information from NMR and crystallographic studies have led to increasingly greater insight into the structural and dynamic determinants of fluorescence decay parameters in the native states of proteins. As our understanding of the fluorescence properties of native proteins has grown, fluorescence spectroscopists have begun to investigate what fluorescence can tell us about the denatured states of proteins as well as the folding/unfolding transitions and pathways. A great deal of progress has been made in the characterization and interpretation of the response of the various fluorescence parameters to protein folding and denaturation. There remain, however, a number of unanswered questions, particularly concerning the structural and dynamic determinants of the fluorescence properties of the denatured states of proteins. Future studies will undoubtedly be aimed toward this goal, and progress in this area will certainly result from systematic comparisons of fluorescence studies with a number of other biophysical and biochemical approaches.

References

1. Garvey, E. P. and Matthews, C. R. (1989) *Biochemistry* **28,** 2083.
2. Banik, U., Saha, R., Mandal, N. C., Bhattacharyya, B., and Roy, S. (1992) *Eur. J. Biochem.* **206,** 15–21.
3. Mann, C. J., Royer, C. A., and Matthews, C. R. (1993) *Prot. Sci.* **2,** 1853–1861.
4. Szabo, A. G., Stepanik, T. M., Wayner, D. M., and Young, N. M. (1985) *Biophys. J.* **41,** 233–244.
5. James, D. R., Demmer, D. R., Steer, R. P., and Verrall, R. E. (1985) *Biochemistry* **24,** 5517–5526.
6. Beechem, J. M. and Brand, L. (1985) *Ann. Rev. Biochem.* **54,** 43–71.
7. Harris, D. L. and Hudson, B. S. (1990) *Biochemistry* **29,** 5275–5284.
8. Eftink, M. R. (1991) *Protein Structure Determination: Methods of Biochemical Analysis,* vol. 35 (Suelter, C. H., ed.), Wiley, New York, pp. 127–205.
9. Royer, C. A., Gardner, J. A., Beechem, J. M., Brochon, J.-C., and Matthews, K. S. (1990) *Biophys. J.* **58,** 363–378.
10. Axelson, P. H., Bajzer, Z., Prendergast, F. G., Cottam, P. F., and Ho, C. (1991) *Biophys. J.* **60,** 650–659.
11. Royer, C. A. (1992) *Biophys. J.* **63,** 741–750.
12. Kim, S.-J., Chowdhury, F. H., Stryjewski, W., Younathan, E. S., Russo, P., and Barkley, M. D. (1993) *Biophys. J.* **65,** 215–226.
13. Kronman, M. J. and Holmes, L. G. (1971) *Photochem. Photobiol.* **14,** 113–134.
14. Grinvald, A. and Steinberg, I. Z. (1976) *Biochim. Biophys. Acta* **427,** 663–678.
15. Janes, S. M., Holtom, G., Ascenzi, P., Brunori, M., and Hochstrasser, R. M. (1987) *Biophys. J.* **51,** 653–660.
16. Gryczynski, I., Eftink, M., and Lakowicz, J. R. (1988) *Biochim. Biophys. Acta* **954,** 244–252.
17. Eftink, M. R., Gryczynski, I., Wiczk, W., Laczko, G., and Lakowicz, J. R. (1991) *Biochemistry* **30,** 8945–8953.
18. Perrin, F. (1926) *J. Phys. Radium* **1,** 390–401.
19. Royer, C. A., Mann, C. J., and Matthews, C. R. (1993) *Prot. Sci.* **2,** 1844–1852.
20. Fernando, T. and Royer, C. A. (1992) *Biochemistry* **31,** 6683–6691.
21. Lane, A. N. and Jardetsky, O. (1987) *Eur. J. Biochem.* **164,** 389–396.
22. Perez, G. M., Howard, P., Weil, A., and Beechem, J. M. (1992) *Biophys. J.* **61,** 2805.
23. Labhardt, A. M. (1986) *Methods Enzymol.* **131,** 126–135.

24. Kuwajima, K., Garvey, E. P., Finn, B. E., Matthews, C. R., and Sugai, S. (1991) *Biochemistry* **30,** 7693–7703.
25. Goldberg, M. E., Semisotnov, G. V., Friguet, B., Kuwajima, K., Ptitsyn, O. B., and Sugai, S. (1990) *FEBS Lett.* **263,** 51–56.
26. Seethaler, N. and Beechem, J. M. (1992) *Biophys. J.* **61,** A179.
27. Beechem, J. M. (1992) *S.P.I.E. Proceedings 1640: Time-Resolved Laser Spectroscopy in Biochemistry III,* 676–680.
28. Semisotnov, G. V., Rodionova, N. A., Razgulyaev, O. I., Uversky, V. N., Gripas, A. F., and Gilmanshin, R. I. (1991) *Bopolymers* **31,** 119–128.
29. Beechem, J. M., Gratton, E., Ameloot, M. A., Knutson, J. R., and Brand, L. (1991) in *Topics in Fluorescence Spectroscopy,* vol. 2 (Lakowicz, J. R., ed.), Plenum, New York, pp. 241–301.
30. Royer, C. A., Hinck, A. P., Loh, S. N., Prehoda, K. E., Peng, X., Jonas, J., and Markley, J. L. (1993) *Biochemistry* **32,** 5222–5232.
31. Amir, D., Levy, D. P., Levin, Y., and Haas, E. (1986) *Biopolymers* **25,** 1645–1658.
32. Amir, D. and Haas, E. (1987) *Biochemistry* **26,** 2162–2175.
33. Amir, D. and Haas, E. (1988) *Biochemistry* **27,** 8889–8893.
34. James, E., Wu, P. G., Stites, W., and Brand, L. (1992) *Biochemistry* **31,** 10,217–10,225.
35. Walbridge, D., Knutson, J. R., and Brand, L. (1987) *Anal. Biochem.* **161,** 467–478.
36. Eriksson, S., Tetin, S., Voss, E., Gratton, E., and Mantulin, W. (1993) *Biophys. J.* **61,** A218.

CHAPTER 4

Ultraviolet Absorption Spectroscopy

Henryk Mach, David B. Volkin, Carl J. Burke, and C. Russell Middaugh

1. Introduction

Four decades ago, ultraviolet absorption spectroscopy played an important and often pivotal role in studies of protein structure and function. A comprehensive discussion of the application of ultraviolet spectroscopy to the study of proteins was provided by Wetlaufer in 1962 *(1)*, which remains a valuable source of information about the technique. The subsequent vigorous growth of protein chemistry was both partially initiated and paralleled by the introduction of powerful new techniques, such as nuclear magnetic resonance, circular dichroism, Fourier-transform infrared, and fluorescence spectroscopies. The widespread use of these methods relegated UV spectroscopy to a secondary technique most often employed for simple concentration determinations or enzymatic assays, although it was still occasionally used in protein structural studies. Nevertheless, ultraviolet spectrophotometry remains a fast, accurate, quantitative, and nondestructive technique with much to offer in studies of protein folding and structural integrity. Moreover, with new instrumental and theoretical improvements, it provides some unique features that have resulted in a moderate resurgence in its use in certain applications. The omnipresence of UV absorption spectrophotometers in biochemical laboratories also suggests that the technique may be used to provide structural information when more specialized instrumentation is not readily available.

From: *Methods in Molecular Biology, Vol. 40: Protein Stability and Folding: Theory and Practice*
Edited by: B. A. Shirley

2. Absorption of Ultraviolet Light

Light consists of perpendicular oscillating magnetic and electric fields whose energy is given by:

$$E = hc/\lambda = h\upsilon \quad (1)$$

where h is Planck's constant, c is the speed, λ is the wavelength, and υ is the frequency of the light. Absorption occurs when the energy of a light photon equals the difference between a ground and excited state of a target molecule. The absorbed energy may be dissipated in the form of heat or radiation of a photon of lower energy (fluorescence or phosphorescence). Under certain conditions, it may produce chemical changes in the absorbing chromophore.

The major determinant of the probability of the absorption of a photon of given energy is the difference between the basal energy configuration of the molecular orbitals and the next highest energetically allowed electronic configuration. The two most commonly observed low energy electronic transitions are of the $n \rightarrow \pi^*$ and $\pi \rightarrow \pi^*$ type (where * denotes an excited state). The discrete nature of these electronic transitions results in the appearance of distinct bands in ultraviolet spectra. However, the energy of various orbitals vary slightly as a consequence of vibrations and rotations about covalent bonds and through energetic interactions with the molecule's immediate environment. The resulting distribution of bond energy is reflected in the significant width and gaussian shape of many absorption bands of molecules in solution. The diffuseness of these bands is further increased when populations of identical chromophores are located in different microenvironments. Thus, the ultraviolet spectra of most molecules in liquids (especially polar solvents) consist of broad band(s), although substructure caused by distinct vibrational transitions or unique microenvironments may be present.

It is well established under ideal conditions that a decrease in the intensity *(I)* of a light beam passing through a layer of sample solution is proportional to the sample concentration *(c)* and the width of the layer (dl):

$$-dI/I_o = kcdl \quad (2)$$

where k is a constant and I_o is the intensity of the incident light. After integration we obtain:

$$\ln(I_o/I) = kcl \quad (3)$$

Thus the absorbance *(A)* is defined as:

$$A = \log(I_o/I) = \varepsilon\, cl \quad (4)$$

where k is now defined as ε, the molar extinction coefficient, *c* is molar concentration, and *l* is pathlength in cm. The above equation is, of course, known as the Beer-Lambert law. In protein chemistry, it is common to replace the molar extinction coefficient with its weight equivalent and this is usually designated with a capital "E." As a rule of thumb, a "typical" 1 mg/mL protein solution usually has an absorbance value on the order of one absorbance unit. It can be shown easily that the maximum molar extinction coefficient of a typical aromatic ring is approx 1×10^5/*M*/cm *(2)*. In practice, however, only a fraction of the photons passing through a molecule is absorbed resulting in a typical extinction coefficient of about 10^4/*M*/cm for a strong absorber. Note that if processes other than pure absorption (e.g., light scattering) are present, the quantity log (I_o/I) is referred to as optical density (OD).

3. UV Spectra of Proteins

The most frequently employed spectral range for proteins is between 250 and 320 nm, a region referred to as the near ultraviolet. All three aromatic amino acid side chains (i.e., tryptophan, tyrosine, and phenylalanine) have prominent absorption bands in the near ultraviolet (Fig. 1). The only other amino acid with an appreciable contribution in this spectral range (typically, a few percent of total absorbance in most proteins) is cysteine. Assuming the absence of metal ions, cofactors, oxidation byproducts, and other absorbing moieties, these four amino acids are the only important chromophores in this region and they, as well as molecules containing them, are expected to behave according to the Beer-Lambert law. In practice, however, deviations from this behavior often occur when:

1. The sample absorbance is too high (i.e., >1.5).
2. The sizes of the molecules in solution are comparable to the wavelength of the incident light and, consequently, a significant portion of the incident light beam is scattered. Furthermore, as particles become larger, the chromophores can begin to optically obscure one another, giving rise to the phenomenon known as absorption flattening.
3. The sample is highly fluorescent.
4. The sample is not chemically stable.

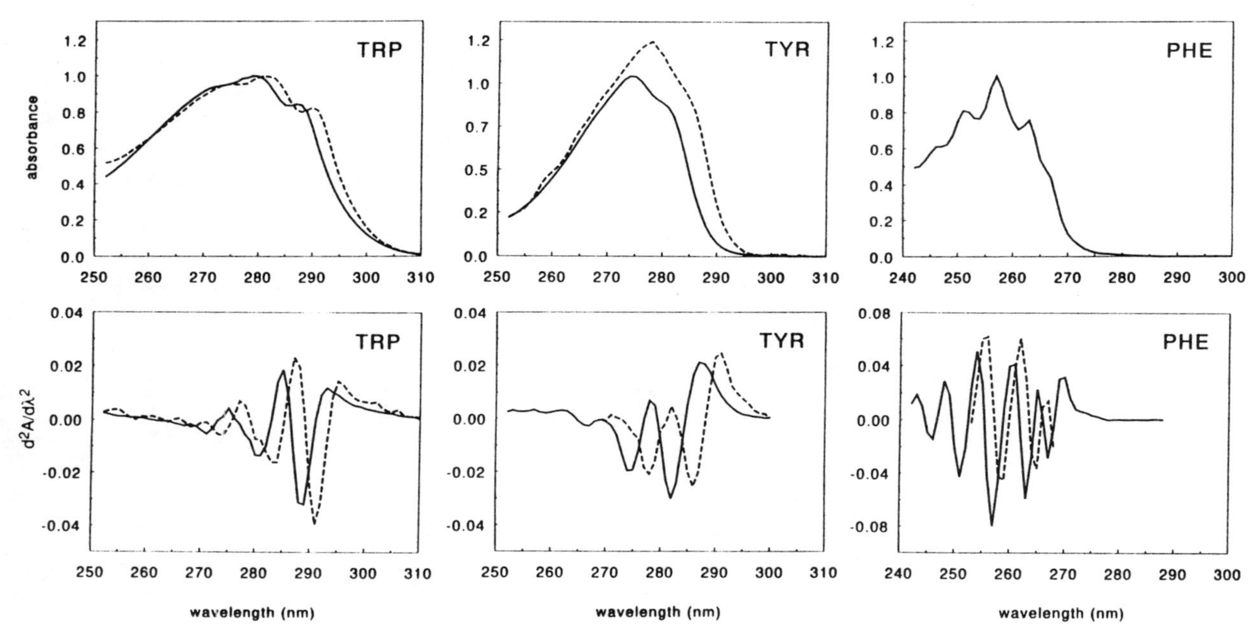

Fig. 1. Upper panels: normal (zero order) spectra of 0.18 m*M* *N*-AcTrpNH$_2$ (Trp), 0.70 m*M* *N*-AcTyrNH$_2$ (Tyr), and 5.1 m*M* *N*-acetyl-L-phenylalanine ethyl ester (*N*-AcPhe OEt). The Trp and Tyr spectra obtained by matrix decomposition *(9)* of a set of 13 globular proteins are also shown (---). Lower panels: The second derivative spectra calculated *(49)* from the normal spectra shown in upper panels. The second derivative spectrum of the second deconvoluted component in the 250–270 nm range, dominated by strong Phe signal, is overlayed with the second derivative spectrum of N-AcPheOEt. Proteins employed were: carbonic anhydrase B, chymotrypsinogen A, α-lactalbumin, lysozyme, ovalbumin, transferrin, IgG, tick anticoagulant peptide, azurin, aldolase, pepsin, trypsin inhibitor, and β-amylase.

In addition, various problems with the measuring technique itself may influence the results, including:

1. A significant amount of stray light.
2. A fast scan rate coupled to an inadequate detector response.
3. Instrumental slitwidth comparable to chromophoric spectral bandwidths.
4. Obstruction of the light pathway, including scratches on optical surfaces and residuals, such as fingerprints or surface-adsorbed solutes.

4. Quantitative Analysis of Proteins

4.1. Concentration Determination

Since almost all proteins possess a well-defined aromatic amino acid composition, absorbance measurements in the near UV provide the most accurate and convenient method of protein concentration determination. Knowledge of the extinction coefficient of a protein at a given wavelength allows its concentration to be determined immediately by a single absorbance measurement at that wavelength. The molar extinction coefficients of proteins are most frequently experimentally determined by an independent concentration determination using methods such as quantitative amino acid analysis, dry weight measurement *(1)*, nitrogen analysis *(3,4)*, and spectral methods *(5,6)*, as well as the Lowry *(7)* and Bradford *(8)* colorimetric techniques. Alternatively, the extinction coefficients of water soluble, globular proteins can usually be calculated with a high degree of accuracy using the average values of the protein resident extinction coefficients of tryptophan, tyrosine, and cysteine residues extracted from a large experimental data set *(9)*:

$$C = A_{280}/\varepsilon l, \quad \varepsilon = 5540 n_{Trp} + 1480 n_{Tyr} + 134 n_{s\text{-}s} \tag{5}$$

where n is the number of each indicated residue in the protein and l is the pathlength in cm. Protein extinction coefficients calculated in this manner generally differ by <2% from values obtained by repetitive quantitative amino acid analysis for globular water-soluble proteins *(9)*. Although similar values have been established for unfolded proteins *(10)*, Eq. (5) allows determination of ε in the native, folded state and direct application to functional protein molecules.

4.2. Correction for Light Scattering

The above procedure is complicated by the presence of any additional spectral components in the near-UV range. When the size of a protein complex becomes comparable to the wavelength of the measuring light (typically, greater than a 10 nm radius), the amount of light scattered by the sample becomes significant, resulting in an elevation of the observed optical density. According to Rayleigh light scattering theory, the intensity of the scattered light is proportional to the inverse fourth power of the incident wavelength. Thus, the presence of such scattering (turbidity) is usually revealed by an apparent absorbance (OD) gradually decreasing toward longer wavelengths in nonabsorbing regions (i.e., >320 nm). This is illustrated by a typical example in the 300–350 nm range in Fig. 2A. A linear plot of the logarithm of the optical density vs the logarithm of the wavelength in this nonchromophoric region can be extrapolated through the 250–300 nm region and subtracted from the experimentally observed spectrum to obtain corrected values *(1,11,12)*. A UV spectrum after such a correction is also illustrated in Fig. 2A. If one needs to correct for light scattering at 280 nm alone for a concentration determination, a simple equation derived from the above log–log relationship can be employed:

$$OD_{280} = 10^{(2.5 \log OD_{320} - 1.5 \log OD_{350})} \tag{6}$$

where OD_{280} denotes the optical density (apparent absorbance, turbidity) at 280 nm and OD_{320} and OD_{350} denote apparent absorbance values at 320 and 350 nm, respectively. In addition, there may be other components in a protein's spectrum that produce broad features in the near-UV region. These can originate from a variety of sources, such as metal ions, nucleic acids, and sulfhydryl-containing compounds. In such cases, this method may yield erroneous results.

An alternative approach for protein concentration determination involves the calculation of second (or higher) order derivative spectra. These spectra (Fig. 1) are sensitive to the fine spectral features of the aromatic amino acids but are not significantly perturbed by the presence of broad underlying spectral bands *(11,13)*. Second derivative peaks also

Fig. 2. *(opposite page)* (**A**) Near-UV spectrum of a 0.1 mg/mL solution of conjugates of polyribosyl ribitol phosphate and outer membrane protein complexes from *Neisseria meningitidis* (PRP-OMPC) (—). The light scattering contribution was removed by subtraction of extrapolated data (---) from a linear plot of log OD vs log λ between 320 and 350 nm and the resultant corrected spectrum is shown as the dashed line. (**B**) The second derivative spectra of the

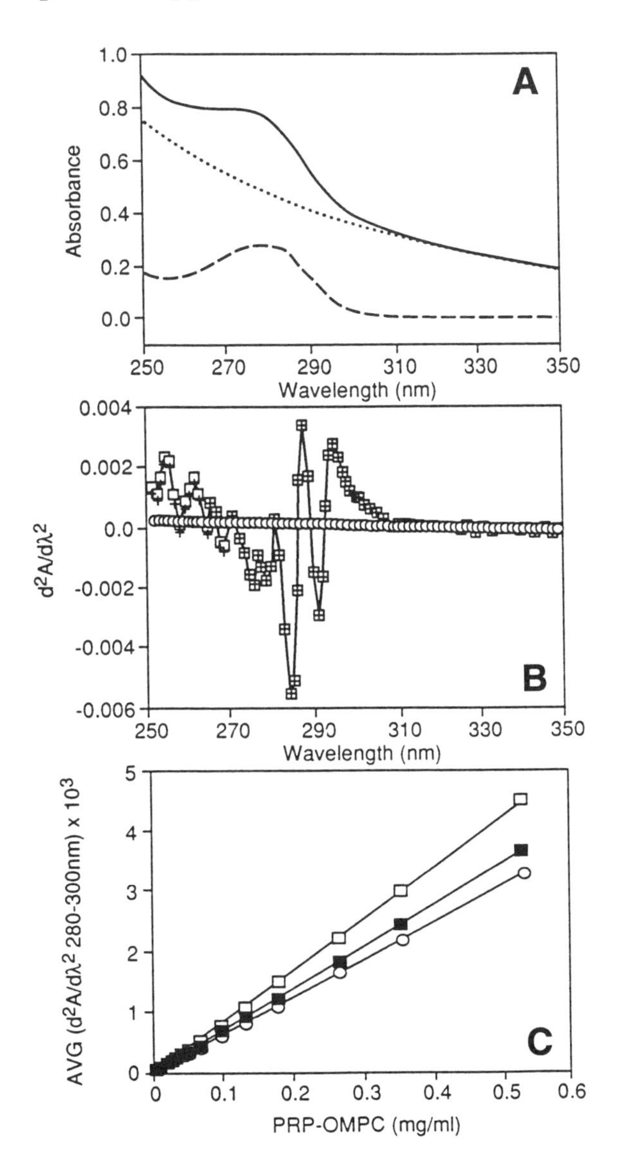

three components shown in (A): original near-UV spectrum (□); light scattering-corrected spectrum (+); and light scattering component (○). (**C**) Linear dependence of the average second derivative peak area in the 280–300 nm range on PRP-OMPC concentration (■). Similar data from highly scattering solutions of aggregated acidic fibroblast growth factor (□) and recombinant hepatitis B surface antigen particles (○) are also shown *(17)*.

obey the Beer-Lambert law and can be integrated to obtain concentration values. An example of a protein second derivative spectrum in the presence of a significant light scattering component is shown in Fig. 2B. As seen in Fig. 2C, the protein concentration determined from integration of the second derivative peaks is in good agreement with the expected values. Other examples of this approach are described in further detail elsewhere *(14–16)*. This method is especially useful when studying protein systems displaying limited aggregation *(17)*.

4.3. Determination of Aromatic Amino Acid Content with Derivative Spectroscopy

The first derivative of the absorbance in the Beer-Lambert (Eq. [4]), is also proportional to the concentration of the sample:

$$dA(\lambda)/d\lambda = d\,[\log I_o(\lambda) - \log I(\lambda)]/d\lambda = 0.43 dI_o(\lambda)/I_o(\lambda)d\lambda - 0.43 dI(\lambda)/I(\lambda)d\lambda = cl\,(d\varepsilon(\lambda)/d\lambda) \quad (7)$$

As indicated above, this also holds for all higher order derivatives although second and fourth derivatives are the most frequently used *(13,18–20)*.

Using various regions of the near-UV spectral range, methods for the quantitative determination of the phenylalanine *(21)*, tyrosine, and tryptophan *(22)* content in proteins have been developed based on these considerations. Employing multiple linear regression, Levine and Federici *(11)* demonstrated that second derivative spectroscopy can be used to measure accurately the aromatic amino acid content in guanidine-HCl denatured protein samples. The strong second derivative peaks of the aromatic amino acids also permit sensitive detection of proteins and other aromatic compounds in the presence of DNA as a consequence of the uniqueness of each component's second derivative spectrum *(23)*. Moreover, since the highly reproducible derivative spectrum potentially provides a unique signature for each protein, the derivative spectra of mixtures containing several proteins can be analyzed by multiple linear regression to find the concentrations of each protein without resorting to actual physical separation of the components *(24)*.

5. Analysis of Protein Unfolding/Refolding by UV Spectroscopy

The wavelength positions of the near-UV bands of the aromatic amino acids in proteins depend on the polarity of their microenvironments *(1)*.

Table 1
Molar Extinction Coefficients
of Selected Biological Chromophores

Chromophore	λ (nm)	ε dm³/*M*/cm)
N-Acetyl-L-tryptophanamide	280	5390
N-Acetyl-L-tyrosinamide	275	1390
N-Acetyl-L-tyrosinamide	280	1185
Tryptophan in proteins	280	5540
Tyrosine in proteins	280	1480
L-cysteine (-S-S-)	280	134
N-Acetyl-L-phenylalanine ethyl ester	257.5	195
DNA (per base)	258	6600
RNA (per base)	258	7400
Cu^{II} in azurin, *P. fluorescens*	781	3200
Fe^{II}-heme in cytochrome C, human	550	27,700
FMN in flavodoxin *C. pasteurianum*	443	9100
FAD in pyruvic dehydrogenase *E. coli*	460	12,700

Wavelengths listed are λ_{max} except for tryptophan and tyrosine in proteins, where a 280 nm wavelength is conventionally used for quantitative analyses.

Consequently, dielectric changes in the immediate vicinity of an aromatic side chain may produce spectral shifts making it difficult to predict the exact UV spectrum of a particular protein from its amino acid composition. However, the use of spectra of "average" protein bound chromophores may still be useful in this regard *(9,25)*. Such spectra of tryptophan and tyrosine side chains in proteins can be determined by matrix analysis of spectra of globular proteins of known amino acid composition (Fig. 1). The normal (zero-order) spectra of tryptophan and tyrosine residues appear to be substantially red shifted compared to the spectra of *N*-acetyl-L-tryptophanamide (*N*-AcTrpNH_2) and *N*-acetyl-L-tyrosinamide (*N*-AcTyrNH_2) in aqueous buffer. In addition, a marked increase in molar absorptivity of tyrosine residues is evident. The molar absorptivity values of Trp and Tyr at 280 nm, obtained previously by the same method, are used for calculation of protein extinction coefficients according to Eq. (5) (Table 1). The second derivative patterns based on these deconvoluted spectra show, as expected, the same extent of shifts, but a peak magnitude increase only occurs in the Trp spectrum. In addition, strong second derivative signals from Phe residues reveals substantial red shift relative to *N*-acetyl-L-phenylalanine ethyl ester (Fig. 1).

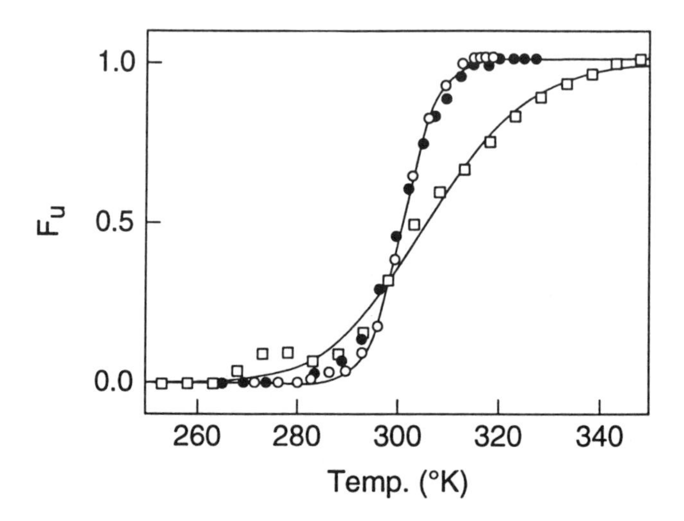

Fig. 3. Thermal denaturation of RNase A in 50% methanol, apparent pH 3.0, as monitored by absorbance at 286 nm (○), fluorescence emission (excitation at 280 nm, emission at 311.5 nm) (●), and circular dichroism (ellipticity at 220 nm) (□) (adapted from Fink and Painter, 1987 *[26]*).

These spectral shifts provide the basis to employ UV spectroscopy to study protein structural transitions.

5.1. Normal (Zero-Order) Spectral Analysis of Protein Unfolding

Since protein UV spectra generally undergo a small shift to lower wavelengths on unfolding owing to the increased polarity of the surrounding medium, one may simply monitor the absorbance at a single wavelength in a region of maximal slope (typically, around 290 nm) to follow this unfolding process. This approach was extensively used in the past to follow alterations in the microenvironments of tryptophan and tyrosine residues with moderate changes usually considered diagnostic of tertiary structure perturbation. For example, Fig. 3 shows unfolding of RNase where the tertiary structure-sensitive fluorescence intensity and near-UV absorption denaturation curves are superimposed. These spectral shifts, however, are usually much smaller than those observed by intrinsic fluorescence or circular dichroism, but can be measured with sufficient precision and accuracy that protein folding/unfolding experiments employing UV spectroscopy can be quite routine.

5.2. Difference Spectroscopy

The principle described above has also been extensively utilized by recording difference spectra, that is, the difference between two protein spectra of identical concentration, one of which is subject to structural perturbation induced by a chaotropic agent or extremes of pH or temperature. Although essentially identical to the former approach, this arrangement minimizes experimental noise. On perturbation, the spectra of aromatic amino acids can manifest shifts in peak positions and/or a change in the molar absorptivity as depicted in Fig. 4A in which the effect of a change in solvent polarity and ionization state on the spectrum of tyrosine is illustrated. After perturbation, the shifted absorption spectrum can be denoted as $\varepsilon' = \varepsilon(\lambda + d\lambda)$ and expressed as a Taylor series *(2)*:

$$\varepsilon' = \varepsilon(\lambda + \Delta\lambda) = \varepsilon(\lambda_o) + \Delta\lambda/1!\,(\delta\varepsilon/\delta\lambda)_{\lambda o} + (\Delta\lambda)^2/(2!)\,(\delta^2\varepsilon/\delta\lambda^2)_{\lambda o} + \ldots \quad (8)$$

If $\Delta\lambda$ is sufficiently small, only the first two terms in the expansion are significant and the difference between the two spectra near λ_o can be approximated by:

$$\varepsilon(\lambda + \Delta\lambda) - \varepsilon(\lambda) = \Delta\lambda(\delta\varepsilon/\delta\lambda) \quad (9)$$

Thus, the difference spectrum will resemble the first derivative spectrum (Eq. [7]). In practice, changes in molar absorptivities are as significant as spectral shifts and, in combination, can result in quite complex difference spectra.

Solvent perturbation difference spectroscopy techniques *(27)* have been extensively used to study the influence of various solutes on the accessibility of protein aromatic side chains. In a typical experiment, a protein solution in the sample beam is supplemented with small amounts of a nondenaturing solvent (e.g., ethylene glycol, ethanol), whereas an identical solution is placed in the reference beam with the protein not in contact with the perturbing solute (Fig. 4). Special double cell (tandem) cuvets are available for this purpose. Since the structure of the protein does not change on addition of small amounts of appropriate solvents, only the accessible chromophores are spectrally altered. By comparison with difference spectra of model aromatic compounds obtained under analogous conditions, the extent of solvent exposure can be estimated *(28)*. Further information can be obtained by the use of solutes of different sizes and by structurally altering the protein with a second solute. Temperature and pH can be employed in a similar manner *(1,27,29)*.

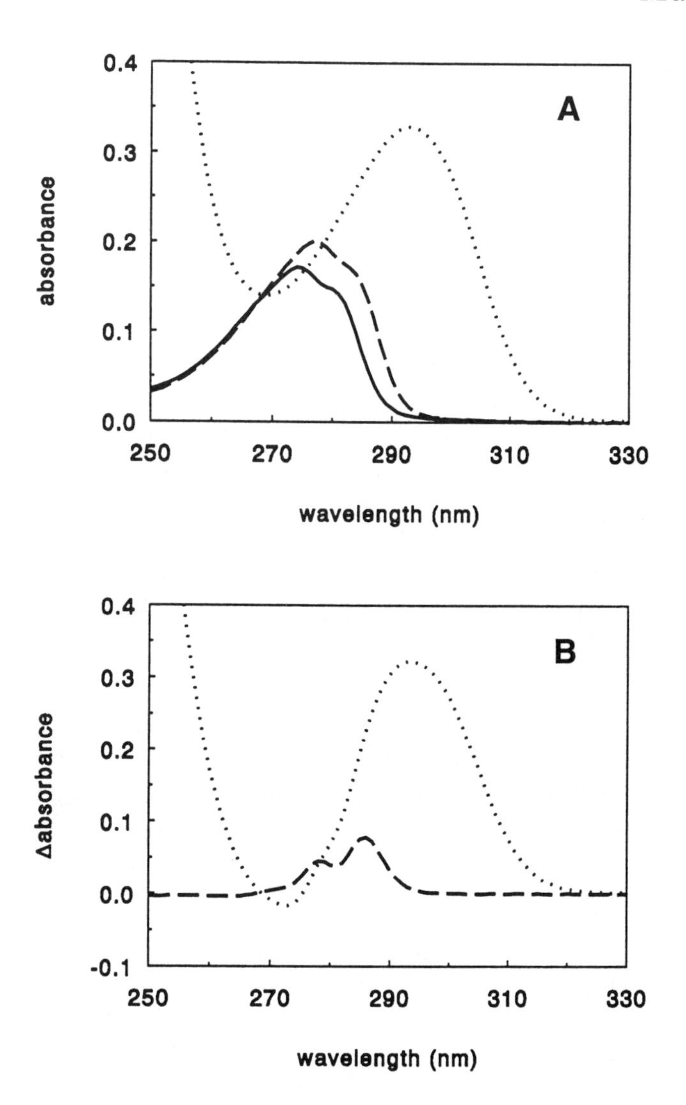

Fig. 4. (**A**) Near-UV spectra of L-tyrosine in aqueous buffer, pH 7 (—), 80% ethylene glycol (---), and aqueous buffer, pH 12 (...). (**B**). Difference between spectra in 80% ethylene glycol (---), aqueous buffer, pH 12 (...), and the spectrum in aqueous buffer, pH 7.

Interactions between associating proteins can also be studied by varying protein concentration employing cuvets of varying pathlength to maintain an optimal absorbance range. For example, the association of insulin *(30)*, lysozyme *(31)*, and hemoglobin *(32)* have all been studied with this method.

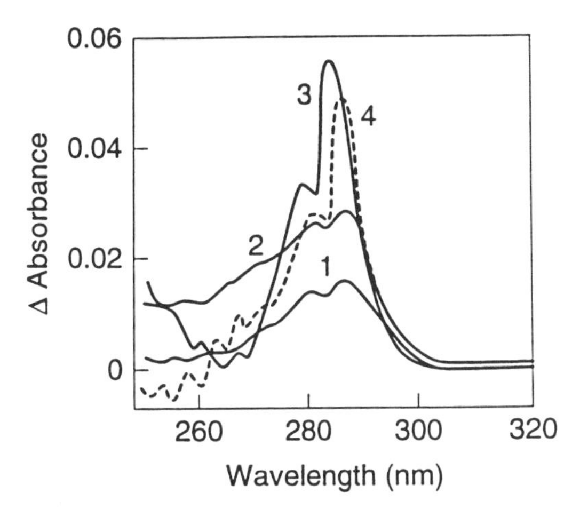

Fig. 5. Solvent perturbation difference spectra of human serum albumin and its respective model compound mixtures obtained with 20% glycerol as perturbant. *1*, native protein, pH 7.6; *2*, acid-isomerized protein, pH 2.1; *3*, disulfide-cleaved protein in 8*M* urea in the presence of 0.02*M* thioglycolate, pH 4.5; *4*, model compound mixture consisting of $1.5 \times 10^{-3}M$ *N*-acetyl-L-tyrosine ethyl ester, $5 \cdot 10^{-5}M$ *N*-acetyl-L-tryptophan ethyl ester, and $2.8 \cdot 10^{-3}M$ phenylalanine ethyl ester, pH 3.5 (adapted from Herskovits and Laskowski, 1962 *[27]*).

Another useful application of UV difference spectroscopy takes advantage of the large shift and increase in molar absorptivity of tyrosine residues on ionization that occurs at pH >10 for solvent exposed tyrosine side chains (Fig. 4) *(33)*. Titration of protein samples frequently permits two populations of tyrosine side chains to be observed: one readily ionizing at pH 10 (solvent exposed) and the other ionizing only when protein structure is disrupted (buried Tyr residues). Furthermore, the sensitivity of the ionization state of Tyr residues to their microenvironment frequently permit individual residues to be recognized by their unique titration behavior *(34)*.

By employing more extreme conditions, UV difference spectroscopy can be applied to the unfolding process itself. Thus, either increasing the concentration of chaotropic agent (e.g., guanidine-HCl, urea), temperature, or altering the pH can be used to induce structural disruption to highly disordered states. In general, changes in each of the distinct difference spectroscopic bands should be superimposable if unfolding transitions reflect two-state behavior *(35)*. Figure 5 illustrates the use of solvent perturbation difference spectroscopy in studying microenvironments of aromatic residues in various forms of human serum albumin *(27)*. In addition, difference spectroscopy can be used in conjunction with spectroscopic methods sensitive to secondary structure content (e.g., CD and FTIR).

5.3. Second Derivative Analysis

The application of derivative spectroscopy to the analysis of protein structural transitions permits one to circumvent many of the limitations encountered with zero-order spectra *(13)*. As indicated above, broad spectral features from sources such as cysteine absorbance, light scattering, and the absorbance of some buffer components and cofactors can be effectively suppressed. On the other hand, spectral features originating from aromatic entities are usually strongly represented (Fig. 1). Although the dependence of UV second derivative spectra on protein conformation have been utilized for some time, the first widely applicable method was only recently described by Ragone et al. *(36)*. These authors employed the ratio between two second derivative peak-to-peak distances to evaluate the average polarity of tyrosine residues in proteins. For example, the technique has been used to detect protein conformational transitions in myoglobin, where subtle structural changes were detected at intermediate denaturant concentrations *(37)*. The development of similar methods for tryptophan residues has proven difficult because of the generally lower degree of solvent exposure of indole rings in proteins and to the decreased sensitivity of their UV bands to solvent polarity. However, least-squares fits of sets of gradually shifted second derivative spectra of *N*-AcTrpNH_2 and *N*-AcTyrNH_2 to experimental second derivative spectra of proteins appear capable of resolving contributions from these residues *(38)*. For example, changes in both zero order and second derivative spectra of the FK506-binding protein (FKBP) are observed on binding of a specific ligand, FK520 (Fig. 6). The corresponding second derivative spectra of the Trp and Tyr components, calculated from the best least-squares fit of gradually shifted model spectra, are shown in Fig. 7. These deconvoluted spectra closely resemble the spectra of the corresponding model compounds (Fig. 1).

A phenylalanine second derivative peak is centered at about 260 nm that is only minimally overlapped by the tryptophan and tyrosine second derivative peaks (Fig. 1). Deconvolution of the Phe contribution permits precise determination of the position of the major negative peak near

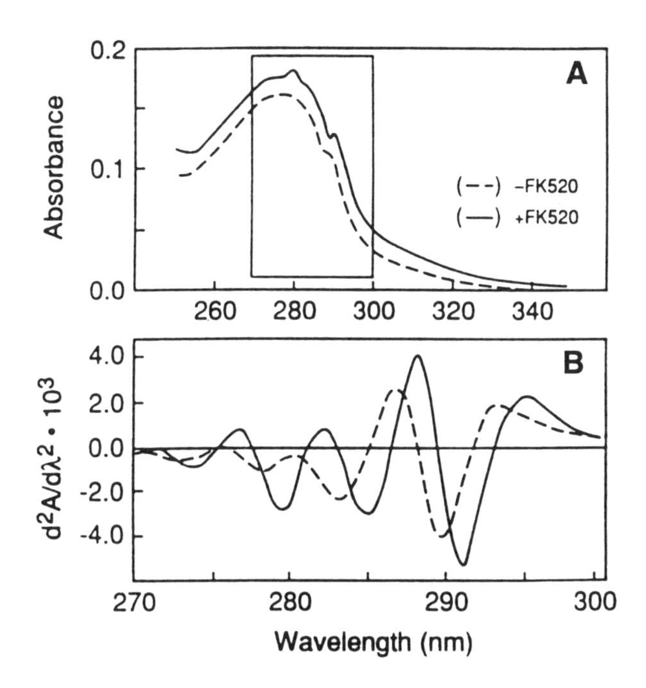

Fig. 6. Normal (**A**) and second-derivative (**B**) spectra of FKBP alone (---) and FKBP-FK520 complex (—). Part of the zero-order spectrum (A) used in the second-derivative calculation (B) is enclosed in a box. Increases of peak magnitudes in both zero-order and second-derivative spectra on binding of the ligand are evident.

258 nm by fitting of a polynomial to second derivative spectra in this region. Subsequent numerical determination of the position of this peak can be accomplished with a precision approaching 0.01 nm *(39)*. As shown in Fig. 8, both the equilibrium and kinetics of unfolding of a protein can be independently monitored by a simple analysis of consecutively acquired derivative spectra in terms of changes in the environment of their Phe and Tyr (and Trp) residues.

In a manner analogous to difference solute-perturbation spectroscopy, derivative spectroscopy can also be used to detect more subtle alterations in the microenvironments of aromatic amino acids. Since relative amplitudes and positions of several derivative peaks are usually compared, this approach has the advantage that baseline problems originating from light scattering can be minimized and that less precise control of concentration is necessary.

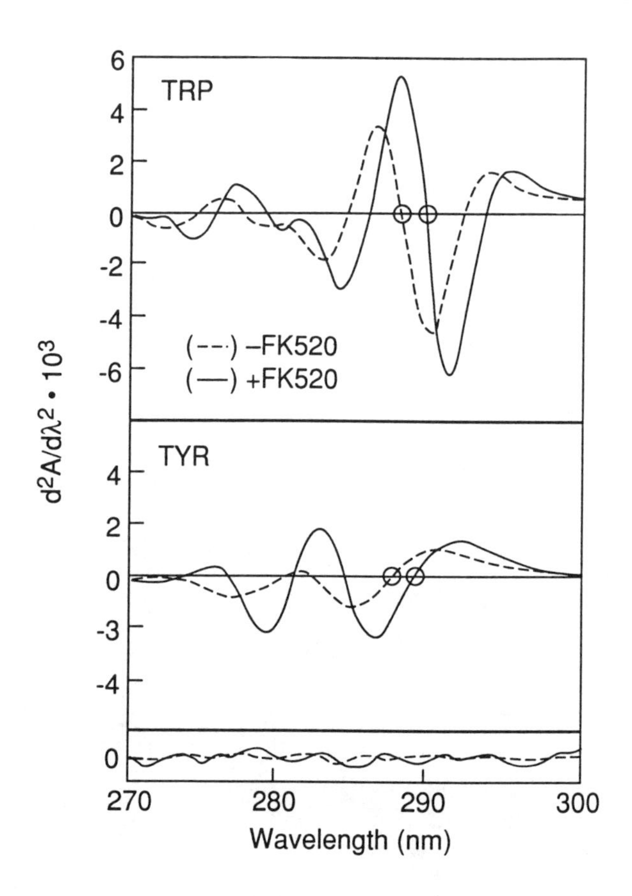

Fig. 7. Deconvoluted spectra of the tryptophan (**Top**) and tyrosine components (**Bottom**) from an FKBP spectrum (---) and the spectrum of an FKBP-FK520 complex (—). The intercepts of the deconvoluted second-derivative spectra with the *x*-axis, used to describe band positions (*see* Fig. 6), are marked with circles. Both tryptophan and tyrosine residues appear to undergo red shifts and an increase in peak magnitude. Residuals of fits to spectra of FKBP-alone (---) and the FKBP-FK520 complex (—) are plotted in the lowest panel employing the same scale.

6. Far UV Spectroscopy

This spectral range is usually considered to extend from 250 nm down to the lower limit of available instrumentation, which is usually between 200 and 180 nm. Nitrogen flushing (to reduce O_2 which absorbs in this region) and optically transparent solutes are necessary to work below 200 nm. Experimental considerations for such experiments are described

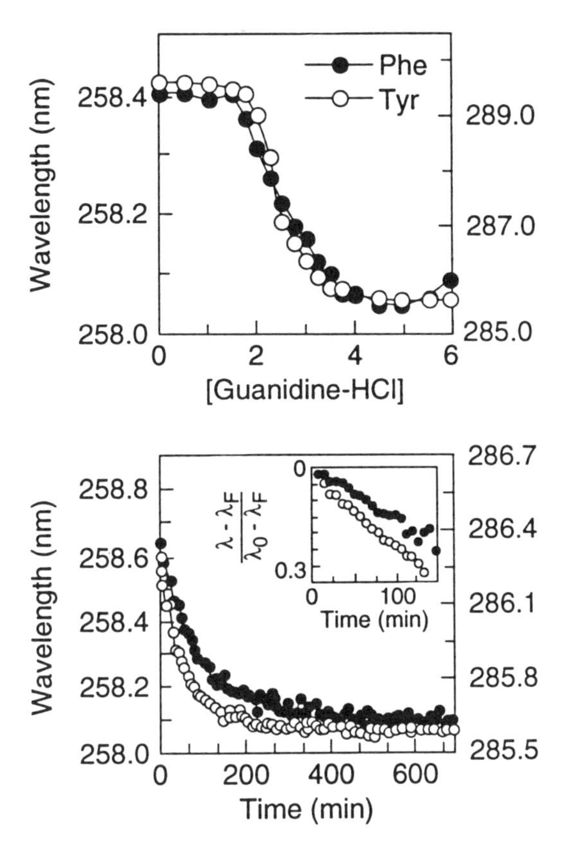

Fig. 8. (**Top**) Plots of the positions of the Phe peak (●) and the positions of the intersection of the second derivative spectrum with the wavelength axis near 288 nm (○) of human serum albumin vs guanidine-HCl concentration. Values from 4–6*M* guanidine-HCl for the Phe curve (●) were determined as an average position of the intersections near 256 and 259 nm *(39)*. (**Bottom**) The kinetics of unfolding of superoxide dismutase (SOD) in 6*M* guanidine-HCl, 50 m*M* sodium cacodylate, pH 6.8 at 30°C. Upper curve (●), the positions of the second-derivative negative peak near 258 nm (Phe). Lower curve (○), the positions of the intersections of the second-derivative spectrum with wavelength axis near 286 nm (Tyr). (**Inset**) plots of In($[\Delta - \Delta_F]/[\Delta_0 - \Delta_F]$) vs time, where Δ is the position of a spectral band on the wavelength axis at a given time, Δ_0 is the position at $t = 0$, and Δ_F is the final (asymptotic) position at long times *(54)*.

in detail by Rosenheck and Doty *(40)*. Unlike the near-UV region where primarily only the three aromatic amino acids of proteins absorb light, virtually all chemical groups absorb in the far-UV. Because of the

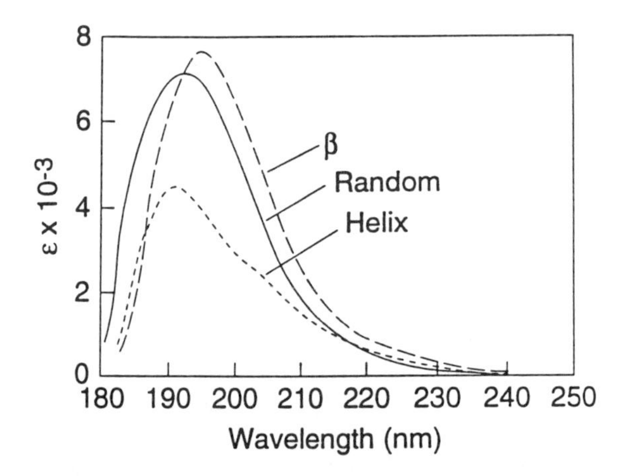

Fig. 9. Far ultraviolet absorption spectra of poly-L-lysine hydrochloride in aqueous solution: random coil; pH 6.0, 25°C; α-helix, pH 10.8, 25°C; β-sheet, pH 10.8, 52°C (adapted fom Rosenheck and Doty, 1961 *[40]*).

plethora of spectral contributions to this region by many common solvents, far-UV spectra have seen only limited application to proteins. The major exception to this statement is the use of peptide bond absorption between 200 and 220 nm for the detection of peptides and proteins in liquid chromatography and capillary electrophoresis. Nevertheless, structural information is contained in this region and it is worth briefly considering further.

It has been known since the early days of protein UV spectroscopy that peptide bonds in the three major types of secondary structure, i.e., α-helix, β-sheet, and random coil, produce distinct spectra in the far-UV region *(41)*. As seen in Fig. 9, these differences seem significant enough to permit them to be used to follow changes in protein secondary structure. As indicated above, however, the presence of spectral contributions from amino acid side chains as well as the intrinsic difficulty in measuring the strong peptide peak itself near 190 nm have discouraged this approach (although, *see* ref. *40*). Side chain contributions in this region can be corrected by a simple subtractive method, but the influences of conformational alterations on any such corrections have yet to be rigorously analyzed. Although the advent of circular dichroism further relegated the use of far-UV absorption as a structural indicator to one of only historical interest, the recent advent of fast scanning and diode-array spectrometers as well as improved data handling capabilities suggest that the UV approach should be reexamined. For example, it should be possible to simultaneously monitor both the near (e.g., 290 nm) and far (e.g., 200 nm) UV, thereby providing a relative measurement of both tertiary and secondary structure, respectively. This could be especially useful in kinetic (including stopped-flow) studies of alternatively folded conformations (e.g., molten globule states) that could, in principle, be performed on-line in a chromatographic mode given multiple pathlengths and carefully selected solvent conditions.

7. Extrinsic Chromophores

In addition to the intrinsic protein chromophores (i.e., the amino acid side chains and the peptide bond), a number of ligands and cofactors display potentially useful UV absorption. The visible absorption bands of such moieties have been employed most often in protein studies, but their UV signals may also be of some utility. The spectra of various protoporphyrin derivatives found in proteins, such as hemoglobin, myoglobin, and the cytochromes, manifest a number of bands in the ultraviolet region. Both zero-order *(42)* and derivative spectra *(43)* have been used to study structural and functional aspects of these chromophores. The signals are often conformationally sensitive and may be altered by changes in the redox state of chelated metals *(44,45)*.

Metal ions are also commonly found directly complexed to proteins. Although calcium and zinc ions are generally transparent, copper and iron exhibit intense, broad bands, as seen in proteins such as superoxide dismutase, azurin, and ferritin. The lack of fine structure in their absorbance spectra, however, allows an essentially complete suppression of their influence by the calculation of second derivative spectra. Metals not normally found in proteins, such as cobalt, terbium, and other lanthanides, can be used to substitute for natural divalent cations and provide intensive and often sharp UV bands. These bands can often provide information about stoichiometry, coordinating side chains, and coordination geometry (for example, *see* ref. *46*). Although not necessarily structurally sensitive themselves, such signals can also be used to follow the formation and destruction of tertiary structure-induced metal binding sites. Although more detailed information concerning metal binding can be provided by nuclear magnetic resonance, electron paramagnetic resonance, and extended X-ray absorption fine structure, the ease and availability of UV-Vis absorption make this a useful technique.

Ultraviolet absorbing nucleic acids are often found associated with a wide variety of proteins. Frequently, the strong near-UV absorbance of the purine and pyrimidine bases (*see* Table 1) dominates the spectra of such complexes. If, however, protein absorbance contributes significantly to the spectrum, the second derivative approach can be used to estimate the relative protein content and follow protein conformational changes *(23)*. In contrast, difference spectroscopy is usually more effective in detecting changes in DNA spectra as a consequence of the well known hypochromic effect *(2)*. It seems fair to say that the utility of high resolution UV spectroscopy, in the context of protein folding and DNA binding, has yet to be fully explored.

8. Instrumental Considerations

The introduction of diode-array detection systems has substantially improved the quality and time of spectral acquisition. Instead of mechanically scanning the wavelength range of interest with monochromatic light, an entire spectrum of light is passed through the sample and, after diffraction by a grating, cast onto an array of photodiodes. Thus, the entire spectrum is typically obtained in less than a second. Moreover, wavelength reproducibility is extremely high since no moving parts are involved. A computer is used to average spectra obtained by multiple measurements and provide noise estimates at each data point. Typically, the light beam has a diameter of <2 mm, allowing spectra to be obtained from samples as small as 50 µL. Conventional scanning spectrophotometers have also been improved considerably with rapid scanning capability that may effectively provide data acquisition times similar to those of diode-array instruments. Although absolute wavelength accuracy and precision may be comprised, frequent calibration with holmium oxide *(47,48)* or benzene vapor can partially offset this disadvantage. Unfortunately, little attention has been paid to far UV measurements below 200 nm and individual modification of many instruments to provide N_2-flushing will be necessary if this region is to be effectively employed.

Rapid improvements in computer speed and memory have made storage and analysis of spectra a trivial task. For instance, derivative spectra were originally obtained by means of special electronic or optical accessories. Savitzky and Golay *(49)*, however, derived equations that allow calculation of derivatives of any order using five or more neighboring data points, thus incorporating smoothing into derivative calculations.

This method provides far superior results when integrated into computer analysis of spectral data. The introduction of powerful computers has also facilitated the use of analytical routines based on matrix algebra. Multiple linear regression, or multicomponent analysis, allow the determination of the relative content of standard spectra in a mixture of up to 12 components *(50,51)*. The more traditional use of absorbance ratios for resolving spectra of mixtures employed only a small fraction of available information and now has been superceded by the former approach. When the spectra of pure components are not known but spectra of several solutions containing different amounts of constituent species can be obtained, singular value decomposition *(52)* can be performed to obtain the spectra of the original components. This technique can also be applied to kinetic data to analyze interconversion between species *(53)*. The use of such programs has become remarkably simple in the user friendly environments provided by modern software. Many computer controlled spectral analyses can be invoked by a single command and completed within seconds. This also allows completely automated thermal unfolding and titration curves of proteins to be obtained with minimal effort.

9. Conclusions

Given the variety of techniques currently available to study protein stability and folding, what does a method that is generally considered to be of low resolution, like UV spectroscopy, have to offer? For example, intrinsic fluorescence is more sensitive and features much larger spectral changes during conformational transitions. Large changes are also often seen in the near and far UV CD, the latter of which also has the additional advantage of providing semiquantitative information about secondary structure content. Nevertheless, recent advances in both instrumentation and data handling capability suggest that UV spectroscopy should be considered as a supplementary method in protein folding studies in certain situations. Potential advantages and unique capabilities of this approach include the following:

1. The availability and convenience of modern UV/VIS absorption spectrophotometers make the approach especially widely available. The precision and accuracy of these instruments allow the intensity and position of peaks to be extremely well defined. Thus, shifts in peak positions of only a few hundredths of a nanometer can be reproducibly measured. In addition, entire spectra can be acquired in a very short period of time, suggesting

that this approach is particularly useful for kinetic studies of protein folding and unfolding. Thus, this offers one method of choice in rapid kinetic (stopped flow) experiments.

2. Rapid scanning monochromators or diode array UV detectors are also now available that permit high resolution spectra to be obtained during elution chromatography. This, in principle, should permit multidimensional studies of protein folding. This idea could be further extended by the use of additional, simultaneous spectroscopic measurements, such as fluorescence, CD, and light scattering using currently available technology.
3. Measurement of a protein's UV spectrum from 190–350 nm gives information about secondary (190–210 nm) and tertiary (250–300 nm) structure as well as aggregation state (light scattering between 300 and 350 nm). Such a high resolution spectrum (typically obtained with different pathlengths for near- and far-UV regions) therefore potentially provides a single comprehensive measure of the structural integrity (stability) of a protein.
4. Detailed derivative spectra of protein can be obtained under conditions that are often present during folding studies (e.g., limited light scattering, presence of strongly absorbing solutes).
5. The method is especially applicable to tyrosine and phenylalanine residues whose intrinsic fluorescence is notoriously insensitive to changes in protein microenvironments. For example, in an ideal protein containing a single tryptophan, tyrosine, and phenylalanine residue, it is possible to simultaneously follow changes in the environment of each residue by monitoring second derivative peaks. Even in the more realistic situation where multiple residues of one or more types of aromatic side chains are present, a very detailed kinetic analysis of populations of residues should be possible by monitoring multiple peaks.

Overall, UV spectroscopy would seem to have much to offer the investigator conducting folding studies of proteins, and may likely see a resurgence of use in the immediate future.

References

1. Wetlaufer, D. (1962) *Adv. Protein Chem.* **17,** 303–390.
2. Cantor, C. R. and Schimmel, P. R. (1980) *Biophysical Chemistry, Part II,* Freeman, San Francisco, pp. 349–408.
3. Johnson, M. J. (1941) *J. Biol. Chem.* **137,** 575–586.
4. Yeh, C. S. (1966) *Microchem. J.* **11,** 229–236.
5. Whitaker, J. R. and Granum, P. R. (1980) *Anal. Biochem.* **109,** 156–159.
6. Scopes, R. K. (1974) *Anal. Biochem.* **59,** 277–282.

7. Lowry, O. H., Rosenbrough, N. J., Farr, A. L., and Randall, R. J. (1951) *J. Biol. Chem.* **193,** 265–275.
8. Bradford, M. M. (1976) *Anal. Biochem.* **72,** 248–254.
9. Mach, H., Middaugh, C. R., and Lewis, R. V. (1992) *Anal. Biochem.* **200,** 74–80.
10. Gill, S. C. and von Hippel, P. H. (1989) *Anal. Biochem.* **182,** 319–326.
11. Levine, R. L. and Federici, M. M. (1982) *Biochemistry* **21,** 2600–2606.
12. Timasheff, S. N. (1966) *J. Colloid Interface Sci.* **21,** 489–497.
13. Fell, A. F. (1983) *Trends Anal. Chem.* **2,** 63–66.
14. Weiser, W. E. and Pardue, H. L. (1983) *Clin. Chem.* **29/9,** 1673–1677.
15. Terada, H., Seki, H., Yamamoto, K., and Kametani, F. (1985) *Anal. Biochem.* **149,** 501–506.
16. Sievert, H.-J. P., Wu, S.-L., Chloupek, R., and Hancock, W. S. (1990) *J. Chromatog.* **499,** 221–234.
17. Mach, H. and Middaugh, C. R. (1993) *BioTechniques* **15,** 240–242.
18. Terada, H., Seki, H., Yamamoto, K., and Kametani, F. (1985) *Anal. Biochem.* **149,** 501–506.
19. Padros, E., Dunach, M., Morros, A., Sabes, M., and Manosa, J. (1984) *Trends Biochem. Sci.* **9(12),** 508–510.
20. Honkawa, T. (1978) *Higher Order Derivative Spectroscopy ADS 112,* Perkin-Elmer, Coleman Instruments Div., Oak Brook, IL.
21. Ichikawa, T. and Terada, H. (1977) *Biochim. Biophys. Acta* **494,** 267–270.
22. Servillo, L., Colonna, G., Balestrieri, C., Ragone, R., and Irace, G. (1982) *Anal. Biochem.* **126,** 251–257.
23. Mach, H., Middaugh, C. R., and Lewis, R. (1992) *Anal. Biochem.* **200,** 20–26.
24. Mach, H., Thomson, J. A., and Middaugh, C. R. (1989) *Anal. Biochem.* **181,** 79–85.
25. Edelhoch, H. (1967) *Biochemistry* **6,** 1948–1954.
26. Fink, A. L. and Painter, B. (1987) *Biochemistry* **26,** 1665–1671.
27. Herskovits, T. T. and Laskowski, J., Jr. (1962) *J. Biol. Chem.* **237,** 2481–2492.
28. Kosen, P. A., Creighton, T. E., and Blout, E. R. (1980) *Biochemistry* **19,** 4936–4944.
29. Donovan, J. W. (1964) *Biochemistry* **3,** 67–74.
30. Lord, R. S., Gubensek, J. A., and Rupley, J. A. (1973) *Biochemistry* **12,** 4385–4392.
31. Holladay, L. A. and Sophianopoulos, A. H. (1972) *J. Biol. Chem.* **247,** 1976–1979.
32. Philo, J. S., Adams, M. L., and Schuster, T. M. (1981) *J. Biol. Chem.* **256,** 7917–7924.
33. Donovan, J. W. (1973) *Methods Enzymol., Part D* **27,** 525–548.
34. Middaugh, C. R., Thomson, J. A., Burke, C. J., Mach, H., Naylor, A. M., Bogusky, M. J., Ryan, J. A., Pitzenberger, S. M., Ji, H., and Cordingley, J. S. (1993) *Protein Sci.* **2,** 900–914.
35. Santoro, M. M. and Bolen, D. W. (1988) *Biochemistry* **27,** 8063–8068.
36. Ragone, R., Colonna, G., Balestrieri, C., Servillo, L., and Irace, G. (1984) *Biochemistry* **23,** 1871–1875.
37. Ragone, R., Colonna, G., Bismuto, E., and Irace, G. (1987) *Biochemistry* **26,** 2130–2134.
38. Mach, H. and Middaugh, C. R. (1994) *Anal. Biochem.,* **222,** 323–331.

39. Mach, H., Thomson, J. A., Middaugh, C. R., and Lewis, R. V. (1991) *Arch. Biochem. Biophys.* **287,** 33–40.
40. Rosenheck, K. and Doty, P. (1961) *Proc. Natl. Acad. Sci. USA* **47,** 1775–1785.
41. Saidel, L. J., Goldfarb, A. R., and Waldman, S. (1952) *J. Biol. Chem.* **197,** 285–291.
42. Zwart, A., van Kampen, E. J., and Zijlstra, W. G. (1986) *Clin. Chem.* **36/2,** 972–978.
43. Lynch, S. R., Sherman, D., and Copeland, R. A. (1992) *J. Biol. Chem.* **267,** 298–302.
44. Zwart, A., Buursma, A., van Kempen, E. J., and Zijlstra, W. G. (1984) *Clin. Chem.* **30/3,** 373–379.
45. Berry, E. A. and Trumpower, B. L. (1987) *Anal. Biochem.* **161,** 1–15.
46. Burke, C. J., Sanyal, G., Bruner, M. W., Ryan, J. A., LaFemina, R. L., Robbins, H. L., Zeft, A. S., Middaugh, C. R., and Cordingley, M. G. (1992) *J. Biol. Chem.* **267,** 9639–9644.
47. Weidner, V. R., Mavrodineanu, R., Mielenz, K. D., Velapoldi, R. A., Eckerle, K. L., and Adams, B. (1985) *J. Res. Nat. Bur. Stand.* **90,** 115–125.
48. *Standard Reference Materials Catalog* (1992/93) US Department of Commerce, Gaithersburg, MD.
49. Savitzky, A. and Golay, J. E. (1964) *Anal. Chem.* **36,** 1627–1639.
50. *AVIV Model 14DS Spectrophotometer Instruction Manual* (1989) AVIV Associates, Inc., Lakewood, NJ.
51. James, G. E. (1980) *Multicomponent Analysis with the HP8450A UV/VIS Spectrophotometer,* Hewlett-Packard Technical Paper, UV-1, Publication No. 23-5953-4751.
52. Noble, B. and Daniel, J. W. (1977) *Applied Linear Algebra,* 2nd ed., Prentice-Hall, Englewood Cliffs, NJ, pp. 323–330.
53. Johnson, W. C., Jr. (1985) in *Physical Optics of Dynamic Phenomena and Processes in Macromolecular Systems* (Sedláček, B., ed.), Walter de Gruyter, Berlin, pp. 493–506.
54. Mach, H., Dong, Z., Middaugh, C. R., and Lewis, R. V. (1991) *Arch. Biochem. Biophys.* **287,** 41–47.

CHAPTER 5

Circular Dichroism

Kunihiro Kuwajima

1. Introduction

Chirality is an important characteristic of protein structure. The secondary structure of a polypeptide backbone, α-helix or β-sheet, is chiral, and the specific tertiary structure of protein side chains is also chiral. Circular dichroism (CD) spectroscopy is a unique optical technique that allows us to detect and quantitate the chirality of the molecular structures, and provides information about both the secondary and the tertiary structures of a protein molecule.

At present, most of the commercially available CD instruments employ piezoelastic birefringence modulators whose modulation frequencies are usually around 50 kHz. CD instruments are thus useful for rapid reaction measurements with a time resolution as short as 0.2 ms *(1–3)*; 10/(modulation frequency) may be taken as the lower limit of the time resolution. The mixing dead time of the stopped-flow method is usually on the order of a few to 10 ms, so that the stopped-flow CD technique is now widely used in studies on the kinetics of structural transitions of proteins and other biological macromolecules *(4–21)*.

The native structure is stabilized cooperatively by various interactions in globular proteins, and the equilibrium unfolding transition is well-approximated as a two-state transition in most cases (Eq. [1]) *(22)*. Figure 1 shows the unfolding transition of lysozyme. The unfolding was induced by a denaturant, guanidine hydrochloride (GuHCl), and measured by the far UV CD band (222 nm) that monitors the backbone secondary structure and by the near UV CD bands (289 and 255 nm) that

From: *Methods in Molecular Biology, Vol. 40: Protein Stability and Folding: Theory and Practice*
Edited by: B. A. Shirley

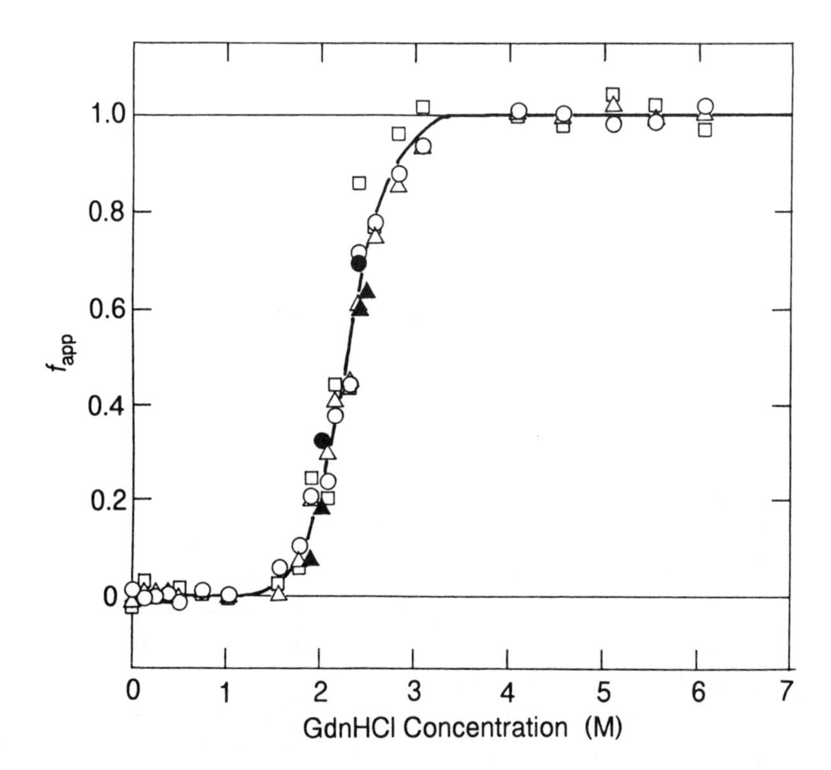

Fig. 1. The GuHCl-induced unfolding transition curve of hen lysozyme measured by CD spectroscopy (pH 1.5 and 4.5°C). The apparent fractional extent of unfolding, f_{app}, was calculated from the ellipticity values at 222 nm (△,▲), 255 nm (○,●) and 287.5 nm (□) using the equation:

$$f_{app} = ([\theta]_N - [\theta])/([\theta]_N - [\theta]_U)$$

where $[\theta]$ represents the observed ellipticity under given conditions, and $[\theta]_N$ and $[\theta]_U$ are the ellipticity values in the N and U states, respectively. The $[\theta]_N$ and $[\theta]_U$ in the transition zone were obtained by extrapolation of the linear dependence of the ellipticity on GuHCl concentration observed in the regions before and after the transition, respectively. Protein concentrations are 0.65 mg/mL for open symbols (△,○,□) and 0.4 mg/mL for filled symbols (▲,●).

monitors the side chain tertiary structure *(23,24)*. The transition curves measured at different wavelengths are coincident with each other, suggesting that unfolding occurs between the two states, the native (N) and unfolded (U) states in the transition zone.

$$N \rightleftharpoons U \quad (1)$$

The above results of lysozyme, however, do not mean that the kinetic refolding of this protein in native condition is also a two-state reaction. The transient accumulation of an intermediate structural state (I) at an early stage in refolding from the U state is now known to be a rather general phenomenon observed in many globular proteins (Eq. [2]) *(25–27).*

$$U \rightleftharpoons I \rightleftharpoons N \quad (2)$$

Kinetic CD measurements are effective in detecting and characterizing such transient intermediates of refolding, and the kinetic CD techniques, including the stopped-flow CD, have been used successfully in recent studies of protein folding. This chapter will thus summarize the practical procedures in these kinetic CD studies.

2. Materials

The use of a denaturant, urea, or GuHCl is almost indispensable in the experimental studies of protein folding. The purity of these denaturants is crucial for CD measurements in the far UV region because the impurities of these compounds strongly absorb the UV light *(28,29).* Absorption spectra of pure and impure samples of aqueous 6*M* GuHCl and a purification procedure for GuHCl have been reported by Nozaki *(29).* The solution of impure urea shows a strong absorbance around 220 nm. At present, highly purified urea and GuHCl useful for the protein folding studies can be purchased commercially. The urea and GuHCl we use are both specially prepared reagent grade from Nacalai tesque, Inc. (Kyoto, Japan). The absorbance values of the 6*M* solutions of these urea and GuHCl tested were 0.6 and 0.5, respectively, at 220 nm and 0.05 for both at 230 nm. The 3*M* solution of "ultrapure" urea from another company, however, showed an absorbance of more than 2 at 220 nm, so the choice of the commercial products of highly purified urea is important. GuHCl stock solutions are stable for months at room temperature. The stock solutions of urea are not so stable, and the decomposition of urea slowly produces cyanate that may react with amino groups of proteins *(28).* Thus, the urea stock solutions (ca. 10*M*) should be freshly prepared and deionized on a mixed-bed ion-exchange column to eliminate cyanate before use. We use Amberlite IR-120B and IRA-402 resin for the mixed-bed column. The concentrations of urea and GuHCl can be determined by refractive index measurements (*28,29*; Chapter 8).

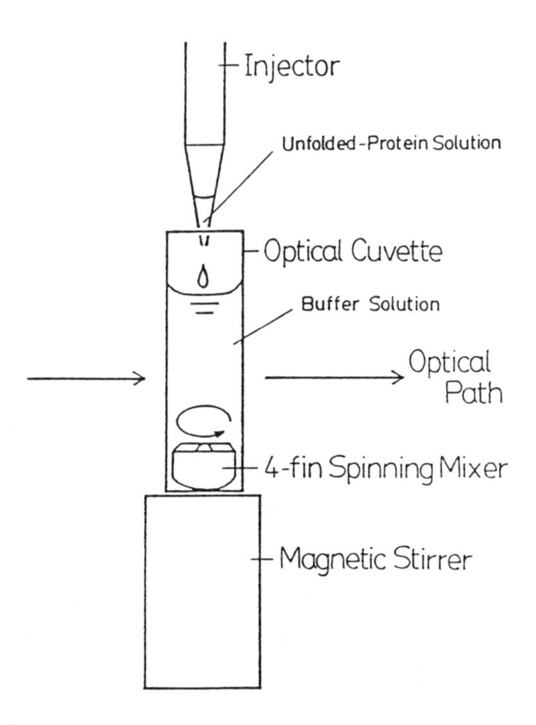

Fig. 2. A simple mixing device for denaturant concentration jump by the use of a quartz cuvet and a magnetic spinner *(23)*.

The protein and buffer solutions are to be freshly prepared in each experiment. We use the following buffer salt and concentration at neutral pH; 10–50 m*M* sodium or potassium phosphate or 50 m*M* sodium cacodylate. Sodium cacodylate was chosen for studies of Ca^{2+}-binding proteins because phosphate forms an insoluble salt with Ca^{2+}. All the solutions for CD measurement should be filtered through a membrane filter (pore size 0.45 μm). It is also important that the solutions used in stopped-flow experiments be degassed to prevent cavitation.

3. Methods

3.1. Kinetic CD Measurements

3.1.1. A Simple Mixing Device

Kinetic CD studies do not always require the stopped-flow technique. Figure 2 shows a device useful for some kinetic CD measurements. It is composed of an injector, a magnetic spinner, and a quartz cuvet with an optical path of 1 cm *(23)*. This type of simple device should not be overlooked. It is possible to realize a mixing dead time of <3 s in denaturant-induced refolding (or unfolding) experiments, if the following requirements are met. First, the dilution of the protein solution must be sufficiently large, usually with a volume ratio of the diluent to the protein solution of at least 20:1. Second, the nozzle of the injector must be in a right position for efficient mixing. For refolding experiments from the U state, the protein solution in concentrated denaturant has a higher density than the diluent, so that it must be injected downward from the nozzle placed just above the surface of the diluent solution, whereas for unfolding experiments, in which the density of the diluent is higher, the nozzle must be immersed in the diluent and placed close to the spinning mixer sitting on the bottom of the cuvet. In the latter case, the nozzle must be made sufficiently thin to prevent diffusion of the solution before injection.

In order to evaluate mixing efficiency and an approximate dead time of the device, a change in the CD signal at 219 nm caused by dilution of a L-(or D-)pantolactone solution with water can be used *(23)*. Because L-(or D-)pantolactone has a CD signal at the wavelength where the secondary structure of a protein is observed, it is useful for calibration of a CD spectropolarimeter for protein structural analysis. We used 0.11% L-pantolactone in 6*M* GuHCl, and this was diluted 20-fold with water. After density fluctuations had disappeared, the CD signal showed a constant value, and the dead time was estimated to be <3 s.

We used this simple mixing device (Fig. 2) for studies on the refolding kinetics of lysozyme and α-lactalbumin *(23,30)*. Typical kinetic traces for the refolding of lysozyme are shown in Fig. 3. Lysozyme was first unfolded at pH 1.5 in 6*M* GuHCl. Refolding was initiated by dilution of the protein solution 20-fold with the diluent solution (0.07*M* NaCl, pH 1.5) at 4.5°C. The kinetic progress curves were measured at three different wavelengths, 222, 250, and 287.5 nm.

3.1.2. Stopped-Flow Method

The stopped-flow method is required when the refolding (or unfolding) reaction of a protein occurs within several seconds. The first measurement of the stopped-flow CD was made in 1974 by Anson and Bayley, who observed CD changes brought about by interactions between DNA and dye molecules in the visible region *(1,2)*. The stopped-flow CD study on the conformational transition of proteins was first made

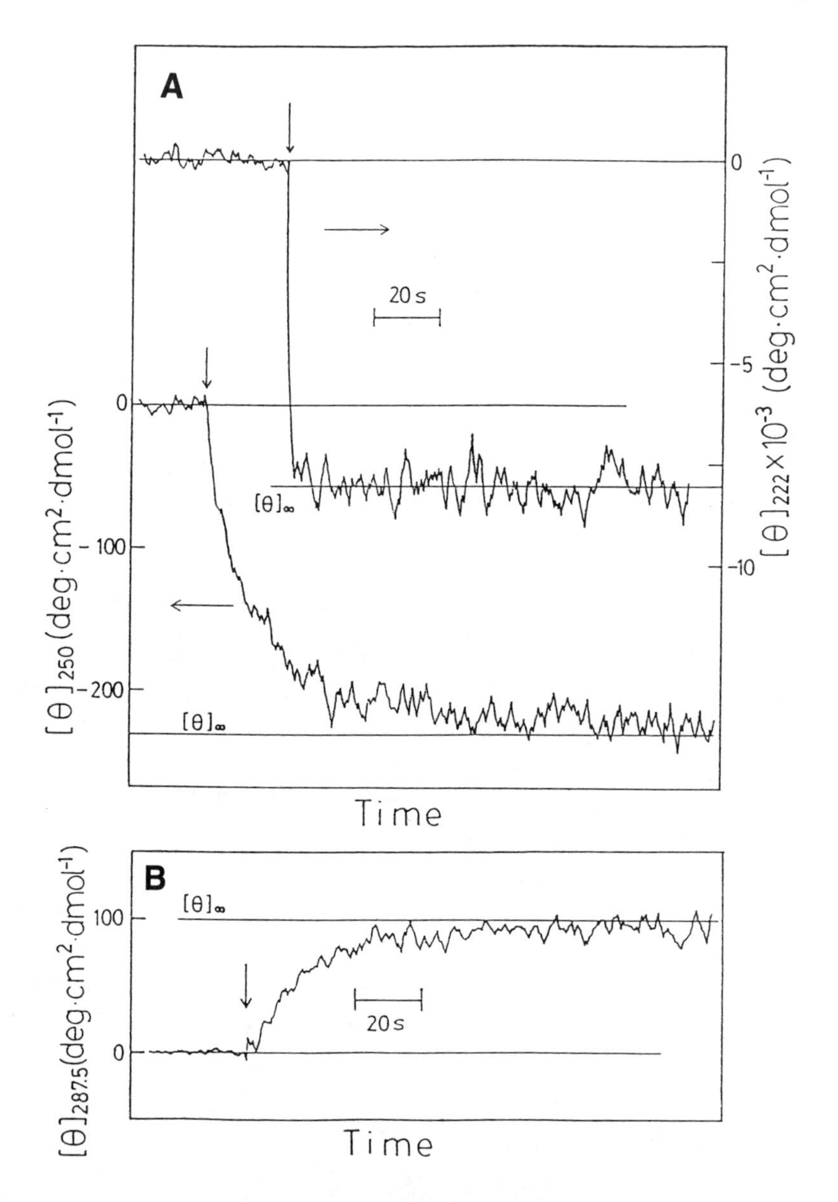

Fig. 3. Kinetic progress curves of refolding of lysozyme measured by the CD ellipticities at different wavelengths (pH 1.58 and 4.5°C) *(23)*. The refolding was initiated by a concentration jump of GuHCl from 6.0 to 0.3*M* in the mixing device of Fig. 2. The wavelengths are: **(A)** 250 and 222 nm and **(B)** 287.5 nm. Protein concentrations are: (A) 0.6 mg/mL for 250 nm and 0.04 mg/mL for 222 nm, and (B) 0.6 mg/mL.

in 1977 by Nitta et al., who measured the near-UV (aromatic) CD changes caused by the acid transition and refolding of α-lactalbumin *(7)*. In 1978, Luchins and Beychok reported the stopped-flow CD observation in the far UV region, which monitored rapid secondary structure changes of hemoglobin caused by its acid transition and refolding *(17)*. In both of the latter two studies, the stopped-flow technique was used to create a pH jump of the solutions.

The stopped-flow CD method has recently been utilized for investigating the refolding reactions from the U state in concentrated urea or GuHCl for a number of globular proteins *(31–45)*. The refolding reactions from the U state are realized by a concentration jump of the denaturant, and special care is required to perform the stopped-flow experiments. First, the jump to a native condition from the U state requires a wide concentration jump (i.e., a high dilution of the protein solution), so that a usual 1:1 stopped-flow mixing device is almost useless, and a mixing apparatus with a high dilution ratio (10:1 or even more) must be used. Second, in order to keep a constant mixing ratio of the two solutions and also to keep stability of the solution after mixing, a mixing apparatus with two driving syringes, in which the solution delivery is controlled by a plunger in each syringe, is superior to a mixing apparatus directly driven by gas pressure; the plungers in the former mixing apparatus may be driven pneumatically, however. The mixing ratio is accurately determined by the ratio of cross-sectional areas of the two syringes. Third, the density and viscosity are very different between both the protein solution in concentrated denaturant and the refolding buffer solution without denaturant, so that the efficiency of mixing of the stopped-flow mixer is crucial for complete mixing of the two solutions. Usually, single mixing is not sufficient and at least double mixing may be required.

Figure 4 shows a block diagram of our stopped-flow CD system and a diagrammatic representation of the stopped-flow mixer and the observation cell *(31)*. We have used this system for studies of refolding intermediates of globular proteins. The structure enclosed by a broken line in the block diagram is a Jasco J-500A CD spectropolarimeter. The stopped-flow mixing apparatus is specially designed by Unisoku Inc. (Osaka, Japan) and is composed of a mixing driver controlled by a three-way magnetic valve, drive syringes, reservoirs, a mixer, and an observation cell. This mixing apparatus is based on a slit-type mixer originally

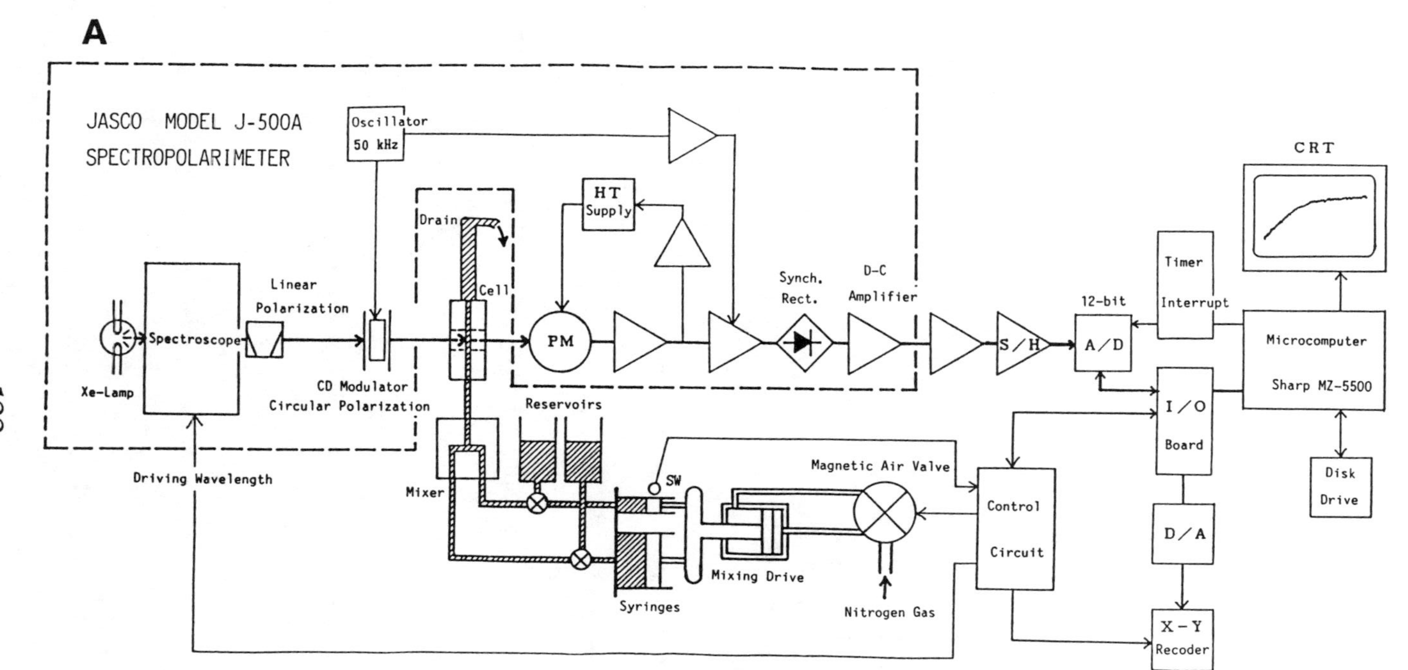

Fig. 4. The block diagram of our stopped-flow CD system **(A)** and the diagrammatic representation of the stopped-flow mixer and the observation cell (B) *(31)*. (A) The part enclosed by a broken line corresponds to a Jasco model J-500A spectropolarimeter.

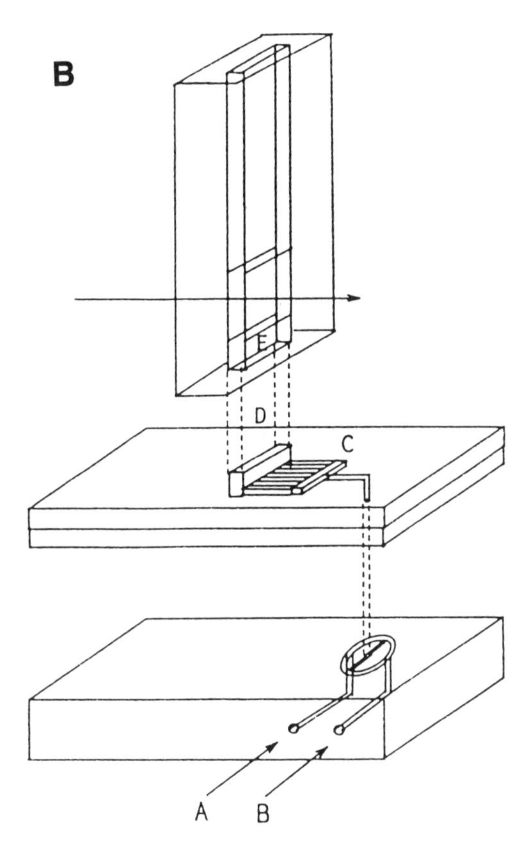

(B) The two solutions from inlets A and B are mixed in a double two-jet mixing part, divided into 8 flow lines at C and then expelled through a slit (D) (1 × 8 mm) to enter a flat observation cell (E) with laminar flow.

designed for stopped-flow X-ray scattering by Kihara and Nagamura *(46)*, and it was modified for CD measurements.

The mixing efficiency of a stopped-flow CD apparatus can be examined by diluting a solution of an optically active compound in concentrated denaturant with a buffer solution without denaturant. Mo et al. reported an example of the mixing artifact that was observed when D-pantolactone in 6*M* urea was diluted tenfold with a buffer solution *(38)*. D-Pantolactone has a negative CD band at 219 nm, and they monitored the CD at 222 nm during the mixing. The artifact disappeared when the viscosities of the two solutions were matched by the addition of sucrose, so they concluded that the artifact was caused by a viscosity difference. This mixing artifact can be eliminated by improving the mixing efficiency of the stopped-flow mixer. However sufficient the mixing efficiency is, there is another artifact (a buoyancy effect) caused by a density difference between the solutions before and after the mixing. This artifact occurs in a time range from several seconds to a few minutes, depending on the design of the mixer, and is very difficult to eliminate. Thus, the simple device depicted in Fig. 2 may be more useful than the stopped-flow apparatus for measurements of slow reactions that occur in a time range longer than a few minutes.

There are a number of methods to determine the dead time of a stopped-flow mixing apparatus *(47–50)*. We have adopted the method reported by Paul et al. *(48)*. In this method, a coupled two-step reaction of 5,5'-dithiobis(2-nitrobenzoic acid) (Ellman's reagent) with excess thioglycerol is used. The reaction produces a progress curve composed of two superimposed exponentials. The reaction is monitored by absorbance changes at 440 nm. In order to measure the absorbance by our CD spectropolarimeter, the direct current component of the photomultiplier output, which is proportional to the transmittance of light, is monitored. The reaction curve is thus first given as the transmittance vs time, and the transmittance-to-absorbance conversion is made in a microcomputer attached to the spectropolarimeter (Fig. 4). Because the ratio of the amplitudes of the two exponential processes is defined by the ratio of the rate constants, the dead time of mixing can be determined with a single mixing experiment. The dead time of our mixing apparatus is 15 ms. It is not difficult to realize a dead time of a few milliseconds in a stopped-flow mixing apparatus. However, in general, the shorter the dead time, the more serious the buoyancy effect, and the worse the mixing efficiency. Thus, the choice of the mixing dead time may be a matter of compromise.

Figure 5 shows a kinetic progress curve of dihydrofolate reductase measured by the ellipticity at 220 nm by our stopped-flow CD system *(36)*. The refolding was initiated by a concentration jump of urea from 4.5 to 0.4*M* at pH 7.8 and 15°C. Both the unfolded protein and the refolding buffer solution contained 10 m*M* potassium phosphate as a buffer salt and 1 m*M* 2-mercaptoethanol.

3.2. Analysis of the Refolding Curves

3.2.1. A Transient Intermediate (I) in Refolding

In Fig. 3, the CD change at 222 nm reflects a change in the backbone secondary structure, and the changes at 250 and 287.5 nm reflect the side

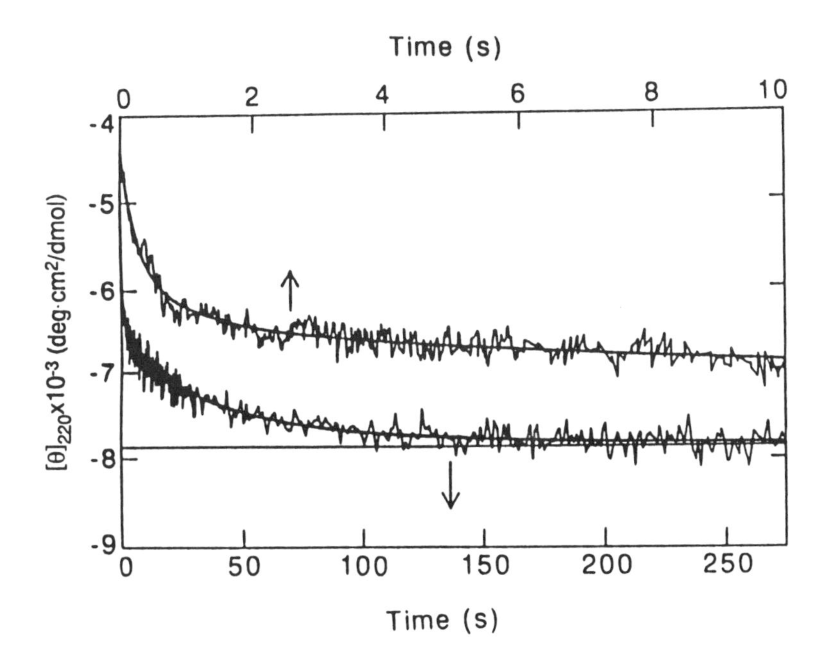

Fig. 5. Kinetic progress curves of *E. coli* dihydrofolate reductase measured by the ellipticity at 220 nm (pH 7.8 and 15°C) *(36)*. The refolding was initiated by a concentration jump of urea from 4.5 to 0.4*M* in the stopped-flow CD system of Fig. 4. The kinetics are shown in different time scales, and a thick solid line represents the theoretical curve by a five-exponential fit *(36)*. The protein concentration was 0.12 mg/mL.

chain tertiary structure. At 250 and 287.5 nm, the total change in the CD specta from the U state to the N state was observed kinetically in refolding. At 222 nm, however, most of the CD change occurs in the burst phase of refolding, that is, within the dead time of the measurement (<3 s). The results clearly demonstrate the transient accumulation of the I state that has folded secondary structure (Eq. [2]), although the equilibrium unfolding transition of the protein is well represented by the two-state model (Fig. 1) *(23,24)*. Chaffotte et al. recently reexamined the refolding kinetics of lysozyme by means of the stopped-flow CD technique *(42)*. The I state has been shown to accumulate within 4 ms. Similarly, the results of Fig. 5 show the formation of the secondary structure in the burst phase of refolding in dihydrofolate reductase *(36)*.

Characterization of the I state is important for elucidation of the folding mechanism of globular proteins. Measurements of the kinetic

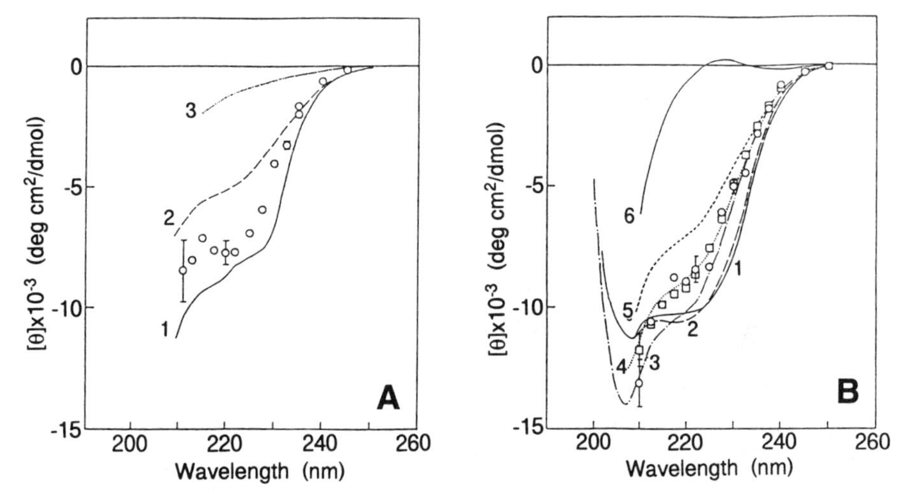

Fig. 6. The CD spectra of the I state for 6 different proteins compared with the spectra of the N and U states and the spectra of the partially unfolded states if present. The CD values obtained by extrapolation to the zero time of the kinetic refolding curves are shown by open circles and squares. **(A)** Lysozyme: 1, the N state; 2, the thermally unfolded state; 3, the U state by GuHCl *(23)*. **(B)** α-Lactalbumin: 1, the N state (Ca^{2+} bound form); 2, the N state (apo form without Ca^{2+}); 3, the acid state (pH 2.0); 4 and 5, the thermally unfolded state; 6, the U state by GuHCl *(23)*. **(C)** Parvalbumin: 1, the N state; 2, the acid state (pH 2.8); 3, the U state by GuHCl *(32)*. **(D)** β-Lactoglobulin: the N state; 2, the U state by GuHCl *(31)*. **(E)** Staphylococcal nuclease: 1, the N state; 2, the acid state (pH 2.1); 3, the U state by urea *(34)*. **(F)** Dihydrofolate reductase: 1, the N state; 2, the U state by urea *(36)*.

progress curves of refolding at various wavelengths will give us the CD spectrum of the I state. Because the formation of the I state occurs in the burst phase much faster than the subsequent folding events, the wavelength dependence of the CD value extrapolated to the zero time of the observed refolding curve may correspond to the CD spectrum of the I state. We have investigated the CD spectra of the I state for a number of globular proteins by the kinetic CD measurements and the spectra were compared with the spectra in the N and U states of the proteins and also with the spectra of partially unfolded states at equilibrium if present *(23,31–36)*. The results are summarized in Fig. 6. In many globular proteins, an appreciable amount of the backbone secondary structure is restored at an early stage of refolding within a millisecond time range.

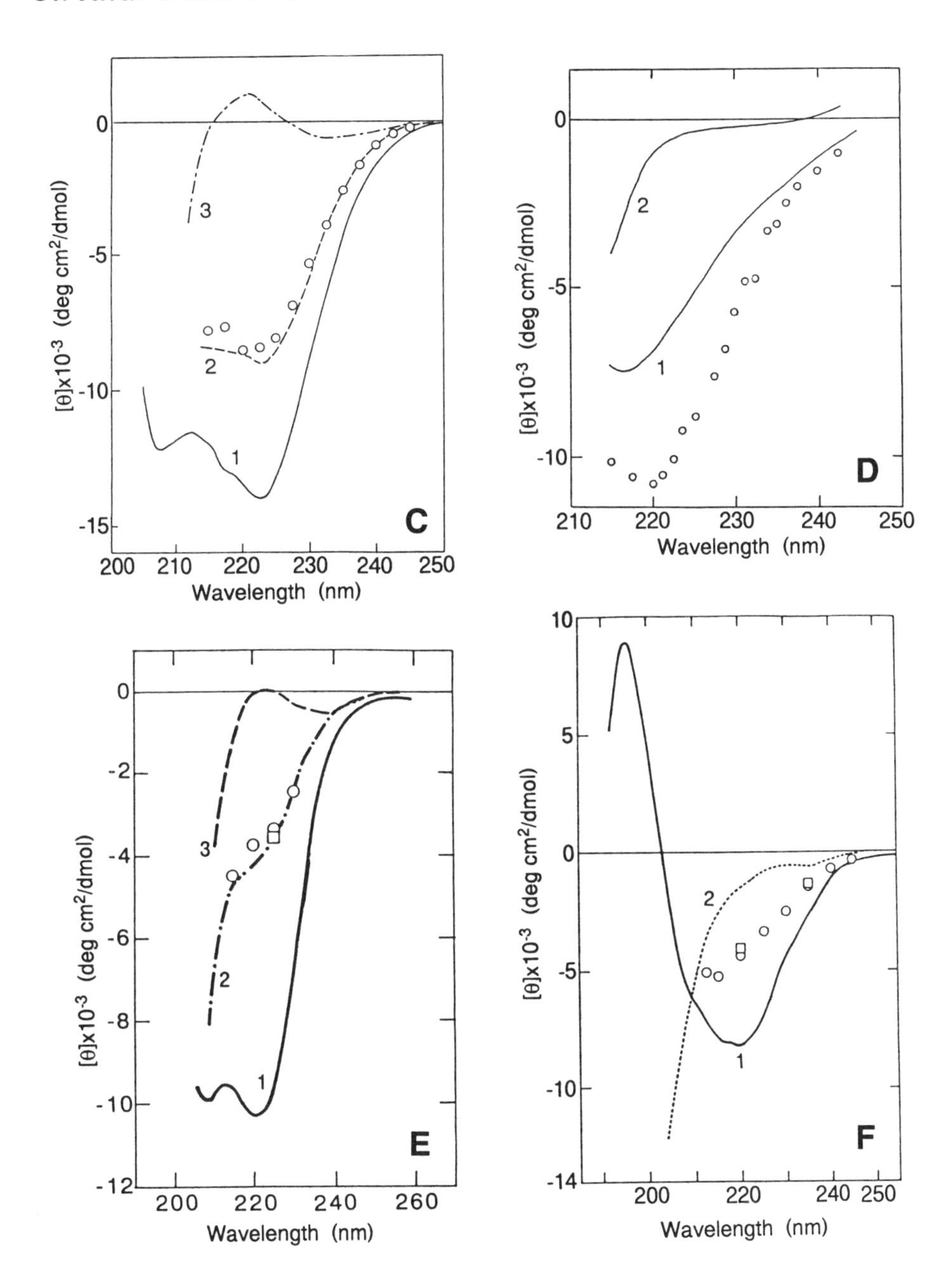

Fig. 6 *(continued)*

How much of the secondary structure is restored in the burst phase, however, largely depends on protein species. In view of the ellipticity change at 222 nm, the I state has secondary structure comparable to that in the N state in lysozyme and α-lactalbumin *(23,42,43)*. Similar observations have also been reported in pepsinogen *(51)* and ribonuclease T1 *(52)*. In β-lactoglobulin, the CD intensity in the I state is larger than in the N state, apparently suggesting that the secondary structure is more extensive in the I state *(31)*. A similar phenomenon was observed in the refolding of lysozyme at neutral pH, but in this case, the increase in the CD intensity occurs in the first observable phase that has a time constant of 16 ms *(42,43)*. It has been suggested that this change in CD may arise from a change in the state of disulfide bonds or from transient formation of some kind of nonnative interactions. In staphylococcal nuclease, on the other hand, the CD value attained in the burst phase of refolding is only 30% of the CD value in the N state *(34)*.

3.2.2. Stability of the I State

The formation of secondary structure in the burst phase of refolding is a very rapid process occurring within a few milliseconds, so that the rapid preequilibrium between U and I is established at a very early stage in the refolding reaction. Therefore, the unfolding transition curve of the I state can be obtained by measuring the refolding reactions at varying concentrations of a denaturant and by investigating the dependence of the burst-phase CD spectrum on the denaturant concentration. In this manner, we have studied the unfolding transitions of the I state for lysozyme, α-lactalbumin, and several other proteins (Fig. 7) *(30–36)*.

When the unfolding from the I state is represented by a two-state transition between I and U, the stabilization free energy, ΔG_{IU}, of the I state can be estimated from the abovementioned unfolding transition curve by the equation:

$$\Delta G_{IU} = G_I - G_U = -RT\ln\,[I]/U] \quad (3)$$

Here, G_I and G_U are the free energies of the I and U states, respectively, and [I] and [U] are the fractions of the respective state, which can be obtained from the unfolding transition curve once the CD values for the pure I and U states have been evaluated. The ΔG_{IU} thus obtained is shown in Fig. 8 as a function of the denaturant concentration and compared with ΔG_{NU} *(25)*.

Most globular proteins, including lysozyme, show a two-state unfolding transition between N and U at equilibrium (Fig. 1). A typical rela-

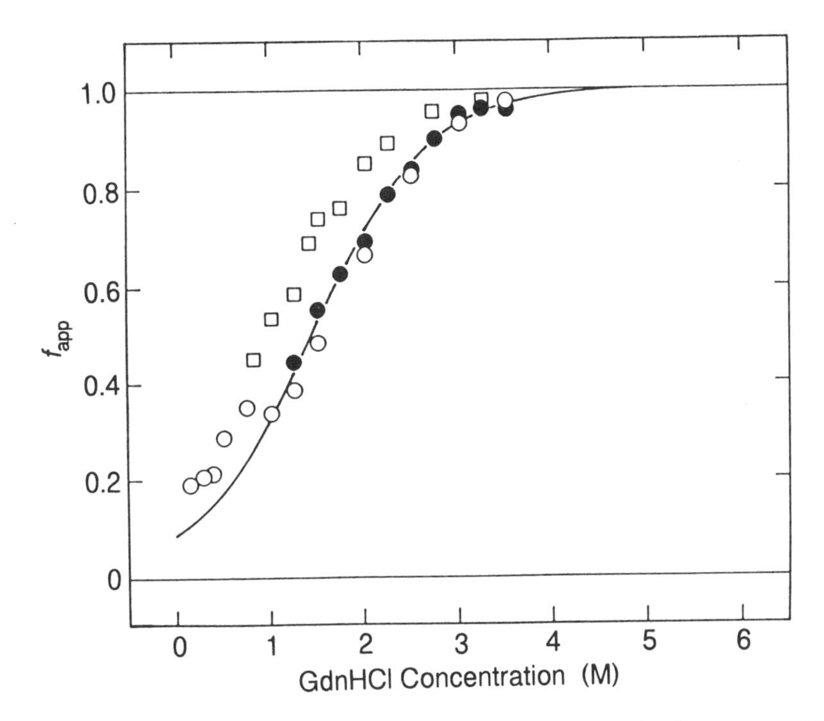

Fig. 7. The unfolding transition curves of the I state for lysozyme (□) (pH 1.5 and 4.5°C) and α-lactalbumin (○,●) (pH 7.0 and 4.5°C) *(30)*. The apparent fractional extent of unfolding, f_{app}, of the I-U transition was obtained from the dependence on GuHCl concentration of the transient CD spectra observed in the burst phase of kinetic refolding from the U state, on the assumption of the two-state transition between I and U. The ellipticity of the pure I state, $[\theta]_I$, is assumed to be identical to $[\theta]_N$. A solid line shows the equilibrium unfolding transition of the MG state of α-lactalbumin at pH 7.0 and 4.5°C, which coincides with the transition of the I state.

tionship between ΔG_{IU} and G_{NU} in these proteins is shown in Fig. 8A. Although the I state is more stable than the U state in the absence of denaturants, the N state is much more stable than the I state, which precludes the population of the I state at equilibrium. During the transition from N to U, the U state becomes more stable than the I state ($\Delta G_{IU} >> 0$) much before the transition midpoint of the $N \rightleftharpoons U$ transition ($\Delta G_{NU} = 0$), so that there is eventually no contribution of the I state in the equilibrium unfolding transition. The I state can be observed only transiently at an early stage of kinetic refolding in native conditions.

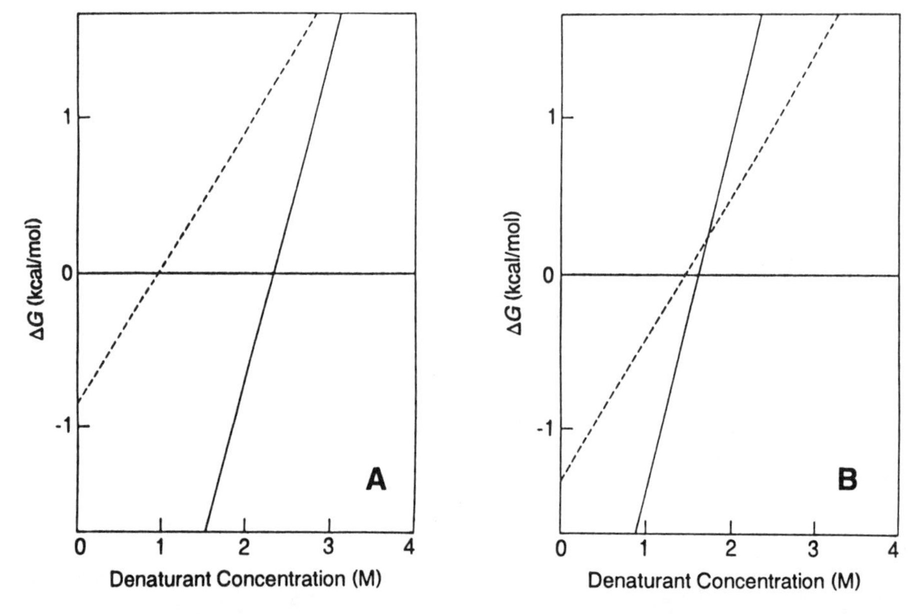

Fig. 8. The stabilization free energy as a function of denaturant concentration for the N state (solid line) and for the I state (broken line) *(25)*. **(A)** The equilibrium unfolding of the N state is represented by a two-state transition. **(B)** A three-state transition involving the MG state will be observed at equilibrium.

3.2.3. Relation to the Molten Globule State

A number of globular proteins, including α-lactalbumin and carbonic anhydrase, are known to show the equilibrium unfolding transition that does not obey the two-state rule but exhibits a compact intermediate that has an appreciable amount of secondary structure *(25–27,53–56; see also* Chapter 15). Previously, such an unfolding transition was thought to be exceptional. The intermediate state has, however, similar characteristics among different proteins, so that the state is thought to be a common state of globular proteins at present. This intermediate structural state of globular proteins is called the molten globule (MG) state *(57; see also* Chapter 15). For α-lactalbumin and several other proteins, the MG state is shown to be identical with the I state observed at an early stage of refolding *(23,25,30,53–56; see also* Chapter 15). Figure 8B shows the plots of ΔG_{IU} and ΔG_{NU} against the concentration of a denaturant for a protein that shows the equilibrium MG state. The stability of the I state is comparable to the stability of the U state at the midpoint of the $N \rightleftharpoons U$ transition (1.5*M* in Fig. 8B), so that there are three states, N, I, and U, in the transition zone (the three-state transition).

Lysozyme and α-lactalbumin are homologous to each other, and their native three-dimensional structures are essentially identical *(58)*. It is, thus, not easy to understand the fact that one of such two proteins shows a typical two-state unfolding, whereas the other shows a three-state unfolding transition. The results of Fig. 8 demonstrate that the difference in the unfolding behavior between the two proteins is brought about by differences in the relative stabilities of the I and N states between the proteins. The unfolding behavior at equilibrium appears to be different between the proteins, but this phenomenon is interpreted in a unified way once the refolding intermediates of the proteins have been detected and characterized by the kinetic CD measurements *(25)*.

4. Notes

1. Table 1 summarizes recent studies of refolding intermediates of globular proteins by means of the stopped-flow CD techniques. The stopped-flow mixing apparatus of Bio-Logic in France, and Unisoku (Osaka) and Jasco (Tokyo) in Japan were used in these studies.
2. The absorption of the UV light by buffer salts or denaturants increases the photomultiplier voltage of a spectropolarimeter and decreases the signal-to-noise ratio in CD measurements. Reliable CD measurements are practically impossible at a photomultiplier voltage higher than 0.6 kV. The choice of the buffer salts that are transparent in a wavelength region interest and the use of highly purified urea or GuHCl are thus important. Phosphate, cacodylate, and Tris buffers are acceptable for measurements in the peptide and aromatic regions, but the buffers that contain carboxyl groups strongly absorb the UV light in the peptide region. When a chelator is to be used, [ethylenebis(oxyethylenenitrilo)]tetraacetic acid (EGTA) is superior to EDTA, because EGTA is more transparent in the peptide region.
3. The shortest limit of the wavelength accessible in the stopped-flow CD experiments of refolding is 212 nm when we use a path length of the observation cell of 1 mm or longer (Table 1). This limitation is caused by strong absorption of the far UV light by the residual denaturant in the solution. However pure the denaturant may be, it absorbs the UV light below 220 nm. The wavelength region may be extended by the use of a shorter path length, and it might be possible to measure the refolding kinetics at a wavelength below 200 nm when using a sufficiently short path length (0.1 mm or even shorter) and a modern spectropolarimeter. CD measurements below 200 nm may be important for quantitative analysis of the secondary structure of folding intermediates *(59,60)*.

Table 1
Stopped-Flow CD Studies of Kinetic Refolding of Globular Proteins

Authors	Protein	Denaturant	CD instrument	Wavelength, nm	Optical path	Mixing apparatus	Dead time	Ref.
Kuwajima, Ikeguchi and Sugai (1985)	α-Lactalbumin and lysozyme	GuHCl	Jasco J-500A	210–300	1 cm	Manual mixing (Fig. 2)	2–3 s	*(23,30)*
Kuwajima et al. (1987)	β-Lactoglobulin and cytochrome *c*	GuHCl	Jasco J-500A	219,222.5 293,420	1 mm and 4 mm	Stopped-flow apparatus (specially designed, Unisoku)	10–20 ms	*(31)*
(1988)	Parvalbumin	GuHCl		215–245				*(32)*
(1990)	Tryptophan synthase β_2 subunit	Urea		225, 285				*(33)*
(1991)	Staphylococcal nuclease	Urea		217–230				*(34,35)*
(1991)	Dihydrofolate reductase	Urea		212–245				*(36)*
Gilmanshin and Ptitsyn (1987)	α-Lactalbumin	Urea	Jasco J-41A	226, 272	3 mm	Stopped-flow apparatus	20 ms	*(37)*
Holtzer et al. (1991)	Tropomyosin	Urea	Jasco J-500A	222	2 mm and 1 cm	Jasco model SFC-5 stopped-flow apparatus	40 ms	*(38,39)*
Goldberg et al. (1992)	C-terminal domain of tryptophan synthase β_2 subunit	GuHCl	Jobin-Yvon CD6	222	5 mm	Bio-Logic model SFM3 stopped-flow apparatus	4 ms	*(40)*
(1992)	Cytochrome *c*	GuHCl		217–289				*(41)*
(1992)	Lysozyme	GuHCl		214–290				*(42)*
Dobson et al. (1992)	Lysozyme	GuHCl	Jobin-Yvon CD6	225–289	—	Bio-Logic model SFM3 stopped-flow apparatus	2 ms	*(43)*
Kiefhaber et al. (1992)	Ribonuclease T1	GuHCl	Jobin-Yvon CD6	225	—	Bio-Logic model SFM3 stopped-flow apparatus	15 ms	*(44)*
Mann and Matthews (1993)	*E. coli* trp aporepressor	Urea	Aviv 62DS	215–240	1.5 mm	Bio-Logic model SFM3	4 ms	*(45)*

Acknowledgments

Most of the kinetic CD studies of the author's group described in this article were done at the Department of Polymer Science, Faculty of Science, Hokkaido University, with which he was previously affiliated. He acknowledges S. Sugai (present address, Soka University), other collaborators, and former students of Hokkaido University.

The author is also grateful to R. L. Baldwin, M. E. Goldberg, A. L. Fink, and C. R. Matthews for sending him their manuscripts before publication.

References

1. Bayley, P. M. and Anson, M. (1974) *Biopolymers* **13,** 401–405.
2. Anson, M. and Bayley, P. M. (1974) *J. Phys. E* **7,** 481–486.
3. Baechinger, H. P., Eggenberger, H. P., and Haenisch, G. (1979) *Rev. Sci. Instrum.* **50,** 1367–1372.
4. Bayley, P. M. (1981) *Prog. Biophys. Mol. Biol.* **37,** 149–180.
5. Pflumm, M., Luchins, J., and Beychok, S. (1986) *Methods Enzymol.* **130,** 519–534.
6. Labhardt, A. M. (1986) *Methods Enzymol.* **131,** 126–135.
7. Nitta, K., Segawa, T., Kuwajima, K., and Sugai, S. (1977) *Biopolymers* **16,** 703–706.
8. Tabushi, I., Yamamura, K., and Nishiya, T. (1978) *Tetrahedron Lett.* 4921–4924.
9. Hatano, M., Nozawa, T., Murakami, T., Yamamoto, T., Shigehisa, M., Kimura, S., Takakuwa, T., Sakayanagi, N., Yano, T., and Watanabe, A. (1981) *Rev. Sci. Instrum.* **52,** 1311–1316.
10. Sano, Y. and Inoue, H. (1979) *Chem. Lett.* 1087–1090.
11. Hasumi, H. (1980) *Biochim. Biophys. Acta* **626,** 265–276.
12. Kihara, H., Takahashi, E., Yamamura, K., and Tabushi, I. (1982) *Biochim. Biophys. Acta* **702,** 249–253.
13. Kawamura-Konishi, Y. and Suzuki, H. (1988) *Biochem. Biophys. Res. Commun.* **156,** 348–354.
14. Kawamura-Konishi, Y., Chiba, K., Kihara, H., and Suzuki, H. (1992) *Eur. Biophys. J.* **21,** 85–92.
15. Takeda, K. (1982) *Bull. Chem. Soc. Japan* **55,** 1335–1339.
16. Takeda, K. (1985) *Biopolymers* **24,** 683–694.
17. Luchins, J. and Beychok, S. (1978) *Science* **199,** 425,426.
18. Leutzinger, Y. and Beychok, S. (1981) *Proc. Natl. Acad. Sci. USA* **78,** 780–784.
19. Erard, M., Burggraf, E., and Pouyet, J. (1982) *FEBS Lett.* **149,** 55–58.
20. Labhardt, A. M. (1984) *Proc. Natl. Acad. Sci. USA* **81,** 7674–7678.
21. Salerno, C., Crifo, C., and Strom, R. (1984) *Eur. J. Biochem.* **139,** 275–278.
22. Tanford, C. (1970) *Adv. Protein Chem.* **24,** 1–95.
23. Kuwajima, K., Hiraoka, Y., Ikeguchi, M., and Sugai, S. (1985) *Biochemistry* **24,** 874–881.
24. Ikeguchi, M., Kuwajima, K., and Sugai, S. (1986) *J. Biochem. (Tokyo)* **99,** 1191–1201.
25. Kuwajima, K. (1989) *Proteins* **6,** 87–103.
26. Kuwajima, K. (1992) *Curr. Opin. Biotechnol.* **3,** 462–467.

27. Matthews, C. R. (1993) *Annu. Rev. Biochem.* **62,** 653–683.
28. Pace, C. N. (1986) *Methods Enzymol.* **131,** 266–280.
29. Nozaki, Y. (1972) *Methods Enzymol.* **26,** 43–50.
30. Ikeguchi, M., Kuwajima, K., Mitani, M., and Sugai, S. (1986) *Biochemistry* **25,** 6965–6972.
31. Kuwajima, K., Yamaya, H., Miwa, S., Sugai, S., and Nagamura, T. (1987) *FEBS Lett.* **221,** 115–118.
32. Kuwajima, K., Sakuraoka, A., Fueki, S., Yoneyama, M., and Sugai, S. (1988) *Biochemistry* **27,** 7419–7428.
33. Goldberg M. E., Semisotnov, G. V., Friguet, B., Kuwajima, K., Ptitsyn, O. B., and Sugai, S. (1990) *FEBS Lett.* **263,** 51–56.
34. Sugawara, T., Kuwajima, K., and Sugai, S. (1991) *Biochemistry* **30,** 2698–2706.
35. Kuwajima, K., Okayama, N., Yamamoto, K., Ishihara, T., and Sugai, S. (1991) *FEBS Lett.* **290,** 135–138.
36. Kuwajima, K., Garvey, E. P., Finn, B. E., Matthews, C. R., and Sugai, S. (1991) *Biochemistry* **30,** 7693–7703.
37. Gilmanshin, R. I. and Ptitsyn, O. B. (1987) *FEBS Lett.* **223,** 327–329.
38. Mo, J., Holtzer, M. E., and Holtzer, A. (1991) *Biopolymers* **31,** 1417–1427.
39. Mo, J., Holtzer, M. E., and Holtzer, A. (1992) *Biopolymers* **32,** 1581–1587.
40. Chaffotte, A. F., Cadieux, C., Guillou, Y., and Goldberg, M. E. (1992) *Biochemistry* **31,** 4303–4308.
41. Eloeve, G. A., Chaffotte, A. F., Roder, H., and Goldberg, M. E. (1992) *Biochemistry* **31,** 6876–6883.
42. Chaffotte, A. F., Guillou, Y., and Goldberg, M. E. (1992) *Biochemistry* **31,** 9694–9702.
43. Radford, S. E., Dobson, C. M., and Evans, P. A. (1992) *Nature* **358,** 302–307.
44. Kiefhaber, T., Schmid, F. X., Willaert, K., Engelborghs, Y., and Chaffotte, A. (1992) *Prot. Sci.* **1,** 1162–1172.
45. Mann, C. J. and Matthews, C. R. (1993) *Biochemistry* **32,** 5282–5290.
46. Nagamura, T., Kurita, K., Tokikura, E., and Kihara, H. (1985) *J. Biochem. Biophys. Methods* **11,** 277–286.
47. Tonomura, B., Nakatani, H., Ohnishi, M., Yamaguchi-Ito, J., and Hiromi, K. (1978) *Anal. Biochem.* **84,** 370–383.
48. Paul, C., Kirschner, K., and Haenisch, G. (1980) *Anal. Biochem.* **101,** 442–448.
49. Brissette, P., Ballou, D. P., and Massey, V. (1989) *Anal. Biochem.* **181,** 234–238.
50. Matsumura, K., Enoki, Y., Kohzuki, H., and Sakata, S. (1990) *Jpn. J. Physiol.* **40,** 567–571.
51. McPhie, P. (1982) *Biochemistry* **21,** 5509–5515.
52. Kiefhaber, T., Quaas, R., Hahn, U., and Schmid, F. X. (1990) *Biochemistry* **29,** 3061–3070.
53. Ptitsyn, O. B. (1987) *J. Prot. Chem.* **6,** 273–293.
54. Ptitsyn, O. B. (1992) in *Protein Folding* (Creighton, T. E., ed.), Freeman, New York, pp. 243–300.
55. Barrick, D. and Baldwin, R. L. (1993) *Biochemistry* **32,** 3790–3796.
56. Barrick, D. and Baldwin, R. L. (1993) *Prot. Sci.* **2,** 869–876.

57. Ohgushi, M. and Wada, A. (1983) *FEBS Lett.* **164,** 21–24.
58. Acharya, K. R., Stuart, D. I., Walker, N. P. C., Lewis, M., and Phillips, D. C. (1989) *J. Mol. Biol.* **208,** 99–127.
59. Johnson, W. C., Jr. (1990) *Proteins* **7,** 205–214.
60. Venyaminov, S. Y., Baikalov, I. A., Wu, C. S. C., and Yang, J. T. (1991) *Anal. Biochem.* **198,** 250–255.

Chapter 6

Infrared Spectroscopy

C. Russell Middaugh, Henryk Mach, James A. Ryan, Gautam Sanyal, and David B. Volkin

1. Introduction

Recently, renewed interest has been generated in the use of infrared spectroscopy to analyze protein structure *(1)*. Although initial applications have emphasized attempts to quantitatively determine secondary structure contents, it is clear that this method offers several unique features applicable to the analysis of protein stability and folding/unfolding processes. These include potentially high resolution, sensitivity, and applicability to a wide variety of physical states of target macromolecules. In this chapter, we will discuss features of vibrational spectroscopy of particular relevance to the study of protein folding with emphasis on the type of information that can be obtained, including potential limitations of the technique.

2. Chromophores

The functionality most frequently employed in studies of protein conformation employing IR spectroscopy is the peptide bond. The vast majority of work has been done in the Amide I region between 1600 and 1700/cm. Absorption of IR light by peptide bonds in this region arises primarily from CO stretching vibrations with minor contributions from CN stretching and CCN deformations *(2,3)*. Deuteration causes small shifts in the position of peaks in this region when D_2O is used as a solvent and a "prime" symbol is used to denote such measurements (Amide I').

From: *Methods in Molecular Biology, Vol. 40: Protein Stability and Folding: Theory and Practice*

The Amide II region (1500–1600/cm) contains signals primarily from N-H bending (and to a lesser extent from C-N stretching) *(2,3)*. This region is also occasionally used for conformational analysis, especially because of its sensitivity to the exchange of protons for deuterons. Amide bands at higher and lower frequency (the Amide A and B and III–VII bands) are of weaker intensity or are less well characterized and have been employed less frequently, but each contains unique features that may make them useful in certain situations.

In most cases, the Amide I band will be the spectral feature of greatest utility for protein folding/unfolding studies and we will therefore emphasize this region in further discussion. Assignments of various secondary structure types to absorption bands between 1600 and 1700/cm is based on a combination of normal mode calculations, analysis of synthetic polypeptides under various environmental conditions, and detailed studies of proteins of known secondary structure *(4)*. Currently accepted assignments are summarized in Table 1. Based on such considerations, α-helices appear to absorb between 1650 and 1658/cm, although bands below 1650/cm have been attributed to α-helix and recent calculations suggest a signal may also be present below 1640/cm *(5,6)*. The assignment of 3_{10}-helices is still uncertain with absorption at both 1639–1640 and 1662–1663/cm proposed *(6)*. The left-handed α-helix has also been suggested to produce a signal in the 1665–1670/cm region *(7,8)*.

Extended chains (β-sheet) usually give one or more absorption bands between 1620 and 1640/cm (or lower) as well as a weaker band between 1670 and 1680/cm *(4)*. Although the complexity of these signals suggests that additional structural information is contained in the number and location of these bands, there is as of yet no consensus on their interpretation. Turns and bends are thought to be reflected in a series of peaks above 1660/cm. Although studies of well defined turns in model peptides suggest that information about the nature of these bends is present, the complexity of this region in actual proteins has inhibited a more definitive analysis *(9)*.

The difficulties of assignment in the Amide I region are probably best illustrated with reference to more random types of structure *(6)*. Unordered polypeptide is generally thought to absorb primarily in the 1640–1650/cm region *(4)*. The unperturbed peptide bond (e.g., *N*-methyl acetamide), however, gives rise to a peak at 1653/cm *(3)*. Furthermore, the appearance of significant absorbance between 1650 and 1660/cm by

Table 1
Summary of Key Assignments in the Amide I Region of Proteins

	Frequency, cm^{-1}
Secondary structure	
α-Helix	1650–1658
	1630–1640 (proposed)
3_{10}-Helix	~1640 (proposed)
	~1662 (α-aminobutyric acid containing peptides)
Left-handed α-helix	1660–1670 (poly-β-benzyl-aspartate)
Poly(L-Pro) helix	~1630
	1630, 1644, 1660 (collagen) *(65)*
Extended chain	1670–1680
	1620–1640 (multiple components)
Turns	1680–1700 (multiple components)
Loops and disordered regions	1645–1655 (multiple components?)
Amino acid side chains	
Asn/Gln/Arg (ionized)	1660–1685
Lys (ionized)/α-NH_3^+	1620–1640
Tyr (ionized)/His (unionized, very weak)/α-COO^-	1590–1605

proteins known to be completely lacking in α-helical structure has also caused peaks in this region to be assigned to random chain conformations *(10,11)*. Although the increased sensitivity of unordered regions to deuteration can sometimes be used to reduce this ambiguity, it is often the case that assignments of bands to helical and unordered conformations are equivocal at best.

Despite these difficulties, the Amide I region after suitable resolution enhancement *(see below)* has been effectively used in the analysis of protein secondary structure content *(4,12–16)*. Both curve fitting and pattern recognition procedures have been employed with some success, although each method has been subject to significant criticism. Although the information from such an analysis is extremely useful in protein folding/unfolding studies, it is not absolutely necessary. Detailed considerations of these types of analyses are reviewed elsewhere *(6)* and considered further below.

It is important to realize that certain amino acid side chains also contribute signals to the Amide I region (Table 1). Both Asn and Gln and ionized Arg absorb in the 1660–1690/cm region, whereas the charged forms of Lys and Tyr have weaker peaks near 1630 and 1600/cm, respectively. An extensive discussion of the IR spectra of amino acid side chains can be found in refs. *17* and *18*. In a typical protein, side chain vibrational spectra can be expected to contribute 15–20% of the total integrated intensity in the Amide I region *(14)*. Therefore, it is perhaps surprising that estimates of protein secondary structure employing this region have been as successful as they appear to be since no correction of this contribution has usually been attempted. This may simply be explained by a constant error in the analysis that is negated by the relative nature of the secondary structure calculations.

The use of the Amide II band in simple structural analysis has proven somewhat more problematic (e.g., *19*). Alpha helices produce a band in the vicinity of 1550/cm and a weaker signal ca. 1520/cm, whereas β-structure gives rise to strong and weak signals around 1530 and 1560/cm, respectively. More random structure generally appears to absorb at approx 1535–1550/cm. In general, however, this region has yet to be subjected to the more comprehensive analysis applied to Amide I spectra. Furthermore, the dramatic reduction or even elimination of this band in the commonly used D_2O solvent has inhibited further interest. We shall continue this prejudice.

All chemical entities having covalent bonds possess vibrational absorption bands. Thus, a variety of other chromophores are potentially available in proteins that could be monitored by FTIR during folding and unfolding studies. These include weak signals from the amino acid side chains in non-Amide regions (e.g., Tyr at 1515/cm, disulfide bonds at 500/cm, SH groups near 2560/cm), phosphate moieties, heme groups, and so on. In addition, ligands with distinctive IR signals, such as cyanide, carbon monoxide, and oxygen, that bind to heme groups have been extensively studied by FTIR *(20)*. To the extent that the IR frequencies and intensities of such bands can be spectrally isolated and measured, and that their spectral characteristics are sensitive to protein conformation, they potentially could be used to monitor states of folding of proteins. As of yet, little work has been done in this area, but the opportunity for such investigations clearly exists.

3. Instrumentation

Infrared analysis of proteins is currently performed almost exclusively with Fourier transform instruments. It is fair to say that the recent resurgence of interest in the technique among biochemists is primarily a result of the availability of such instrumentation with their greatly improved sensitivity, signal-to-noise ratios, wavenumber precision, and built-in data handling capability greatly exceeding the performance of their dispersive counterparts *(21)*. Although an appropriate FTIR spectrometer may not yet be found in the immediate environment of many biochemists, most large research institutions (especially chemistry and analytical departments) will have such an instrument. The only particular instrumental requirements are moderate resolution (1–2/cm), a liquid-nitrogen cooled MCT (mercury-cadmium-telluride)-detector for enhanced sensitivity, and appropriate sampling devices and data handling capability as described below. Less sensitive deuterated triglycine sulfate (DTGS) detectors can also be employed but require much longer data acquisition times. All of this is currently available at a reasonable cost in a user friendly format from a variety of manufacturers. Recently, the potential for kinetic analysis has been enhanced by the availability of step scanning instruments that can obtain tens to hundreds of spectra per second *(22)*.

4. Physical State of Samples and Appropriate Instrument Geometries

One of the major advantages of FTIR analyses of protein folding is the ability of the technique to examine proteins in a wide variety of physical states. Solutions, suspensions, gels, and solids as well as protein adsorbed onto various interfaces can all be investigated. This is in marked contrast to CD, fluorescence, NMR, and many other commonly used methods in which the range of states of the test sample are more restricted. This very flexible use of FTIR is particularly facilitated by the availability of a number of different types of sampling geometries. Four of these appear to be of particular utility and are summarized diagrammatically in Fig. 1.

The majority of protein FTIR experiments are still conducted with conventional transmission geometries (Fig. 1A). The main considerations in the use of these methods concern both the nature of the material of which the windows are constructed and the path length. With regard to the former, the windows must be both water insoluble and transparent in the spectral regions of interest. Most commonly, demountable or fixed path length cells constructed of CaF_2 or related halide salts are employed. Choice of the proper optical material can provide IR transparency through all of the regions commonly of interest (4000–600/cm). When

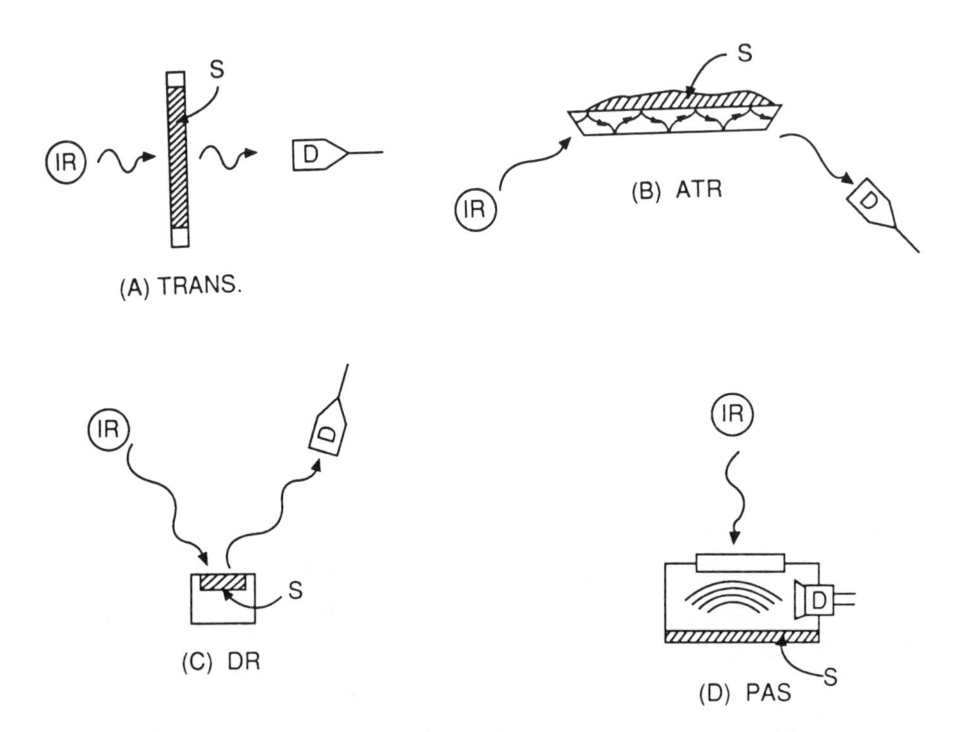

Fig. 1. Diagrams of the four major sampling geometries used in studies of proteins: **(A)** transmittance; **(B)** attenuated total reflectance; **(C)** diffuse reflectance; and **(D)** photoacoustic spectroscopy. The infrared radiation source is indicated by "IR," the radiation path by the wavy line, the sample by "S," and the detector by "D." In the case of PAS, the detector is a microphone. For an extended discussion of each method, *see* ref. *21*.

water is used as a solvent, path lengths on the order of a few μm (e.g., 1–10) are usually employed. When, as is often the case, D_2O is used, longer path lengths, generally 50–100 μm, are chosen, consequently permitting much lower protein concentrations to be examined. Although path lengths can be experimentally determined for quantitative work, this is not usually necessary in protein folding studies where relative changes in band intensity and positions are the primary parameters of interest.

A second approach that has shown increasing protein applications involves various attenuated total reflectance (ATR) configurations (e.g., ref. *23*). In this method, IR light is passed through a prism that has a refractive index greater than the sample that is positioned in direct contact with the prism's surface(s) (Fig. 1B). An evanescent wave is produced that penetrates partially (a small fraction of the wavelength of the incident light) into the sample, dropping off exponentially from the surface of the internal reflectance element (IRE). The light path can exhibit a number of reflections depending on the dimensions and refractive index of the IRE, with the total effective path length the product of the penetration depth and the number of reflections (e.g., four in Fig. 1B). The IRE is often rectangular in cross-section with the sample placed in intimate contact with one surface of the plate, but since about 1982 cylindrical IREs have become available that have been optimized for solution measurement *(24)*. The effective path length for most IREs is in the range of a few microns making them particularly useful for aqueous samples.

The major advantages of ATR methods include their wide applicability to a variety of physical states (solutions, suspensions, solids, films, and gels such as membrane preparations, and so on), high reproducibility, and ease of use. Unfortunately, these advantages are often offset by a major disadvantage *(25)*. Since the molecules being probed are just those near or on the surface of the IRE, surface adsorbed protein may constitute the primary origin of the spectral features observed. This spectrum may or may not represent native protein. Data have been presented both in support and in opposition to the presence of structurally unperturbed proteins at the surface of the IRE in ATR measurements *(23,25,26)*. At the very least, however, the potential for protein structural alteration on surface adsorption exists in ATR measurements. Thus, it is imperative that a solution spectrum be obtained and compared to ATR data before analyses of spectra. Furthermore, the even greater likelihood of unique interactions of partially or totally unfolded proteins with the IRE suggests great caution in the use of this approach in folding experiments.

A third and probably less useful geometry is that of diffuse reflectance, sometimes referred to as "DRIFTS" when used in conjunction with FTIR instrumentation. In this method, the detector collects IR light diffusely reflected by a usually solid sample (Fig. 1C). In this approach radiation travels some limited distance through the sample before it is scattered and/or reflected. This results in enhancements of weaker signals and frequency shifts that can be corrected by what is known as the Kubelka-Munk equation, permitting direct comparison to conventional absorption spectra. Diffuse reflectance has been used to examine proteins adsorbed onto the surface of various matrices with some success

(27). Another technique that permits both solids and liquids to be examined is photoacoustic spectroscopy. This technique takes advantage of the fact that incident modulated radiation will cause alternate heating and contraction of a gas (i.e., sound) in direct contact with an absorbing sample. This sound can be detected by a sensitive microphone in the sample chamber and then be used to obtain an absorption spectrum (Fig. 1D). Unfolding studies of solid proteins employing both near- and mid-range PAS FTIR have been performed (e.g., *28,29*), but the inherent sensitivity and signal-to-noise ratio of this technique are somewhat less than those of other FTIR approaches and it is therefore unclear whether PAS has much to offer for investigations of the type of interest here.

FTIR also offers the user the option of obtaining spectra through a microscope employing either transmission or reflectance geometries. Spatial resolution on the order of a few microns can be achieved with excellent signal-to-noise ratios. Thus, for example, protein inclusion bodies, organized protein arrays in membranes, and macromolecules adsorbed onto the surface of individual particles can all be spectrally analyzed by FTIR microscopy *(30,31)*. Furthermore, individual protein fibers can be investigated using this technology *(32)*.

Finally, FTIR can also be used with polarized radiation. Preferential absorption of light occurs when the electric dipole moment of the incident radiation is polarized parallel to the absorption transition dipole moment of the absorbing species. This principle has been extensively used in studies of fibrous proteins since maximum Amide I absorbance will occur when the incident polarization direction is parallel to the axis of an α-helix with minimum absorbance observed in the Amide II band. Reverse behavior is expected for β-sheets. Use of this approach requires oriented samples, however, a situation not usually applicable to protein folding/unfolding studies. An example of how this approach can be profitably employed, however, is illustrated by studies of large tension-induced conformational transitions in spider silk proteins *(32)*.

5. Data Analysis

Traditionally, infrared spectroscopists presented data in the form of transmittance spectra. In contrast, IR spectra of proteins are now almost always initially displayed as zero order absorbance spectra, i.e., a plot of absorbance vs frequency in wavenumbers. Such spectra in the Amide I region generally consist of a single broad peak with perhaps the presence of one or more shoulders hinting at the existence of the multiple components of interest resulting from the heterogeneity of peptide band environments. Two methods are commonly applied to enhance the resolution of these FTIR spectra. The first involves calculation of the *n*th order derivative from the absorption spectrum, with first, second, and fourth derivatives favored in FTIR spectroscopy. This approach is discussed in more detail in this volume in Chapter 4. The second involves a band-narrowing procedure known as "Fourier self-deconvolution" (FSD) and is lucidly explicated in several recent reviews *(33–36)*. The operator needs to select a single postulated bandwidth for all underlying components (i.e., σ = the width at half-height of the unresolved bands) and a "resolution enhancement factor" (K). The resulting "resolution-enhanced" spectrum is rapidly calculated by instrument software. Values of σ = 10–20/cm and K = 1.2–2.0 for 2/cm resolution spectra are usually employed in the analysis of Amide I spectra with interactive optimization employed to obtain the final values. This is done by enhancing resolution by variation of the two parameters until "over-deconvolution" occurs. The maximum effective resolution after deconvolution cannot be higher than the resolution of the instrument used to record the data. Both derivative and Fourier deconvolution also amplify the noise in the absorption spectrum. This is manifested by the presence of apparent signal in regions of the spectrum that should be optically transparent. Great care must be taken to avoid this artifact since it could result in the attribution of spectral components to structural features when, in fact, no such relationship exists. Proper use of this procedure is described in more detail elsewhere *(33–36)*. Examples of deconvoluted spectra of a primarily α-helix containing protein (the hepatitis B surface antigen) and a β-sheet rich protein (IgG) as well as one displaying a mixture of both secondary structure types (a 40 kDa fragment of *Pseudomonas* exotoxin) are illustrated in Fig. 2A–C, respectively.

One of the major advantages to FSD is that it should not perturb the integrated intensities of actual absorption signals that are responsible for the observed, broad Amide I composite band. Therefore, the FSD spectrum can be used to quantitatively analyze underlying components. Methods based on either curve fitting or pattern recognition have been used in this regard. In the curve fitting methods, Gaussian and/or Lorentzian functions are used to represent characteristic component absorption bands and a nonlinear least squares method is employed to

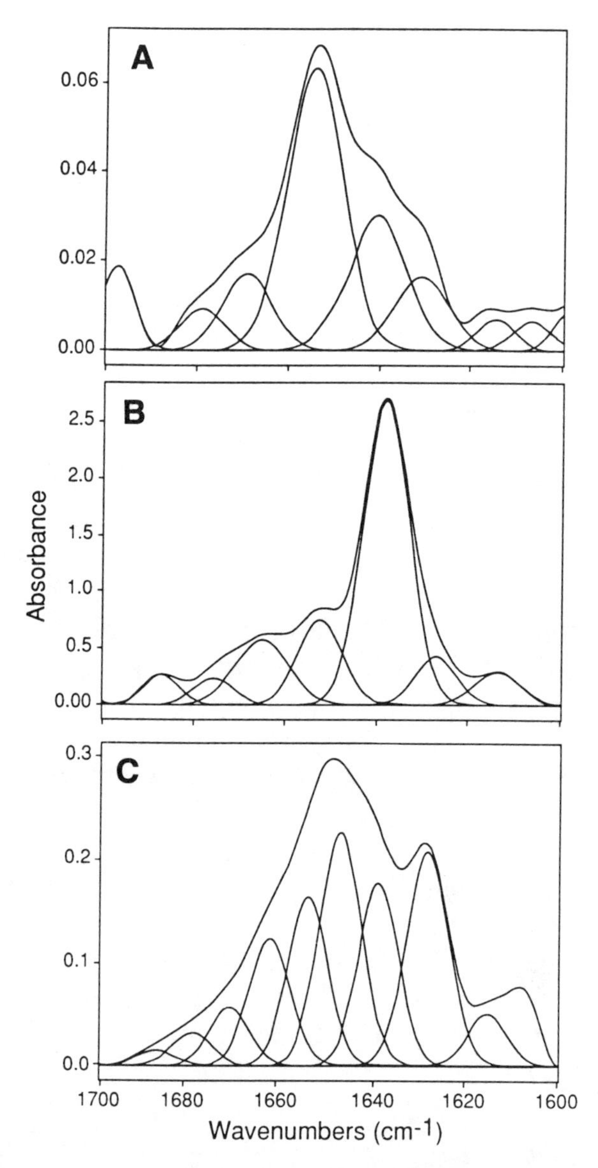

Fig. 2. Fourier self-deconvoluted Amide I' spectra of a **(A)** primarily α-helical protein (the hepatitis B surface antigen), **(B)** a β-sheet rich protein (an IgG monoclonal immunoglobulin), and **(C)** a protein displaying a mixture of secondary structure types (a 40 kDa fragment of *Pseudomonas* exotoxin). Spectra were obtained in D_2O at protein concentrations of 1–10 mg/mL with a Digilab FTS-60 spectrometer. Deconvoluted spectra were obtained using a bandwidth

obtain best fit values of position, height, and width for each band *(4)*. The number of bands is based on identifiable features in the FSD spectrum with second derivative peaks often used as initial guesses of position. Examples of the results of such an analysis for three proteins are shown in Fig. 2. The ultimate accuracy of this procedure is critically dependent on somewhat problematic peak assignment as discussed above and correction for side chain absorption is preferred although currently only occasionally performed *(14,37)*. Furthermore, the assumption of identical extinction coefficient for each secondary structure type and the inherent problem of uniqueness of fit when so many components are involved (usually 7–10 in the 1600–1700/cm region) must further temper the use of the type of analysis in absolute secondary structure estimation *(6,38)*. Despite these reservations, significant success has been reported for the method. Importantly, however, this analysis can be used to quantitatively describe specific changes in the secondary structure of proteins subjected to environmental perturbation (e.g., temperature, pH, denaturing agent) in terms of changes in the areas of individual component peaks.

Pattern recognition methods employ a set of reference Amide I spectra of proteins of known crystal-structure to fit the Amide band of a subject protein *(12,13,15)*. This method is directly analogous to well known procedures used in the analysis of CD and Raman spectra *(39,40)*. Importantly, pattern recognition based methods avoid both the assignment problem and the amplification of noise inherent in spectral deconvolution algorithms, but retain the assumption of identical extinction coefficients for each secondary structure type. This method is, of course, totally dependent on the reference database, which is still relatively limited. Nevertheless, these procedures are probably to be preferred for absolute estimation of secondary structure contents. At present, however, they would appear to offer little obvious advantage in the analysis of protein folding. Therefore, either deconvolution or pattern-based analysis of the

---

of 20/cm and a resolution enhancement factor of 2.0. Curve fitting was performed with mixed Lorentzian/Gaussian functions as described previously *(11)*. Note the strong α-helical peak at 1654/cm in (A), the large peak owing to β structure at 1638/cm in (B), and the presence of both peaks in (C). *See text* for further discussion of assignments.

Amide I band can be used profitably in this regard. A critical comparative analysis of both procedures has recently appeared and we recommend that this review be carefully consulted prior to attempting detailed analysis of changes in Amide I spectra in terms of alterations in protein secondary structure *(6)*.

6. Protein Stability Analysis

The major question concerning the potential use of FTIR (in particular, of the Amide I band) in analyzing the stability of proteins is the ability of the technique to detect small changes in protein structure. In our experience, if Amide I spectra of high quality (good signal-to-noise ratio) can be obtained, it should easily be possible to detect changes in the integrated intensity of individual (deconvoluted) bands of >5%. Under ideal conditions (i.e., S/N > 1000:1), a precision of perhaps 2–3% may be observed. What are the requirements for obtaining such spectra? First, no interfering absorbtivity must appear in the spectral region of interest (i.e., 1600–1700/cm). Several specific problems are frequently encountered in this regard. By far the most perplexing quandary concerns the absorption bands of liquid and gaseous water that appear in this region. In the case where experiments are performed in water, the large bending vibration of this solvent is present near 1640/cm. Great care must be taken that the necessary solvent subtraction does not introduce artifactual spectral features. This is usually accomplished by use of an internal standard such as the 600/cm libration band or 2125/cm association band of water *(41)*. Because such studies usually also require relatively high protein concentrations, they are probably not ideally suited for studies of protein folding and unfolding and D_2O is the preferred solvent. The use of D_2O as a solvent does not eliminate the water problem, however, because water vapor produces a series of sharp, characteristic peaks in the same region. These can be substantially, if not entirely, eliminated by flushing the instrument with dry nitrogen or air. If this is not completely successful, however, these water vapor peaks must be removed by careful subtraction and examination of an empty region of the spectrum as well as demonstration of the complete absence of the characteristic water vapor bands at 1684, 1670, 1662, 1653, 1646, and 1617/cm in the Amide I region itself *(16)*. These considerations are especially critical when FSD or derivative procedures are employed since water contributions could be substantially magnified by these analysis procedures and confused with actual amide features. The presence of small quantities of water in D_2O must also be considered with subtractive correction of characteristic HOD signals usually the simplest solution.

A second problem concerns the potential presence of solutes that absorb in the Amide I window. Many commonly employed buffers have little or no absorption in this region (e.g., phosphate) but some compounds containing primary amines or peptide moieties are to be avoided. A potentially major problem occurs with proteins and peptides that are purified by HPLC in which trifluoroacetate is employed. This agent produces a very strong band near 1673/cm *(42)*, which could be erroneously identified as absorption from extended chain structure. Trace amounts of trifluoroacetate can be very difficult to dissociate from polypeptides (dialysis may be insufficient) and an ion-exchange step may be necessary to affect complete removal. Alternatively, this band can be subtracted from the spectrum after deconvolution, but protein spectral features in this same region will also be lost, thereby limiting a more comprehensive analysis. Many protein stability studies involve the use of a structural perturbant. Two common methods employ alterations in temperature and pH. These approaches are directly applicable to FTIR studies with little modification of standard procedures employed with other spectroscopic monitoring techniques (e.g., CD, fluorescence, and so on). In contrast, the two most commonly used denaturing agents, urea and guanidine hydrochloride (GuHCl), are not very useful since they absorb strongly in the Amide I region, especially at the high concentrations generally necessary to achieve significant protein structural alteration. Fortunately, a number of chaotropic salts are available that have potencies similar to urea and GuHCl on a molar basis. The monovalent salt LiBr is especially useful since it is completely devoid of infrared absorbance. Furthermore, $LiClO_4$ absorbs only weakly between 1600 and 1700/cm and can therefore be employed. The latter salt is particularly useful since its great UV transparency also permits its application to far UV CD analysis and direct comparisons between CD and FTIR measurements.

7. Folding and Unfolding of Proteins

In general, studies of protein unfolding using FTIR can be performed by methods similar to those employed for other spectroscopic techniques. As discussed above, the native state is most conveniently analyzed by a curve-fitting analysis of the Amide I band to define secondary structure content and this state then perturbed in the desired manner. As structural alterations are incrementally induced by increases in temperature, extremes of pH, or solute addition, equilibrium is permitted to occur and a new spectrum is then obtained. Changes in the integrated areas under individual bands derived from total deconvolution or secondary structure contents from pattern recognition methods can then be plotted as a function of temperature, pH, or solute concentration. If appropriate, these data can be fit to standard two state models or stable intermediates resolved by their unique spectral properties. Preliminary studies suggest that intermediates such as molten globule states are indeed detectable by this method and display unique secondary structural features *(42a)*.

The Amide I FTIR spectra of protein unfolded states are currently rather poorly defined. In general, however, the following generalizations seem appropriate, employing a wide variety of structural perturbations including temperature, pH, pressure, and solutes:

1. A substantial loss in the integrated intensity of bands arising from α-helix and/or β-sheet;
2. The appearance of a strong new band between 1610 and 1620/cm; and
3. An increase in a weaker band between 1680 and 1690/cm.

Partially deconvoluted spectra as a function of temperature that illustrate these phenomena for two different proteins are shown in Fig. 3. These new signals have generally been assigned to intermolecular β-sheet formation as the protein aggregates. The reader is referred to some representative original reports on which these generalizations are based *(43–52)*, but it is clear that analyses of nonnative states of proteins by Amide I FTIR are in their infancy. It should be possible, however, to describe the unfolding of a protein in terms of its secondary structure by this method in more detail than an analogous investigation using CD.

Recent developments suggest that FTIR has the potential to become a much more powerful tool for the study of protein conformation. When a ^{12}C-amide (carbonyl) is replaced with a ^{13}C-substituted amide group, the Amide I frequency is decreased by 35–45/cm *(53)*. This provides the ability to specifically label a particular segment of secondary structure within a protein that can be spectrally resolved. Although as of this writing the method has only been applied to peptides *(54)*, a mixture of two different proteins *(53)* and individual residues in a fibrillar protein *(55)*, this

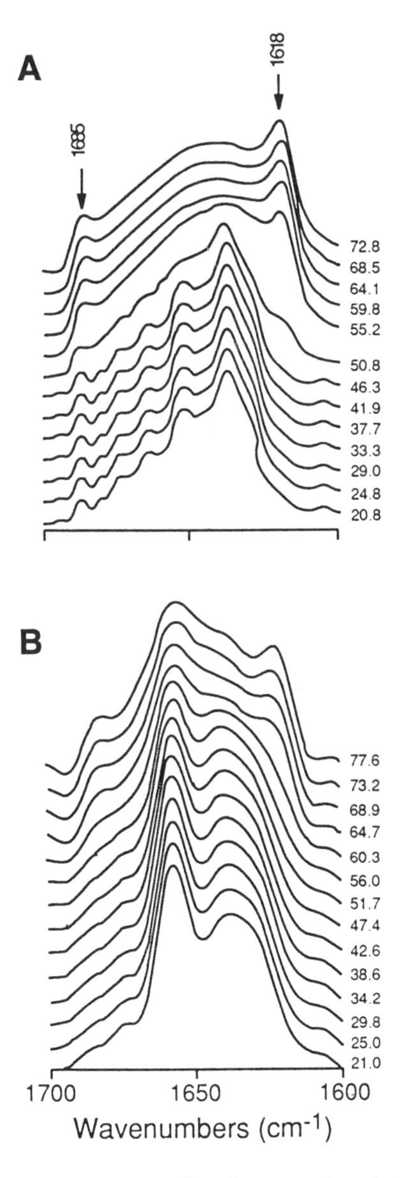

Fig. 3. Effect of temperature on the deconvoluted Amide I' band of **(A)** a globular protein (chymotrypsinogen A) and **(B)** a membrane protein (acetylcholine receptor-rich membranes). The numbers on the right side of each spectrum indicate the temperature (°C) at which the data were obtained. Spectra are reproduced by permission from refs. *52* and *51*, respectively, where the experimental conditions of the measurements are described.

approach should allow the formation and disruption of specific regions of secondary structure to be monitored.

Another potentially useful application of FTIR to protein folding studies involves monitoring isotopic exchange (e.g., *56*). This approach takes advantage of the slower exchange rates of hydrogen-bonded amide groups present in regular secondary structure. When a small peptide is placed in D_2O, NH groups are rapidly converted to ND moieties, which results in a lowering of the frequency of the Amide II band by approx 100/cm. Much smaller effects are seen in the Amide I carbonyl stretching vibrations (shifts of <10–15/cm). Although either peak can be used to follow the rate of isotope exchange, the large decrease in Amide II intensity between 1400–1500/cm has made measurement of this signal preferable. When similar experiments are performed with proteins, a wide distribution of exchange rates is found, with a significant decrease in the rate of exchange observed for both buried and intramolecularly hydrogen-bonded amides. Thus, simple plots of Amide II signal intensity as a function of time after initiation of isotope exchange in the presence and absence of structural perturbation (temperature, pressure, pH, solute, and so on) can be constructed and analyzed in terms of the relative exposure of the various classes of amides. The experiment can also be reversed by extensively exchanging the protein in D_2O (usually at low pH to enhance exchange) and then following reappearance of signal in the 1500–1600/cm region. As described above for the amide carbonyl group, additional isotopic substitution can be used to enhance the resolution of this approach. In this case, ^{15}N can be exchanged for ^{14}N, producing a large shift (e.g., 25/cm) in the Amide II band when D_2O is used as a solvent *(53)*.

Although the kinetics of protein folding and unfolding have yet to be extensively examined by FTIR, this seems a most promising area for future exploration. The method has been used, however, to follow changes at short times (ns to ms) in chromophoric proteins such as bacteriorhodopsin (e.g., *57*) and relevant technologies appear to be emerging from this and related work. In particular, the commercial introduction of step-scan FTIR instrumentation with temporal resolution in the submillisecond time range *(22)* suggests that such studies only await accompanying development of appropriate stopped-flow sample handling capability to permit protein folding and unfolding reactions to be monitored by FTIR. Whether sufficient signal-to-noise ratios can be obtained to actually visualize deconvoluted amide spectra at early times in folding/unfolding pathways remains to be demonstrated, but seems highly probable. It is anticipated that by the time of publication of this book, the first versions of such studies should have appeared. Unfortunately, the dead-time for any such instrument will probably be similar to analogous apparatuses for stopped-flow UV, CD, fluorescence, and so on (on the order of a few milliseconds), which will prohibit accumulation of data at times (submillisecond) prior to the formation of the majority of secondary structure in early molten globule intermediates.

8. Conclusions

The analysis of protein folding employing the IR Amide I (or Amide I') band has the potential to provide more detail on the state of secondary structure than CD spectroscopy. It may also be possible to combine the two methods to obtain a maximum amount of information *(58)* as well as employ vibrational circular dichroism to further enhance resolution *(59)*. Although similar information is obtainable employing Raman and NMR spectroscopy, the higher protein concentrations required for these methods and lower signal-to-noise ratio of the former make the FTIR approach more facile. The availability of a wide variety of sampling geometries with FTIR based methods also permits proteins to be examined in virtually any physical state. For example, membrane proteins and proteins adsorbed onto highly scattering surfaces are easily analyzed by these methods *(60–64)*. Nevertheless, the use of FTIR spectroscopy to study protein folding is in its infancy. As such, several problems described above need to be emphasized *(6)*. The assignment of various types of secondary structure to unique Amide frequencies remains tentative. Despite this, the very complex spectra that result from a deconvolution or derivative analysis of the Amide I band strongly tempt the investigator to attempt a detailed analysis of changes in individual spectral components in terms of distinct alterations in secondary structure. Although such efforts are certainly worthwhile, especially when employed in a comparative manner, the eventual use of site-specific deuteration procedures seems far more likely to produce definitive results in both equilibrium and kinetic studies. In addition to these concerns, potential problems arising from incompletely compensated water peaks, nonprotein absorbance and deconvolution artifacts must all be kept in mind. Nevertheless, FTIR spectroscopy seems on the verge of providing important information concerning the folding and unfolding of proteins that is complementary to more commonly employed methods.

Note added in proof: A description of an FTIR study has recently appeared in which it is demonstrated that differential responses to temperature of secondary structural elements (as reflected in the deconvoluted Amide I and Amide I' bands) of Ribonuclease T_1 can be resolved (Fabian, H., Schultz, C., Naumann, D., Landt, O., Hahn, U., and Saenger, W. [1993] *J. Mol. Biol.* **232,** 967–981).

References

1. Haris, P. I. and Chapman, D. (1992) *Trends Biochem. Sci.* **17,** 328–333.
2. Krimm, S. (1987) *Vib. Spectra Struct.* **16,** 1–72.
3. Bandekar, J. (1992) *Biochim. Biophys. Acta* **1120,** 123–143.
4. Byler, D. M. and Susi, H. (1986) *Biopolymers* **25,** 469–487.
5. Torii, H. and Tasumi, M. (1992) *J. Phys. Chem.* **97,** 92–98.
6. Surewicz, W. K., Mantsch, H. H., and Chapman, D. (1993) *Biochemistry* **32,** 389–394.
7. Hashimoto, M. and Arakawa, S. (1967) *Bull. Chem. Soc. Japan* **40,** 1698–1701.
8. Middaugh, C. R., Thomson, J. A., Burke, C. J., Mach, H., Naylor, A. M., Bogusky, M. J., Ryan, J. A., Pitzenberger, S. M., Ji, H., and Cordingley, J. S. (1993) *Prot. Sci.* **2,** 900–914.
9. Mantsch, H. H., Perczel, A., Hollosi, M., and Fasman, G. D. (1993) *Biopolymers* **33,** 201–207.
10. Prestelski, S. J., Byler, D. M., and Liebman, M. N. (1991) *Biochemistry* **30,** 133–143.
11. Middaugh, C. R., Mach, H., Burke, C. J., Volkin, D. B., Dabora, J. M., Tsai, P. K., Bruner, M. W., Ryan, J. A., and Marfia, K. E. (1992) *Biochemistry* **31,** 9016–9024.
12. Lee, D. C., Haris, P. I., Chapman, D., and Mitchell, R. C. (1990) *Biochemistry* **29,** 9185–9193.
13. Sarver, R. W., Jr. and Krueger, W. C. (1991) *Anal. Biochem.* **194,** 89–100.
14. Kalnin, N. N., Baikalov, I. A., and Venyaminov, S. Y. (1990) *Biopolymers* **30,** 1273–1280.
15. Doussea, F. and Pezolet, M. (1990) *Biochemistry* **29,** 8771–8779.
16. Dong, A., Huang, P., and Caughey, W. S. (1990) *Biochemistry* **29,** 3303–3308.
17. Chirgadze, Y. N., Fedorov, O. V., and Trushina, N. P. (1975) *Biopolymers* **14,** 679–694.
18. Venyaminov, S. Y. and Kalnin, N. N. (1990) *Biopolymers* **30,** 1243–1257.
19. Venyaminov, S. Y. and Kalnin, N. N. (1990) *Biopolymers* **30,** 1259–1271.
20. Maxwell, J. C. and Caughey, W. S. (1978) *Methods Enzymol.* **56,** 302–323.
21. Griffiths, P. R. and de Haseth, J. A. (1986) *Fourier Transform Infrared Spectroscopy,* Wiley, New York.
22. Palmer, R. A. (1993) *Spectroscopy* **8,** 26–36.
23. Singh, B. R. and Fuller, M. P. (1991) *Appl. Spectrosc.* **45,** 1017–1021.

24. Marley, N. A., Gaffney, J. S., and Cunningham, M. M. (1992) *Spectroscopy* **7,** 44–53.
25. Jackson, M. and Mantsch, H. H. (1992) *Appl. Spectrosc.* **46,** 699–701.
26. Swedberg, S. A., Pesek, J. J., and Fink, A. L. (1990) *Anal. Biochem.* **186,** 153–158.
27. Katzenstein, G. E., Vrona, S. A., Wechsler, R. J., Steadman, B. L., Lewis, R. V., and Middaugh, C. R. (1986) *Proc. Natl. Acad. Sci. USA* **83,** 4268–4272.
28. Martin, B. L., Wu, D., Tabatabai, L., and Graves, D. J. (1990) *Arch. Biochem. Biophys.* **276,** 94–101.
29. Sadler, A. J., Horsch, J. G., Lawson, E. Q., Harmatz, D., Brandau, D. T., and Middaugh, C. R. (1984) *Anal. Biochem.* **138,** 44–51.
30. Steadman, B. L., Thompson, K. C., Middaugh, C. R., Matsuno, K., Vrona, S., Lawson, E. Q., and Lewis, R. V. (1992) *Biotech. Bioeng.* **40,** 8–15.
31. Matsuno, K., Lewis, R. V., and Middaugh, C. R. (1991) *Arch. Biochem. Biophys.* **291,** 349–355.
32. Dong, Z., Lewis, R. V., and Middaugh, C. R. (1991) *Arch. Biochem. Biophys.* **284,** 53–57.
33. Kauppinen, J. K., Moffatt, D. J., Mantsch, H. H., and Cameron, D. G. (1981) *Anal. Chem.* **53,** 1454–1457.
34. Susi, H. and Byler, D. M. (1986) *Methods Enzymol.* **130,** 290–311.
35. Mantsch, H. H., Moffatt, D. J., and Casal, H. L. (1988) *J. Mol. Struct.* **173,** 285–298.
36. Moffatt, D. J. and Mantsch, H. H. (1992) *Methods Enzymol.* **210,** 192–200.
37. Garcia-Quintana, D., Garriga, P., and Manyosa, J. (1993) *J. Biol. Chem.* **268,** 2403–2409.
38. Wilder, C. L., Friedrich, A. D., Potts, R. O., Daumy, G. O., and Francoeur, M. L. (1992) *Biochemistry* **31,** 27–31.
39. Johnson, W. C., Jr. (1992) *Methods Enzymol.* **210,** 426–447.
40. Williams, R. W. (1983) *J. Mol. Biol.* **166,** 581–603.
41. Dousseu, F., Therrien, M., and Pezolet, M. (1989) *Appl. Spectrosc.* **43,** 538–542.
42. Surewicz W. K. and Mantsch, H. H. (1989) *J. Mol. Struct.* **214,** 143–147.
42a. Mach, H., Ryan, J. A., Burke, C. J., Volkin, D. B., and Middaugh, C. R. (1993) *Biochemistry* **32,** 7703–7711.
43. Muga, A., Arrondo, J. R. R., Bellon, T., Sancho, J., and Bernabeu, C. (1993) *Arch. Biochem. Biophys.* **300,** 451–457.
44. Kirsch, J. L. and Koenig, J. L. (1989) *Appl. Spectrosc.* **43,** 445–451.
45. Jackson, M., Haris, P. I., and Chapman, D. (1989) *Biochim. Biophys. Acta* **998,** 75–79.
46. Schlereth, D. D. and Mantele, W. (1993) *Biochemistry* **32,** 1118–1126.
47. Yamamoto, T. and Tasumi, M. (1988) *Can. J. Spectro.* **33,** 133–137.
48. Sosnick, T. R. and Trewhella, J. (1992) *Biochemistry* **31,** 8329–8335.
49. Wong, P. T. T. and Hermans, K. (1988) *Biochim. Biophys. Acta* **956,** 1–9.
50. Jackson, M. and Mantsch, H. H. (1988) *Biochim. Biophys. Acta* **1118,** 139–143.
51. Fernandez-Ballester, G., Castresana, J., Arrondo, J.-L. R., Ferragut, J. A., and Gonzalez-Ros, J. M. (1992) *Biochem. J.* **288,** 421–426.
52. Ismail, A. A., Mantsch, H. H., and Wong, P. T. T. (1992) *Biochim. Biophys. Acta* **1121,** 183–188.

53. Haris, P. I., Robillard, G. T., van Dijk, A. A., and Chapman, D. (1992) *Biochemistry* **31,** 6279–6284.
54. Tadesse, L., Nazarbaghi, R., and Walters, L. (1991) *J. Am. Chem. Soc.* **113,** 7036,7037.
55. Halverson, K. J., Sucholeiki, I., Ashburn, T. T., and Lansbury, P., Jr. (1991) *J. Am. Chem. Soc.* **113,** 6701,6702.
56. Zhang, Y.-P., Lewis, R. N. A. H., Hodges, R. S., and McElhaney, R. N. (1992) *Biochemistry* **31,** 11,572–11,578.
57. Sasaki, J., Maeda, A., Kato, C., and Hamaguchi, H. (1993) *Biochemistry* **32,** 867–871.
58. Sarver, R. W., Jr. and Kreuger, W. C. (1991) *Anal. Biochem.* **199,** 61–67.
59. Pancoska, P., Wang, L., and Keiderling, T. A. (1993) *Protein Sci.* **2,** 411–419.
60. Buchet, R., Varga, S., Seidler, N. W., Molnar, E., and Martonosi, A. N. (1991) *Biochim. Biophys. Acta* **1068,** 201–216.
61. Braiman, M. S. and Rothschild, K. J. (1988) *Ann. Rev. Biophys. Chem.* **17,** 541–570.
62. Cornell, D. G., Dluhy, R. A., Briggs, M. S., McKnight, C. J., and Gierasch, L. M. (1989) *Biochemistry* **28,** 2789–2797.
63. van Straaten, J. and Peppas, N. A. (1991) *J. Biomater. Sci. Polymer Edn.* **2,** 113–121.
64. Giroux, T. A. and Cooper, S. L. (1991) *J. Colloid Interface Sci.* **146,** 179–194.
65. George, A. and Veis, A. (1991) *Biochemistry* **30,** 2372–2377.

CHAPTER 7

Identifying Sites of Posttranslational Modifications in Proteins Via HPLC Peptide Mapping

Kenneth R. Williams and Kathryn L. Stone

1. Introduction

The intent of this chapter is to provide realistic procedures for identifying sites of posttranslational modifications in proteins. Particular emphasis will be directed toward identifying degradative covalent changes, such as deamidation and oxidation, that are thought to be associated with protein "aging" and to influence protein stability. The first issue that arises is how one goes about detecting the more than 100 different posttranslational modifications that have already been identified in proteins *(1,2)*. In light of this diversity, those techniques are most useful that can detect any, or at least most, possible modifications. Although a technique like isoelectric focusing can detect modifications, such as phosphorylation and deamidation, that result in altering the net charge, this technique would be less likely to detect sulfoxide formation. The two techniques that come closest to being able to rapidly detect most possible modifications are comparative HPLC peptide mapping and mass spectrometry, which can be used to determine accurate masses on both the intact protein and its HPLC-separated peptides. Since electrospray mass spectrometry (ESMS) can routinely determine the mass of a 50 kDa protein within 0.01% *(3)*, any modification that altered the mass of this protein by more than ±5 atomic mass units (amu) could be detected. Hence, assuming that the mass of the nonmodified protein is known, ESMS of the intact protein should succeed in rapidly detecting phospho-

From: *Methods in Molecular Biology, Vol. 40: Protein Stability and Folding: Theory and Practice*
Edited by: B. A. Shirley

rylation (+80) and oxidation (+16), but would, for instance, fail to detect deamidation (+1). Providing that a sample of the nonmodified protein is available, comparative HPLC peptide mapping provides an extremely sensitive means of detecting most posttranslational modifications and of rapidly isolating peptides containing these alterations. The ability of this approach to detect even subtle modifications is further enhanced by either on-line ESMS or off-line ESMS or laser desorption mass spectrometry (LDMS). Although ESMS usually requires a larger amount of sample (i.e., 0.5–50 pmol as compared to 0.1–10 pmol for LDMS *[3]*), its mass accuracy is about fivefold better than that of LDMS (i.e., ±0.005–0.01% for ESMS as compared to ±0.01–0.05% for LDMS *[3]*). Hence, it is likely that ESMS, but not LDMS, would be able to detect the deamidation of a typical 15-mer tryptic peptide. The advantage of mass spectrometry is that in many cases, this approach not only detects posttranslational modifications, but also helps to identify them. Hence, if comparative peptide mapping of the modified vs nonmodified protein detected an alteration in the relative mobility of a particular peptide, whose mass was then demonstrated to be +16 in the case of the modified peptide, one of the most likely explanations for this difference would be methionine oxidation.

Since SDS polyacrylamide gel electrophoresis (PAGE) is the current method of choice for purifying the 50–100 pmol amounts of protein needed for comparative peptide mapping, those procedures for generating internal peptides from proteins that have been separated by SDS-PAGE are often the most useful. Several approaches may be taken to obtain peptides from SDS-PAGE separated proteins. These include, but are not limited to:

1. Digest the protein in the gel in the presence *(4)* or absence of SDS *(5,6)* and then diffuse the resulting peptides out of the gel;
2. Blot the protein onto nitrocellulose *(7,8)* or PVDF *(8)* and then digest it *in situ;* and
3. Blot the protein onto PVDF, cleave it *in situ* with cyanogen bromide, elute the resulting peptides, and then subject them to further digestion with trypsin *(9)*.

Each of these procedures has its own advantages and disadvantages with the impetus toward deriving better "in gel" techniques deriving from the observation that electroblotting is often not quantitative. In this chapter, techniques will be described for carrying out comparative HPLC peptide mapping on proteins isolated either in solution or in SDS polyacrylamide gels.

2. Materials

2.1. SDS-PAGE

1. Laemmli gels *(10)* are prepared using 7–15% polyacrylamide with the percent polyacrylamide being determined by the size of the protein.
2. Gel stain: 0.1% Coomassie blue in 10% acetic acid, 50% methanol, 40% H_2O (1.0 g Coomassie blue + 100 mL acetic acid + 500 mL methanol + 400 mL H_2O).
3. Gel destain: 10% acetic acid, 50% methanol, 40% H_2O (100 mL acetic acid + 500 mL methanol + 400 mL H_2O).

2.2. Amino Acid Analysis

1. Acid for hydrolysis: 6*N* HCl, 0.2% phenol containing 2 nmol/100 µL norleucine as an internal standard (50 mL concentrated HCl + 200 µL phenol + 100 µL 20 m*M* norleucine + H_2O to a total vol of 100 mL).
2. Na-S sample dilution buffer: 2% sodium citrate, 1% hydrogen chloride, 0.5% thiodiglycol, 0.1% benzoic acid, pH 2, in water.

2.3. Enzymatic Digestion of Proteins

1. Enzymatic digestion is carried out with either sequencing grade trypsin, chymotrypsin, or endoproteinase Glu-C (Protease V8 from *Staphylococcus aureus*) from Boehringer Mannheim (Mannheim, Germany) or with lysyl endopeptidase (#129-02541, Achromobacter Protease I from *Achromobacter lyticus*) from Wako Pure Chemical Industries, Ltd. (Osaka, Japan).
2. Trypsin and chymotrypsin stock solutions (0.1 mg/mL) are prepared by dissolving 100 µg enzyme (as purchased from Boehringer Mannheim) in 1 mL 1 m*M* HCl.
3. Protease V8 is prepared by dissolving 50 µg enzyme in 500 µL 50 m*M* NH_4HCO_3.
4. Lysyl endopeptidase (as purchased) is dissolved in 2 m*M* Tris-HCl, pH 8.0, to make a 0.1 mg/mL stock solution.
5. Pepsin (purchased from Sigma, St. Louis, MO) is dissolved in 5% formic acid (Baker, Phillipsburg, NJ) at a concentration of 0.1 mg/mL.
6. 8*M* urea, 0.4*M* NH_4HCO_3 buffer used for digesting proteins in solution is prepared using Pierce (Oug-Beigerland, The Netherlands) Sequanal Grade urea (4.8 g) and Baker ammonium bicarbonate (0.316 g) in 10 mL H_2O.
7. The in gel digestion buffer, 0.1*M* NH_4HCO_3, is prepared using Baker ammonium bicarbonate (0.79 g in 100 mL H_2O).
8. 45 m*M* dithiothreitol (DTT) for protein reduction: 6.9 mg dissolved in 10 mL H_2O.
9. 100 m*M* iodoacetic acid (IAA) for alkylation: 185.9 mg dissolved in 10 mL H_2O.
10. 0.05% SDS/5 m*M* NH_4HCO_3 for protein dialysis: 5 mL 1% SDS and 39.5 mg NH_4HCO_3 in 100 mL H_2O.

2.4. HPLC Separation of Peptides

1. HPLC system: Peptide separations should be carried out on either a Hewlett Packard (Wilmington, DE) 1090*M* HPLC system equipped with a diode array detector and a 250 μL injection loop or on any other HPLC system capable of delivering reproducible gradients over the flowrate range extending from 0.15–0.5 mL/min. Fractions are collected automatically via peak detection into capless Eppendorf tubes (positioned in 13 × 100 mm test tubes) using an Isco (Lincoln, NE) Foxy fraction collector with an Isco Model 2150 Peak Separator or comparable system.
2. A Vydac C-18 reverse-phase column (25 cm, 5 μ particle size, 300 Å pore size, Separations Group) or other C-18 reverse-phase column capable of providing a similar degree of resolution is recommended (*see* Note 9). In general, amounts of protein digests that are <250 pmol are separated on 2.1 ID columns and amounts in the 250 pmol to 10 nmol range are separated on 4.6 mm ID columns.
3. pH 2 Mobile phase:
   a. Buffer A: 0.06% TFA (3 mL 20% TFA/H_2O/L of H_2O).
   b. Buffer B: 0.052% TFA/80% acetonitrile (2.7 mL 20% TFA, 800 mL CH_3CN, 200 mL H_2O).
4. pH 6.0 Mobile phase:
   a. Buffer A: 5 m*M* potassium phosphate, pH 6.0 (12 mL 0.5*M* KH_2PO_4 in a total vol of 1200 mL H_2O).
   b. Buffer B: 1 m*M* potassium phosphate, pH 6.0, 80% (v/v) CH_3CN (200 mL buffer A and 800 mL acetonitrile).

3. Methods

3.1. SDS-PAGE

The percent acrylamide gel used for protein purification is determined by the size of the protein. In general, proteins >100 kDa are electrophoresed in 7–10% polyacrylamide gels, whereas smaller proteins are electrophoresed in 10–15% polyacrylamide gels. Staining of the gel should be carried out for 60 min at room temperature and destained for several hours. After destaining, the protein of interest (along with a blank section of gel equal in size to that containing the protein of interest) is excised from the gel using a razor blade and tweezers and frozen at –20°C in an Eppendorf tube.

3.2. Amino Acid Analysis

1. Remove 10–15% of the sample (or of the gel slice containing the sample) and hydrolyze *in vacuo* in 100 μL (or, in the case of gel samples, 200 μL)

6*N* HCl, 0.2% phenol (containing 2 nmol/100 µL norleucine as an internal standard) at 115°C for 16 h.
2. Following hydrolysis, solution samples should be dried in a SpeedVac and, in the case of gel hydrolysates, the supernatant transferred to a second tube and then dried in a SpeedVac.
3. The dried hydrolysate is then dissolved in Na-S sample dilution buffer and run on a Beckman (Fullerton, CA) Model 7300 Amino Acid Analyzer using ion exchange separation of the amino acids and post column ninhydrin detection.

3.3. Enzymatic Digestion of Proteins

3.3.1. Digestion of Proteins in Solution

3.3.1.1. Trypsin, Chymotrypsin, Lysyl Endopeptidase, or Protease V8 Digestion

Enzymatic digestion of proteins requires that a reasonable level of care be exercised in terms of final sample preparation (*see* Note 1). Proteins that are isolated in solution and that contain <~0.1 mmol monovalent salt and that are free of detergents and glycerol can often simply be dried in a SpeedVac. Higher levels of salts and many detergents, such as SDS, can be removed from the sample by adding 1/9 vol of 100% TCA, incubating on ice for 30 min, centrifuging, and then washing the protein pellet with 100 µL cold acetone. In general, we recommend that the glycerol concentration be lowered to below 15% and that, if possible, the protein concentration be increased to at least 100 µg/mL prior to TCA precipitation. An alternative approach that may be taken with samples that contain high concentrations of salts and SDS is to first dialyze them vs 0.05% SDS, 5 m*M* NH_4HCO_3 to remove the salt. After dialysis, the sample may then be dried in a SpeedVac prior to adding 50 µL water and 450 µL cold acetone. After incubating the sample at –20°C for at least an hour and centrifuging, the pellet is then washed with 100 µL cold acetone to remove the SDS *(5)*. If necessary, the acetone extraction may be repeated one or more times to remove even relatively large amounts of detergent.

Following removal of any detergent that might have been present and lowering of the salt concentration, the sample is ready for enzymatic digestion. Trypsin is often the enzyme of choice for comparative mapping since it cleaves with high specificity at the COOH-terminal side of lysine and arginine. Although bonds involving acidic amino acids are cleaved slowly, the only bonds that are extremely resistant to trypsin cleavage are those involving lysine-proline and arginine-proline linkages. Another advantage of trypsin (and chymotrypsin) is that it can readily digest proteins that are insoluble in the 2*M* urea, 0.1*M* NH_4HCO_3 digestion buffer. Chymotrypsin has significantly less specificity than trypsin in that it cleaves at the COOH-terminal side of tryptophan, tyrosine, and phenylalanine residues with additional cleavages occurring after some leucine, methionine, and other amino acids containing hydrophobic amino acid side chains. Another commonly used enzyme is lysyl endopeptidase, which cleaves specifically after lysine. This enzyme has an advantage in that it will produce larger fragments than trypsin. Finally, Protease V8 cleaves after glutamic acid residues in either ammonium bicarbonate (pH 8) or ammonium acetate (pH 4) buffers and cleaves after both aspartic acid and glutamic acid residues in phosphate buffer (pH 7.8).

All enzyme stocks should be divided into 100 µL aliquots and frozen at –20°C. Stocks of trypsin and chymotrypsin are stable under these conditions for at least 6 mo whereas the lysyl endopeptidase is stable for up to 2 yr (based on the manufacturer's recommendations). According to the manufacturer, endoproteinase Glu-C is stable for approx 1 mo at –20°C. All enzyme stocks should be discarded once thawed.

The digestion procedure that follows can be used with any of the above enzymes (assuming that the buffer is changed in the case of Protease V8).

1. Dissolve the dried or precipitated protein in 20 µL 8*M* urea, 0.4*M* NH_4HCO_3 and then remove a 10–15% aliquot for amino acid analysis. If the analysis indicates there is sufficient protein to digest (*see* Note 2), proceed with step 2, otherwise, additional protein should be prepared to pool with the sample.
2. Check the pH of the sample by spotting 1–2 µL on pH paper. If necessary, adjust the pH to 7.5–8.5.
3. Add 5 µL 45 m*M* dithiothreitol and incubate at 50°C for 15 min to reduce the protein (*see* Note 3 regarding the necessity of this step).
4. After cooling to room temperature, alkylate the protein by adding 5 µL 100 m*M* iodoacetic acid and incubating at room temperature for 15 min.
5. Dilute the digestion buffer with H_2O so that the final digest will be carried out in 2*M* urea, 0.1*M* NH_4HCO_3.
6. Add the enzyme in a 1/25, enzyme/protein [w/w] ratio (*see* Note 4 for exceptions to the 1:25 [w/w] guideline).
7. Incubate at 37°C for 24 h.
8. Stop the digest by freezing, acidifying the sample with TFA, or injecting onto a reverse-phase HPLC system.

3.3.1.2. Pepsin Digestion

Although the very broad specificity of pepsin hinders its routine use for comparative peptide mapping, it is applicable in the case of otherwise intransigent proteins as well as for further digesting relatively small peptides and, particularly, for studies directed at identifying disulfide bonds (*see* Note 10). Under acidic conditions, pepsin cleaves proteins at a wide variety of peptide bonds. Although it cleaves preferentially between adjacent aromatic or leucine residues, pepsin also cleaves at either the NH_2- or COOH-terminal side of any amino acid except proline. A typical digestion procedure follows:

1. Dissolve the dried protein in 100 μL 5% formic acid.
2. Add pepsin at a 1:50, enzyme/protein (w/w) ratio.
3. Incubate the sample at room temperature for 1–24 h with the time of incubation being dependent on the desired extent of digestion.
4. Dry the digest in a SpeedVac prior to dissolving in 0.05% TFA and injecting onto a reverse-phase HPLC system.

3.3.2. Digestion of Proteins in SDS Polyacrylamide Gels

Although the in-gel digestion procedure has succeeded with as little as 25 pmol protein (Fig. 1), in general, we recommend that at least 50–100 pmol protein be subjected to SDS-PAGE and that the density of protein in the gel band that is to be digested be above ~0.075 μg/mm^3. Generally, the latter requires that at least 1–2 μg of the protein of interest be loaded in a single lane of a 0.75 mm thick SDS polyacrylamide gel. The protein of interest, along with a blank, is then excised and 10–15% quantitated by amino acid analysis as outlined in Section 3.2. Notes 5–8 contain further discussion relating to the amounts and concentrations of protein that are required and of the results that might be expected from the in-gel digestion procedure. The actual procedure follows:

1. Swell the intact gel band(s) by soaking in 500 μL 0.2*M* NH_4HCO_3 per ~1 cm long gel piece while shaking (on a tilt platform) for 1 h at room temperature.
2. After removing the wash solution, measure the total length of the gel band(s) and then remove 10–15% for hydrolysis and amino acid analysis.
3. If amino acid analysis indicates sufficient protein is present, cut the gel bands (including the equal size "blank" gel) into 1 × 2 mm pieces.
4. Add sufficient (i.e., 500 μL or more) 0.2*M* NH_4HCO_3 to cover the gel pieces and to permit efficient shaking on the tilt table.

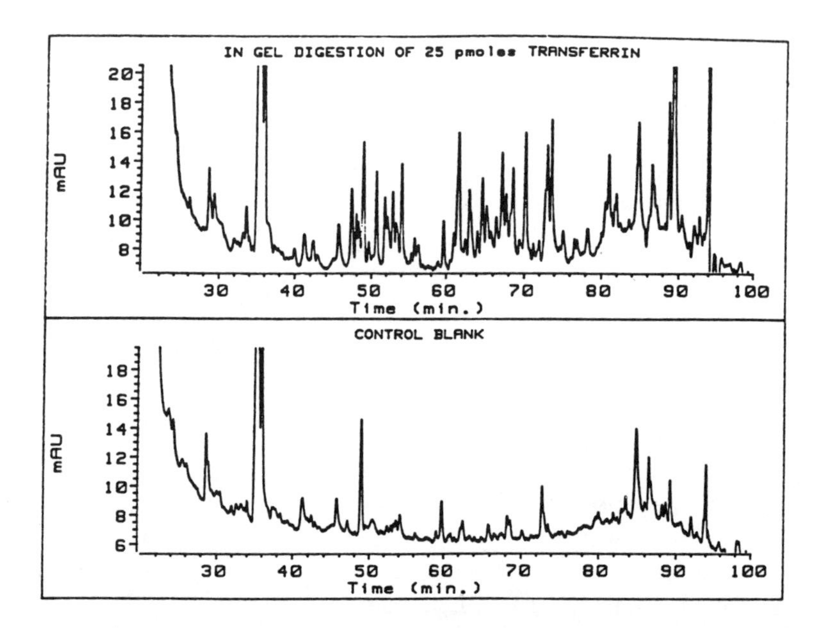

Fig. 1. In gel lysyl endopeptidase digestion of transferrin. Following SDS-PAGE of 25 pmol of transferrin in a 12.5% polyacrylamide gel, the gel was stained and destained as described in Section 2. The protein band was then excised, along with a control blank, and digested in the gel with lysyl endopeptidase as described in Section 3. Peptides were chromatographed on a Vydac C-18, 2.1 × 250 mm reverse-phase column that was eluted at a flowrate of 150 µL/min as described in Section 2. A comparison of the transferrin digest (top profile) with the control digest (bottom profile), which was carried out on a blank section of gel, indicates the digest proceeded well. However, our preliminary studies indicate the overall success rate of the current in gel procedure at the 25 pmol level is considerably below the 94–100% level that can be readily achieved at the >100 pmol level (*see* Table 1 for the latter data).

5. Shake at room temperature for 1 h and remove the wash.
6. Check the pH of the wash. If the pH is below 8.0, repeat steps 4 and 5.
7. Add sufficient 0.2*M* NH_4HCO_3 to cover the gel pieces.
8. Add 5 µL 45 m*M* dithiothreitol and then incubate at 50°C for 20 min.
9. Remove samples from the incubator, cool to room temperature, and add 5 µL 100 m*M* iodoacetic acid.
10. Incubate at room temperature in the dark for 20 min.
11. Estimate the approximate total volume of the sample, including the swelled gel, by comparing to tubes containing known volumes of water.

12. Calculate the amount of trypsin (chymotrypsin or Protease V8) that would be needed to give a final enzyme concentration of 4.0 μg/mL or, for lysyl endopeptidase, 12 μg/mL. Calculate the resulting w/w ratio of substrate/enzyme.
    a. In the case of trypsin (chymotrypsin or Protease V8) digests, if the calculated w/w ratio is >25, add sufficient enzyme so that the w/w ratio is 25.
    b. If the calculated w/w ratio is <25, add trypsin (chymotrypsin or Protease V8) enzyme to a final concentration of 4.0 μg/mL.
13. Incubate at 37°C for 24 h.
14. Add 500 μL 2*M* urea, 0.1*M* NH_4HCO_3 to the gel pieces and shake at room temperature to extract the peptides.
15. After ~6 h, transfer the supernatant to another Eppendorf tube.
16. Add 500 μL 0.1*M* NH_4HCO_3 to the gel pieces and shake overnight at room temperature.
17. Transfer the supernatants to the corresponding Eppendorf tubes.
18. SpeedVac dry the supernatants.
19. Redissolve the dried supernatant in 160 μL of H_2O.
20. Check the volume and adjust to 210 μL by adding water prior to reverse-phase HPLC.

3.4. HPLC Separation of Peptides

The TFA/acetonitrile buffer system described in Section 2.4. is an almost universal reverse phase solvent system because of its low ultraviolet absorbance, high resolution, and excellent solubilizing properties. By adding a slightly higher percentage of TFA (0.06%) to buffer A than to buffer B (0.052%), the baseline at 210 nm can be readily balanced to allow very high sensitivity runs *(11)*. The gradient that we have generally used is:

Time	*%B*
0–60 min	2–37.5%
60–90 min	37.5–75%
90–105 min	75–98%

In the case of extremely complex digests (i.e., tryptic digests of proteins that are above about 100 kDa), the gradient times given above may be doubled. Peptides that are incompletely resolved may be rechromatographed under the same conditions on a C-8 Aquapore column (1.0 × 250 mm). This can most conveniently be done by diluting the peak twofold (or more) with 8*M* urea, 0.4*M* NH_4HCO_3 prior to injecting onto the Aquapore column. Alternatively, peptides isolated from the Vydac C-18 column can be rechromatographed on the same column using the pH 6 buffer system. The differing selectivity of the Aquapore column and of the pH 6 buffer system often bring about significant further purification of peptides that were originally isolated from a Vydac C-18 column developed in the TFA buffer system *(11)*. Further discussion regarding optimizing the reproducibility, resolution, and sensitivity of reverse-phase HPLC may be found in Note 9. Finally, Note 10 discusses more general aspects of identifying and isolating peptides that contain post-translational modifications.

4. Notes

1. In many instances large losses occur during the final purification steps when the protein concentrations are invariably lower. Hence, although ultrafiltration or dialysis of a 5 mg/mL crude solution of a partially purified enzyme may lead to nearly 100% recovery of activity, similar treatment of a 25 µg/mL solution of the purified protein might well lead to significant, if not total, loss of activity because of nonspecific adsorption. Similarly, the effectiveness of organic and acid precipitation procedures often decreases substantially as the final protein concentration is decreased below about 100 µg/mL. Whenever possible, therefore, the final purification step should be arranged such that the resulting protein solution is as concentrated as possible and, ideally, can simply be dried in a SpeedVac prior to enzymatic digestion. In this regard it should be noted that a final NaCl concentration of 1*M* does not significantly effect the extent of trypsin digestion. When it is necessary to carry out an organic or acid precipitation to remove salts or detergents, the protein should first be dried in a SpeedVac (in the 1.5 mL tube in which it will ultimately be digested) prior to redissolving or suspending in either a minimum volume of water (in the case of an acetone precipitation) or in 10% TCA (in the case of an acid precipitation/extraction) so as to increase the protein concentration and minimize loss. Two common contaminants that are extremely deleterious to enzymatic cleavage are detergents (as little as 0.005% SDS will noticeably decrease the rate of tryptic digests carried out in the presence of 2*M* urea *[5]*) and ampholines. Since detergent removal is often associated with protein precipitation and since many detergents (such as SDS) form large micelles that cannot be effectively dialyzed, it is usually preferable to extract the detergent from the protein (that has been dried in the tube in which it will be digested) rather than to dialyze it away from the protein. In the case of ampholines, our experience is that even prolonged dialysis extending over several days with 15,000 Dalton cutoff membrane is not

sufficient to decrease the ampholine concentration to a level that permits efficient trypsin digestion. Rather, the only effective methods that we have found for complete removal of ampholines are TCA precipitation or hydrophobic chromatography.

2. One of the most common causes of "failed" digests is that the amount of protein being subjected to digestion has been overestimated. Often this is because of the inaccuracy of dye-binding and colorimetric assays. For this reason we recommend that an aliquot of the sample be taken for hydrolysis and amino acid analysis prior to digestion. The aliquot for amino acid analysis should be taken either immediately prior to drying the sample in the tube in which it will be digested or after redissolving the sample in 8*M* urea, 0.4*M* NH_4HCO_3. Although up to 10 μL of 8*M* urea is compatible with ion exchange amino acid analysis, this amount of urea may not be well tolerated by PTC amino acid analysis. Hence, in the latter case the amino acid analysis could be carried out prior to drying and redissolving the sample in urea. Although it is possible to succeed with less material, to ensure a high probability of success we recommend that a minimum of 50–100 pmol protein be digested. Typically, 10–15% of this sample would be taken for amino acid analysis. In the case of a 25 kDa protein, the latter would correspond to only 0.125–0.188 μg protein being analyzed. When such small amounts of protein are being analyzed it is important to control for the ever-present background of free amino acids that are in buffers, dialysis tubing, plastic tubes and tips, and so on. If sufficient protein is available, aliquots should be analyzed both before (to determine the free amino acid concentration) and after hydrolysis. Alternatively, an equal volume of sample buffer should be hydrolyzed and analyzed and this concentration of amino acids should then be subtracted from the sample analysis.
3. Since many (probably most) native proteins are resistant to enzymatic cleavage, it is usually best to denature the protein prior to digestion. Although some proteins may be irreversibly denatured by heating in 8*M* urea (as described in the above protocol), this treatment is not sufficient to denature transferrin. In this instance, prior carboxymethylation, which irreversibly modifies cysteine residues, brings about a marked improvement in the resulting tryptic peptide map *(5)*. Another advantage of carboxymethylating the protein is this procedure enables cysteine residues to be identified during amino acid sequencing. Cysteines have to be modified in some manner prior to sequencing to enable their unambiguous identification. Under the conditions that are described in Section 3., the excess dithiothreitol and iodoacetic acid do not interfere with subsequent digestion. Although carboxymethylated proteins are usually relatively insoluble,

the 2*M* urea that is present throughout the digest is frequently sufficient to maintain their solubility. However, even in those instances where the carboxymethylated protein precipitates following dilution of the 8*M* urea to 2*M*, trypsin and chymotrypsin will usually still provide complete digestion. Often, the latter is evidenced by clearing of the solution within a few minutes of adding the enzyme. If carboxymethylation is insufficient to bring about complete denaturation of the substrate, an alternative approach is to cleave the substrate with cyanogen bromide (1,000-fold excess over methionine, 24 h at room temperature in 70% formic acid). The resulting peptides can then either be separated by SDS-PAGE (since they usually do not separate well by reverse-phase HPLC) or, preferably, they can be enzymatically digested with trypsin or lysyl endopeptidase and then separated by reverse-phase HPLC. If this approach fails, the protein may be digested with pepsin, which, as described above, is carried out under very acidic conditions, or may be subjected to partial acid cleavage *(13)*. However, the disadvantage of these latter two approaches is that they produce an extremely complex mixture of overlapping peptides. Finally, extensive glycosylation (i.e., typically >10–20% by wt) can also hinder enzymatic cleavage. In these instances it is usually best to remove the carbohydrate prior to beginning the digest.

4. Every effort should be made to use as high substrate and enzyme concentration as possible to maximize the extent of cleavage. Although the traditional 1:25, w/w ratio of enzyme to substrate provides excellent results with mg amounts of protein, it will often fail to provide complete digestion with low µg amounts of protein. For instance, using the procedures outlined above, this w/w ratio is insufficient to provide complete digestion when the substrate concentration falls below about 20 µg/mL *(12)*. The only reasonable alternative to purifying additional protein is to either decrease the final digestion volume below the 80 µL value used above or to compensate for the low substrate concentration by increasing the enzyme concentration. The only danger in doing this, of course, is the increasing risk that some peptides may be isolated that are autolysis products of the enzyme. Assuming that only enzymes such as trypsin, chymotrypsin, lysyl endopeptidase, and Protease V8 are used, whose sequences are known, it is usually better to risk sequencing a peptide obtained from the enzyme (which can be quickly identified via a database search) than it is to risk incomplete digestion of the substrate. Often, protease autolysis products can be identified by comparative HPLC peptide mapping of an enzyme (i.e., no substrate) control and by subjecting candidate HPLC peptide peaks to laser desorption mass spectrometry prior to sequencing. The latter can be extremely beneficial both in ascertaining the purity of candidate peptide

peaks as well as in identifying (via their mass) expected protease autolysis products. In order to promote more extensive digestion, we have sometimes used enzyme:substrate mole ratios that approach unity. If there is any doubt concerning the appropriate enzyme concentration to use with a particular substrate concentration, it is usually well worth the effort to carry out a control study (using a similar concentration and size standard protein) where the extent of digestion (as judged by the resulting HPLC profile) is determined as a function of enzyme concentration.

5. As in the case of in solution digests, care must be exercised to guard against sample loss during final purification. Whenever possible, SDS (0.05%) should simply be added to the sample prior to drying in a SpeedVac and subjecting to SDS-PAGE. Oftentimes, however, if the latter procedure is followed, the final salt concentration in the sample will be too high (i.e., above ~1*M*) to enable it to be directly subjected to SDS-PAGE. In this instance the sample may either be concentrated in a SpeedVac and then precipitated with TCA (as described above) or it may first be dialyzed to lower the salt concentration. If dialysis is required, the dialysis tubing should be rinsed with 0.05% SDS prior to adding the sample, which should also be made 0.05% in SDS. After dialysis vs a few m*M* NH_4HCO_3 containing 0.05% SDS, the sample may be concentrated in a SpeedVac and then subjected to SDS-PAGE (note that samples destined for SDS-PAGE may contain several percent SDS).
6. Although most estimates of protein amounts are based on relative staining intensities, our data suggest that, in general, there is at least a 5–10-fold range in the relative staining intensity of different proteins. Obviously, when working in the 50–100 pmol range, such a range could well mean the difference between success and failure. Hence, we routinely subject an aliquot of the SDS-PAGE gel (usually 10–15% based on the length of the band) to hydrolysis and ion exchange amino acid analysis prior to proceeding with the digest. Because these analyses will often contain <0.5 μg protein, it is important that a "blank" section of gel about the same size as that containing the sample be hydrolyzed and analyzed as a control to correct for the background level of free amino acids that are usually present in polyacrylamide gels. Based on samples taken from 20 different gels submitted by users of the W. M. Keck Foundation Biotechnology Resource Laboratory (Yale University, New Haven, CT), the background typically ranges up to about 0.2 μg and averages about 0.05 μg in these 10–15% aliquots. Although amino acid analyses on gel slices are complicated by this background level of free amino acids and by the fact that some amino acids (i.e., glycine, histidine, methionine, and arginine) usually cannot be quantitated following hydrolysis of gel slices, these estimates are still con-

siderably more accurate than are estimates based on relative staining intensities. In those instances where amino acid analysis indicates <50 pmol protein, the stained band may be stored frozen while additional material is purified.

7. In general, the sample should be run in as few SDS-PAGE lanes as possible to maximize the substrate concentration and to minimize the total gel volume present during the digest. Whenever possible, a 0.5–0.75 mm thick gel should be used and at least 1–2 μg of the *protein of interest* should be run in each gel lane so the density of the protein band is at least 0.075 μg/mm^3. As shown in Table 1, the in-gel procedure has a success rate of at least 94% when 100 pmol or more protein is digested. This rate decreases to about 81% when 51–100 pmol protein is subjected to SDS-PAGE (Table 1). Although we have succeeded when as little as 25 pmol transferrin had been subjected to SDS-PAGE (Fig. 1), our preliminary data suggest the overall success rate of in-gel digests is substantially decreased as the amount of protein is reduced below the 50–100 pmol range. Although several enzymes (i.e., trypsin, chymotrypsin, lysyl endopeptidase, and Protease V8) may be used with the in-gel procedure, nearly all our experience has been with trypsin. In general, we recommend using a minimum final enzyme concentration of 4 μg/mL. The latter means that in the case of a 50 pmol digest of a 25 kDa protein that is carried out in 100 μL, the final weight (and molar) ratio of substrate to enzyme will be ~3. When larger amounts of protein are available, we recommend increasing the final enzyme concentration so as to maintain the traditional 1:25 w/w ratio.
8. To ensure as complete digestion as possible, we recommend that the sample be reduced and carboxymethylated prior to in-gel digestion and that when extensive (i.e., >10–20% by wt) glycosylation is present, that the sample be deglycosylated prior to cleavage. Often the latter can be accomplished immediately prior to SDS-PAGE, which thus prevents loss (owing to insolubility) of the deglycosylated protein and effectively removes the added glycosidases.
9. Three factors that play a critical role in comparative HPLC peptide mapping are reproducibility, resolution and, when the amount of protein is limiting, sensitivity. In the following paragraphs each of these topics will be briefly discussed. For a more complete discussion the reader is referred to Mant *(14)*.

   Comparative peptide mapping requires that successive chromatograms of digests of the same protein be sufficiently similar that they can be overlayed onto one another with little or no detectable differences. In general, the latter requires that average peak retention times not vary by more than about ±0.20% *(11)*. Assuming the digests were carried out under identical conditions, problems with regard to reproducibility often relate to the inability of

Table 1
Effect of Sample Load on 71 Recent "In Gel" Digests Carried Out in the W. M. Keck Facility at Yale University

	51–100 pmol		101–200 pmol		201–300 pmol	
Parameter	Range	Mean or #	Range	Mean or #	Range	Mean or #
Number of proteins	—	16	—	31	—	24
Mass of protein, kDa	25–108	65	30–195	71	12–350	76
Amount of protein digested, pmol	55–94	75	102–200	149	203–300	244
Density of protein band[a], μg/mm^3	.02–.15	.068	.028–.26	.11	.04–.48	.24
Initial yield[b]	3.7–45.3	14 ± 12	2.4 – 55.5	13 ± 11	.92 – 24.1	9.0 ± 6
Results of sequencing						
Peaks sequenced/protein	1–5	2.0	1–5	1.9	1–6	2.4
≥4 Residues called, %	—	77	—	89	—	78
Mixture, %	—	10	—	3.2	—	10
No sequence, %	—	13	—	11	—	12
Residues called/peptide sequenced	6–20	12.2	6–26	12.1	5–22	11.3
Overall success rate, %[c]	—	81	—	94	—	100

[a]Data on the density of the protein band are based on hydrolysis/amino acid analysis of 10–15% of the stained band.

[b]Calculated as the yield of Pth-amino acid sequencing at the first or second cycle compared to the amount of protein that was digested. Since initial yields of purified peptides directly applied onto the sequencer are usually about 50%, the actual percent recovery of peptides from the above digests is probably twice the percent initial yields that are given above. The standard deviation is given after the average yield.

[c]As judged by the ability to sequence a sufficient number of internal residues to either identify the protein via database searches or to allow oligonucleotide probes to be synthesized to enable cDNA cloning studies to proceed. The total number of positively identified residues for those proteins that were scored as a success varied from 12–58, with the minimum being 12 residues. The latter consisted of two stretches of 7 and 5 residues, respectively, in two different peptides.

the HPLC pumps to deliver accurate flowrates at the extremes of the gradient range. That is, to accurately deliver a 99% buffer A/1% buffer B composition at an overall flowrate of 0.15 mL/min requires that pump B be able to accurately pump at a flowrate of only 1.5 µL/min. The latter is well beyond the capabilities of many conventional HPLC systems. Although reproducibility can be improved somewhat by restricting the gradient range to 2–98%, as opposed to 0–100% buffer B, the reproducibility of each HPLC system will be inherently limited in this regard by the ability of its pumps to accurately deliver low flowrates. Obviously, some HPLC systems that provide reproducible chromatograms at an overall flowrate of 0.5 mL/min might be unable to do so at 0.15 mL/min *(11)*. Similarly, minor check valve, piston seal, and injection valve leaks that go unnoticed at 0.5 mL/min, might well account for reproducibility problems at 0.15 mL/min.

The ability of HPLC to discriminate peptides that contain minor post-translational modifications is critically dependent on resolution, which, in turn, depends on a large number of parameters, including the flowrate, gradient time, column packing, and dimensions as well as the mobile phase *(11,15)*. Studies with tryptic digests of transferrin suggest that, within reasonable limits, gradient time is a more important determinant of resolution than is gradient volume. In general, a total gradient time of ~100 min seems to represent a reasonable compromise between optimizing resolution and maintaining reasonable gradient times *(15)*. Optimal flowrates depend on the inner diameter of the column being used, which, in turn, is dictated by the amount of protein that has been digested. In general, we find that amounts of protein digests in the 50–250 pmol range are best chromatographed at 0.15 mL/min on 2.0–2.1 mm ID columns, whereas larger amounts are best chromatographed at ~0.5 mL/min on 3.9–4.6 mm ID columns. Amounts of digests that are below ~50 pmol can probably be best chromatographed on 1 mm ID columns developed at flowrates in the range of 25–75 µL/min. Unless precautions are taken to minimize dead volumes, significant problems may, however, be encountered in terms of automated peak detection/collection and post column mixing as flowrates are lowered much below 0.15 mL/min *(11)*. Typically, the use of flowrates in the 25–75 µL/min range requires that fused silica tubing be used between the detector and the fraction collector and that a low volume flow cell (i.e., 1–2 µL) be substituted for the standard flow cell in the UV detector.

Several commercially available C-18, reverse-phase supports provide high resolution. These include (but certainly are not limited to) Alltech (Deerfield, IL) Macrosphere, Vydac, Waters' Delta Pak *(12,15)*, and Reliasil. Although under the conditions tested, the resolution obtained on a Brownlee (Applied Biosystems Inc., Foster City, CA) Aquapore C-8 column was somewhat less than that on some other supports *(12,15),* the Aquapore column appeared to have significantly different selectivity and hence provides a valuable means of further purifying peptides that may have been isolated on one of these other supports *(11).* In general, best results appear to be obtained on 300 Å pore size, 5 µ particle size supports. In addition, since peptide resolution has been shown to be directly related to column length *(11,12,15),* whenever possible, the 25 cm versions of these columns should be used. One caveat in regard to the latter is that we have found that a 15 cm Delta-Pak C-18 column provides similar resolution to that obtained on a 25 cm Vydac C-18 column *(15).*

Although the low ultraviolet absorbance, high resolution, and excellent solubilizing properties of the 0.05% TFA/acetonitrile, pH 2.2 buffer system have made it the almost universal mobile phase for reverse-phase HPLC, there are occasions when a different mobile phase might be advantageous. Hence, the differing selectivity of the 5 m*M* pH 6.0 phosphate system *(11)* makes this a valuable mobile phase for detecting posttranslational modifications (such as deamidation) that may be more difficult to detect at the lower pH of the TFA system (where ionization of side chain carboxyl groups would be suppressed). In addition, changing the mobile phase provides another approach for further purifying peptides that were originally isolated in the TFA system.

The sensitivity of HPLC is dictated primarily by the volume in which each peak is eluted. Although sensitivity can be increased by simply decreasing the flowrate (while maintaining a constant gradient time program) eventually, the linear flow velocity on the column will be reduced to such an extent that optimal resolution will be lost. At this point the column diameter needs to be decreased so that an optimal linear flow velocity can be maintained at a lower flowrate. General guidelines for selecting flowrates and column diameters that optimize both resolution and sensitivity have been given *(see above).* In general, the sensitivity of detection is increased as the wavelength is decreased with the practical limit being about 210 nm. As noted in Section 3.4., high sensitivity HPLC requires that the baseline be "balanced" by adding a slightly higher percentage of TFA to buffer A than to buffer B. Although we generally use 0.06% TFA in buffer A compared to 0.052% TFA in buffer B, minor alterations in these compositions can be made accurately following the running of a blank run with each new set of buffers. Finally, an important determinant of sensitivity (that is often overlooked) is the path length of the flow cell. For instance, an HP1090 equipped with a 0.6 cm path length cell provides (at the same flowrate) a threefold increase in sensitivity over that afforded by a Michrom UMA System equipped with a 0.2 cm path length cell.

10. Provided that samples of both the modified and unmodified protein are available, comparative HPLC peptide mapping provides an extremely facile means of rapidly identifying peptides that contain posttranslational modifications. In the case of proteins that have been expressed in *E. coli,* the latter can often serve as the unmodified control since relatively few posttranslational modifications occur in this organism. Certainly the first attempt at comparative HPLC tryptic peptide mapping should be with enzymes such as trypsin or lysyl endopeptidase that have high specificity and the digests should be separated using acetonitrile gradients in 0.05% TFA. Although elution position (as detected by absorbance at 210 nm) provides a sensitive criterion to detect subtle alterations in structure, the value of comparative HPLC peptide mapping can be further enhanced by multiwavelength monitoring and, especially, by on-line or off-line mass spectrometry of the resulting peptide fractions. If comparative peptide mapping fails to reveal any significant changes, it is often worthwhile running the same digest in the pH 6.0 phosphate buffered system that was mentioned above. At this higher pH, some changes, such as deamidation of asparagine and glutamine, produce a larger effect on elution position than at pH 2.2 where ionization of the side chain carboxyl groups would be suppressed. Another possible reason for failing to detect differences on comparative HPLC is that the peptide(s) containing the modifications are either too hydrophilic to bind or too hydrophobic to elute from reverse-phase supports. Hence, in addition to trying a different HPLC solvent system, another approach that may be taken to expand the capabilities of comparative HPLC peptide mapping is to try a different proteolytic enzyme, such as chymotrypsin or Protease V8. Finally, the failure to observe any difference on comparative HPLC peptide mapping may result from loss of the posttranslational modification during either the cleavage or the subsequent HPLC. Assignment of disulfide bonds is one example where this can be a problem. Even if the reduction and carboxymethylation steps are deleted from the method that is outlined above (so that native disulfide bonds are left intact), disulfide interchange may occur during enzymatic cleavage, which is typically carried out at pH 8. This problem can be addressed by either going to shorter digestion times *(16)* or by carrying out the cleavage under acidic conditions where disulfide interchange is less likely to occur. For this reason, pepsin (which is active in 5% formic acid) digests are often used for isolating peptides containing disulfide bonds. Providing that the control sample is reduced, comparative HPLC peptide mapping can be used to rapidly isolate disulfide linked peptides. If the sequence of the protein of interest is known then laser desorption mass

spectrometry (before and after reduction of the disulfide linked peptide) can be used to identify the two peptides that are disulfide bonded.

References

1. Krishna, R. G. and Wold, F. (1993) in *Methods in Protein Sequence Analysis* (Imahori, K. and Sakiyama, F., eds.), Plenum, New York, pp. 167–171.
2. Harris, R. (1993) *ABRF News* **4(3),** 8–10.
3. Williams, K. R. and Carr, S. A. (1995) in *Encyclopedia of Molecular Biology and Biotechnology* (Meyers, R. A., ed.), VCH, New York, in press.
4. Kawasaki, H., Emori, Y., and Suzuki, K. (1990) *Anal. Biochem.* **191,** 332–336.
5. Stone, K. L. and Williams, K. R. (1993) in *A Practical Guide to Protein and Peptide Purification for Microsequencing,* 2nd ed. (Matsudaira, P. T., ed.), Academic, New York, pp. 43–69.
6. Rosenfeld, J., Capdevielle, J., Guillemot, J., and Ferrara, P. (1992) *Anal. Biochem.* **203,** 173–179.
7. Aebersold, R. H., Leavitt, J., Saavedra, R. A., Hood, L. E., and Kent, S. B. (1987) *Proc. Natl. Acad. Sci. USA* **84,** 6970–6974.
8. Fernandez, J., DeMott, M., Atherton, D., and Mische, S. M. (1992) *Anal. Biochem.* **201,** 255–264.
9. Stone, K. L., McNulty, D. E., LoPresti, M. B., Crawford, J. M., DeAngelis, R., and Williams, K. R. (1992) in *Techniques in Protein Chemistry III* (Angeletti, R., ed.), Academic, New York, pp. 23–34.
10. Laemmli, U. K. (1970) *Nature* **227,** 680.
11. Stone, K. L., Elliott, J. I., Peterson, G., McMurray, W., and Williams, K. R. (1990) in *Methods in Enzymology* (McCloskey, J., ed.), Academic, New York, pp. 389–412.
12. Stone, K. L., LoPresti, M. B., and Williams, K. R. (1990) in *Laboratory Methodology in Biochemistry: Amino Acid Analysis and Protein Sequencing* (Fini, C., Floridi, A., Finelli, V. N., and Wittman-Liebold, B., eds.), CRC, Boca Raton, FL, pp. 181–205.
13. Sun, Y., Zhou, Z., and Smith, D. (1989) in *Techniques in Protein Chemistry* (Hugli, T., ed.), Academic, New York, pp. 176–185.
14. (1991) in *High Performance Liquid Chromatography of Peptides and Proteins: Separation, Analysis and Conformation* (Mant, C. T. and Hodges, R. S., eds.), CRC, Boca Raton, FL.
15. Stone, K. L., LoPresti, M. B., Crawford, J. M., DeAngelis, R., and Williams, K. R. (1991) in *High Performance Liquid Chromatography of Peptides and Proteins: Separation, Analysis and Conformation* (Mant, C. T. and Hodges, R. S., eds.), CRC, Boca Raton, FL, pp. 669–677.
16. Glocker, M. O., Arbogast, B., Schreurs, J., and Deinzer, M. L. (1993) *Biochemistry* **32,** 482–488.

CHAPTER 8

Urea and Guanidine Hydrochloride Denaturation Curves

Bret A. Shirley

1. Introduction

Proteins are synthesized as a linear chain of amino acids. In order to become biologically active, a protein must fold and adopt one out of an enormous number of possible conformations. This conformation, which we will refer to as the native state, exists in solution as a very compact, highly ordered structure. This native state structure results from a delicate balance between large and opposing forces. In order to form the native state, the forces that favor the unfolded state (mainly conformational entropy) must be overcome by the covalent and noncovalent interactions favoring the folded protein (*see* Chapter 1). Under physiological conditions the native (folded) and the denatured (unfolded) states of a protein are in equilibrium. The free energy change, ΔG, for the equilibrium reaction

$$\text{Native (N)} \rightleftharpoons \text{Denatured (D)} \tag{1}$$

is referred to as the conformational stability of a protein. The determinants of native state stability in aqueous solutions are the amino acid sequence of the protein as well as the variable conditions of pH, temperature, and the concentration of salts and ligands *(1,2)*. Although the native conformation is essential for activity, the conformational stability is remarkably low. The native state of most naturally occurring proteins is only about 5–15 kcal/mol more stable than its unfolded conformations *(3)*.

From: *Methods in Molecular Biology, Vol. 40: Protein Stability and Folding: Theory and Practice*
Edited by: B. A. Shirley

Denaturation curves using urea and guanidine hydrochloride are a convenient method for estimating the conformational stability of a protein. Under physiological conditions, the equilibrium of most proteins so greatly favors the native state that measuring the equilibrium constant for the folding/unfolding reaction is difficult. However, in the presence of urea or guanidine hydrochloride the folding/unfolding equilibrium will be shifted so that an equilibrium constant can be accurately measured, and an extrapolation can be made to physiological conditions.

This chapter will describe how urea and guanidine hydrochloride denaturation curves are determined and analyzed. The value derived from these experiments that will be of most interest is the free energy change for the folding/unfolding reaction at 25°C in the absence of denaturant. This value will be designated $\Delta G(H_2O)$ and will be referred to as the conformational stability of a protein. The terms *denaturation* and *unfolding* will be used interchangeably throughout the chapter and should be considered equivalent.

2. Materials

2.1. Preparation of Urea and Guanidine Hydrochloride Stock Solutions

Urea and guanidine hydrochloride can both be purchased in highly pure forms from a number of sources (e.g., Schwarz-Mann Biotech [Cleveland, OH] or United States Biochemical [Cleveland, OH]). Urea stock solutions can be prepared by weight and their molarity checked by refractive index measurements. Molarities calculated by these two methods should agree within 1%. It should be noted that solutions of urea slowly decompose to form cyanate and ammonium ions *(4)*. Since cyanate ions are capable of chemically modifying the amino groups of a protein *(5)*, urea stock solutions should be prepared and used within one day. By contrast, solutions of guanidine hydrochloride are stable for months. Since guanidine hydrochloride is a hygroscopic compound, its molarity is most conveniently determined by refractive index measurements after the solution has been prepared. Table 1 gives useful information for preparing urea and guanidine hydrochloride stock solutions.

A general method for preparing ~100 mL of a ~10*M* urea stock solution is the following:

1. Place a 150 mL beaker containing a stir bar onto an accurate top loading balance and tare.

Table 1
Urea and Guanidine Hydrochloride Stock Solutions

Property	Urea	GuHCl
Mol wt	60.056	95.533
Solubility, 25°C	10.49*M*	8.54*M*
d/d_o[a]	$1 + 0.2658W + 0.0330W^2$	$1 + 0.2710W + 0.0330W^2$
Molarity[b]	$117.66(\Delta N) + 29.753(\Delta N)^2 + 185.56(\Delta N)^3$	$57.147(\Delta N) + 38.68(\Delta N)^2 - 91.60(\Delta N)^3$
Grams of denaturant/ gram of water to prepare		
6*M*	0.495	1.009
8*M*	0.755	1.816
10*M*	1.103	—

[a] W is the weight fraction of denaturant in the solution, d is the density of the solution and d_o is the density of water *(6)*.

[b] ΔN is the difference in refractive index between the denaturant solution and water (or buffer) at the sodium D line. The equation for urea solutions is based on data from Warren and Gordon *(7)*, and the equation for GuHCl solutions is from Nozaki *(8)*.

Table 2
Example of the Preparation of a Urea Stock Solution

1. Tare a 150 mL beaker containing a stir bar on a top loading balance with an accuracy of ± 0.02 g.
2. Add ~60 g urea to the beaker and weigh (59.91 g).
3. Add 0.649 g of MOPS buffer (sodium salt), 1.8 mL of 1*M* HCl and ~52 mL of distilled water and weigh the solution again (114.65 g).
4. Allow the urea to dissolve with stirring and check the pH. If necessary, add a weighed amount of 1*M* HCl to adjust to pH 7.0.
5. Prepare a 30 m*M* MOPS buffer solution, pH 7.0.
6. Determine the refractive index of the urea stock solution (1.4173) and the buffer (1.3343). Therefore, $\Delta N = 1.4173 - 1.3343 = 0.0830$.
7. Calculate the urea molarity from ΔN using the equation in Table 1: $M = 10.08$.
8. Calculate the urea molarity based on the recorded weights. The density is calculated with the equation given in Table 1: Weight fraction of urea $(W) = 59.91/114.65 = 0.5226$; therefore $d/d_o = 1.148$. Therefore volume of the solution = 114.65/1.148 = 99.88 mL. Therefore urea molarity = 59.91/60.056/0.09988 = 9.99*M*.
9. Since the molarities calculated in steps 7 and 8 differ by <1%, the solution can be used to determine a urea unfolding curve.

2. Add to the beaker ~60 g of urea, the desired amount of buffer, and ~50 mL of water.
3. Cover the beaker and allow the urea to dissolve with gentle stirring.
4. Adjust the solution to the final desired pH with acid or base.
5. Take a final weight of the solution.
6. Prepare ~100 mL of a buffer solution at the same pH and buffer concentration as the urea stock solution.
7. Measure the refractive index of both the buffer and urea stock solutions.
8. Use the information in Table 1 to calculate the concentration of the urea stock solution both by weight and refractive index.

Table 2 gives a specific example of the preparation and concentration determination for urea stock solution at pH 7 containing 30 m*M* MOPS.

3. Methods

3.1. Following Unfolding

In principle, any physical technique that is capable of distinguishing the native and denatured states of a protein can be used to monitor the unfolding transition. Biological activity measurements, immunochemical techniques, hydrodynamic methods, such as viscosity, and the spectroscopic techniques of circular dichroism (Chapter 5), nuclear magnetic

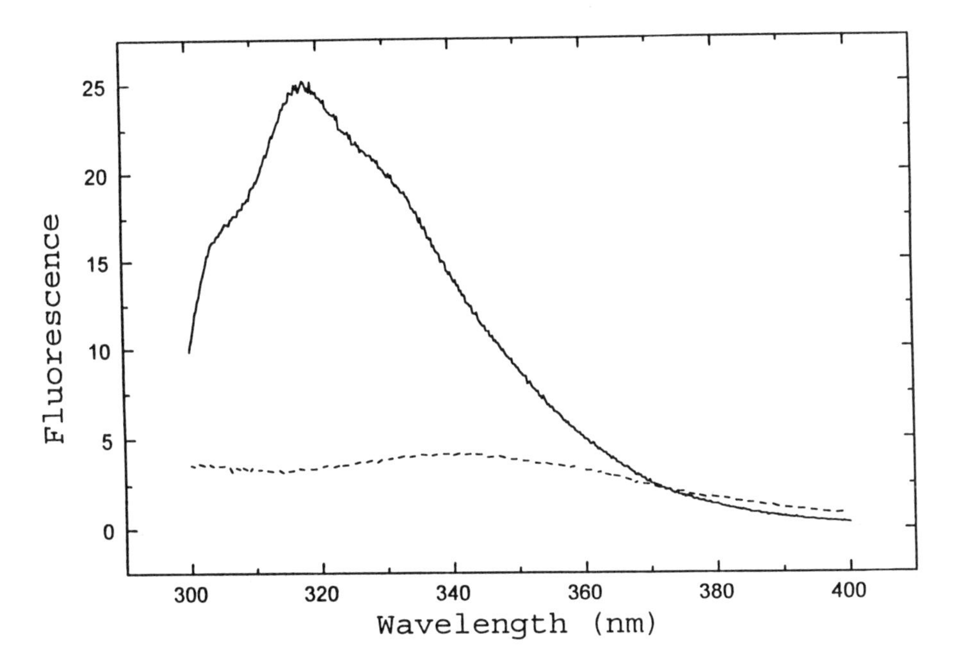

Fig. 1. Intrinsic fluorescence emission spectra upon excitation at 278 nm for folded (——) and urea unfolded (– – – –) RNase T1. (Figure supplied courtesy of Lanette Fee, Texas A&M University.)

resonance, fluorescence (Chapter 3) and ultraviolet difference spectroscopy (Chapter 4) can all be used to follow unfolding. Spectroscopic techniques are most often used and will be discussed exclusively here.

There are several considerations in choosing a spectroscopic technique to follow unfolding. The initial step in deciding on a technique is to determine the spectra of the native and denatured states of the protein. It is important to pick a technique and a wavelength where the physical property being monitored will differ significantly for the native and denatured states of the protein. As an example, the fluorescence spectra for native and denatured ribonuclease T1 are shown in Fig. 1. When ribonuclease T1 unfolds, there is a five- to sixfold decrease in fluorescence intensity when emission is monitored at 320 nm after excitation at 278 nm.

The sensitivity of the technique being used is of importance, especially when there are limited amounts of protein available. The urea denaturation of ribonuclease T1 requires <1 mg of protein when unfolding is followed with fluorescence, but as much as 100 mg may be required to determine a similar curve if circular dichroism were used. In general, fluorescence requires far less protein than other spectroscopic techniques.

In choosing a spectroscopic technique it is sometimes important to know the properties of the protein that will be monitored. Some techniques depend on global changes in protein structure, whereas others depend only on local effects. Fluorescence and absorption depend on the change in environment of the chromophoric amino acid side chains (tyrosine and tryptophan) and hence are sensitive to changes in tertiary structure. Circular Dichroism measurements below 230 nm depend largely on secondary structure, whereas those above 250 nm depend mainly on the change in environment of the aromatic amino acids *(9)*. These are important considerations when attempting to gain information about the completeness of the transition or the mechanism of folding (*see* Section 3.5.). It should also be noted that difference spectroscopy is more difficult to use than other spectroscopic techniques since twice as many solutions will have to be prepared.

3.2. Equilibrium and Reversibility

Before measurements are made it is imperative that the folding/unfolding reaction be at equilibrium. Depending on the protein and the conditions, the time required to reach equilibrium can range from milliseconds to days. The temperature at which measurements are taken can have a large effect on the equilibration time. For ribonuclease T1 the time required to reach equilibrium decreases from hours to minutes when the temperature is increased from 20–30°C. To establish the time required to reach equilibrium, the chosen observable parameter y (e.g., fluorescence intensity) should be measured as a function of time until no further changes are observed (the time required to reach equilibrium generally decreases as the denaturant concentration or temperature increases).

It is also necessary to establish the reversibility of the unfolding reaction. This can be achieved by allowing the reaction to reach equilibrium at a denaturant concentration sufficient to completely denature the protein, and then diluting the protein to conditions where it is folded. For a reversible reaction, the value of y measured after complete denaturation and refolding should be identical to the value measured directly.

Proteins that contain free sulfhydryls present special problems because the formation of nonnative disulfide bonds can lead to irreversibility. For proteins containing only free sulfhydryls, a reducing agent such as

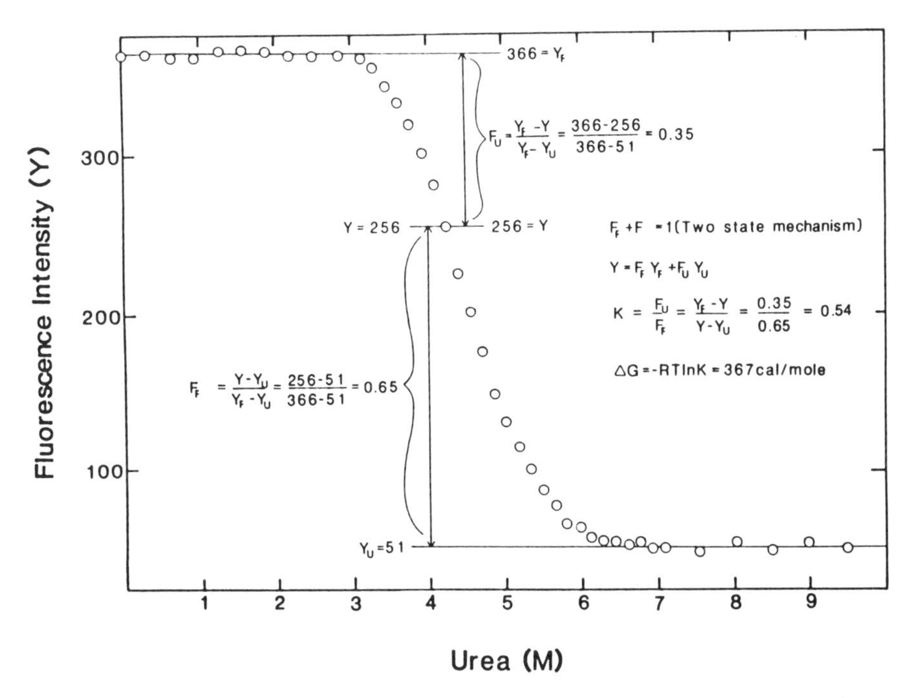

Fig. 2. The analysis of a urea denaturation curve of ribonuclease T1 (30 m*M* MOPS, pH 7.0, 25°C) assuming a two-state mechanism. The fluorescence intensity at 320 nm was measured after excitation at 278 nm. Reprinted with permission from ref. *11*.

dithiothreitol can be added to prevent disulfide bond formation. For those proteins that have both disulfide bonds and free sulfhydryls, disulfide interchange can be minimized by working at low pH *(10)*.

3.3. Determining Urea and Guanidine Hydrochloride Denaturation Curves

Each point of a urea or guanidine hydrochloride denaturation curve is determined with a separate solution. If sufficient care is taken, experimental solutions can be prepared such that the limiting factor in defining a denaturation curve will be the accuracy of the measurement of the physical property used to follow unfolding (*see* Note 1).

A separate solution was used to determine each point for the urea denaturation curve in Fig. 2. Experimental solutions are prepared volumetrically from protein, urea, and buffer stock solutions. The urea con-

Table 3
Determination of a Urea Denaturation Curve

Urea, mL[a]	Buffer, mL[b]	Protein, mL[c]	Urea, *M*[d]	*y*[e]
0.20	2.80	0.20	0.62	362
0.30	2.70	0.20	0.94	364
0.40	2.60	0.20	1.25	365
0.50	2.50	0.20	1.56	369
0.60	2.40	0.20	1.87	368
1.25	1.75	0.20	3.90	301
1.30	1.70	0.20	4.06	282
1.35	1.65	0.20	4.21	256
1.40	1.60	0.20	4.37	227
1.45	1.55	0.20	4.53	203
1.50	1.50	0.20	4.68	177
1.55	1.45	0.20	4.83	150
1.60	1.40	0.20	5.00	133
1.65	1.35	0.20	5.15	115
1.70	1.30	0.20	5.31	101
2.20	0.80	0.20	6.87	51
2.25	0.75	0.20	7.02	51
2.40	0.60	0.20	7.49	50
2.55	0.45	0.20	7.96	53
2.70	0.30	0.20	8.42	49

[a]9.99*M* urea stock solution (30 m*M* MOPS, pH 7).
[b]Buffer stock solution (30 m*M* MOPS, pH 7).
[c]0.01 mg/mL ribonuclease T1 stock solution (30 m*M* MOPS, pH 7).
[d]Urea molarity = [9.99][(mL urea)/(total volume of experimental solution in mL)].
[e]Fluorescence intensity (278 nm excitation, 320 nm emission) was measured with a Perkin-Elmer MPF 44B spectrofluorometer. Solutions were incubated at 25°C for 6 h before measurements were taken *(12)*.

centration of each solution is determined by varying the amount of urea and buffer stock solution added. After measurements have been made, it is advisable to measure the pH of at least the solutions in the transition region to insure that a constant pH has been maintained.

In order to accurately analyze a denaturation curve, the transition region and the pre- and posttransition baselines must be sufficiently defined. It is therefore advisable to determine at least five points in each of these regions. Table 3 gives an example of the preparation and measurement of experimental solutions for a typical denaturation curve of ribonuclease T1.

3.4. Analysis of Denaturation Curves

The thermodynamic parameter that is most readily determined from urea or guanidine hydrochloride denaturation curves is the free energy change, ΔG_U, for the folding/unfolding reaction. The classic method for analyzing denaturation curves assumes a two-state mechanism for the folding/unfolding reaction. This method assumes that at equilibrium only the fully folded or native protein and the fully unfolded or denatured protein are present at significant concentrations. Of course, this assumption cannot be valid since intermediates must necessarily exist along the pathway between native and denatured proteins. However, for many proteins the concentration of intermediates at equilibrium has been found to be small relative to the concentrations of the native and denatured proteins. Disregarding these intermediates greatly simplifies evaluation of the thermodynamic parameters, and the error introduced by this assumption is small. The two-state assumption must be validated for each protein under a given set of conditions. This point is discussed further in Section 3.5.

By assuming a two-state mechanism, only the folded and unfolded forms of the protein are present at significant concentrations and

$$f_F + f_U = 1 \tag{2}$$

where f_F and f_U represent the fraction of total protein in the folded and unfolded conformations, respectively. The observed value at any point on the transition curve is given by

$$y = y_F f_F + y_U f_U \tag{3}$$

where y is any observable parameter chosen to follow unfolding, and y_F and y_U represent the values of y characteristic of the folded and unfolded protein. The values of y_F and y_U for any point in the transition region are obtained by extrapolation of the pre- and posttransition baselines, which is generally achieved by a least squares analysis *(13)* (*see* Note 2). Combining Eqs. (2) and (3) yields

$$f_U = (y_F - y)/(y_F - y_U) \tag{4}$$

The equilibrium constant K_U, and the free energy change, ΔG_U, for the folding/unfolding reaction can be calculated using

$$K_U = f_U/(1 - f_U) = f_U/f_F = (y_F - y)/(y - y_U) \tag{5}$$

Table 4
Analysis of a Urea Denaturation Curve

Urea, M	y^a	f_U	K_U	ΔG_U, cal/mol[b]
3.90	301	0.21	0.26	800
4.06	282	0.27	0.36	600
4.21	256	0.35	0.53	370
4.37	227	0.44	0.79	140
4.53	203	0.52	1.08	–40
4.68	177	0.60	1.50	–240
4.83	150	0.69	2.19	–460
5.00	133	0.74	2.87	–620
5.15	115	0.79	3.93	–810
5.31	101	0.84	5.35	–990

[a]These data are the points in the transition region of Fig. 2. For each point, $y_F = 366$ and $y_U = 51$.
[b]A least squares analysis of this data fit to Eq. 7 yields: $\Delta G(H_2O)$ = 5.7 kcal/ mol, m = 1280 (cal/mol/M), and $[\text{urea}]_{1/2} = 4.50M$ urea. These parameters describe the solid line in Fig. 3.

and

$$\Delta G_U = -RT\ln K_U = -RT\ln[(y_F - y)/(y - y_U)] \tag{6}$$

where R is the gas constant (1.987 calories/deg/mol) and T is the absolute temperature (K). The equilibrium constant can be measured most accurately near the midpoint of a denaturation curve, and for values of K_U outside the range 0.1–10 ($0.1 \leq f_U \leq 0.9$) the error becomes substantial. As an example, the analysis of a urea denaturation curve of ribonuclease T1 assuming a two-state mechanism is shown in Fig. 2 and Table 4.

To obtain an estimate of $\Delta G(H_2O)$ from these studies, accurately measurable values of the equilibrium constant, K_U, are obtained under denaturing conditions, and an attempt is made to extrapolate back to zero denaturant concentration. When data such as those in Fig. 2 are analyzed as described above, ΔG_U is generally found to vary linearly with denaturant concentration. The simplest and at present most widely used model for estimating $\Delta G(H_2O)$ is the linear extrapolation model. This model assumes that the linear dependence of ΔG_U on denaturant concentration observed in the transition region continues to zero concentration of denaturant. A least squares analysis can be used to fit the data to an equation of the form

$$\Delta G_U = \Delta G(H_2O) - m[\text{denaturant}] \tag{7}$$

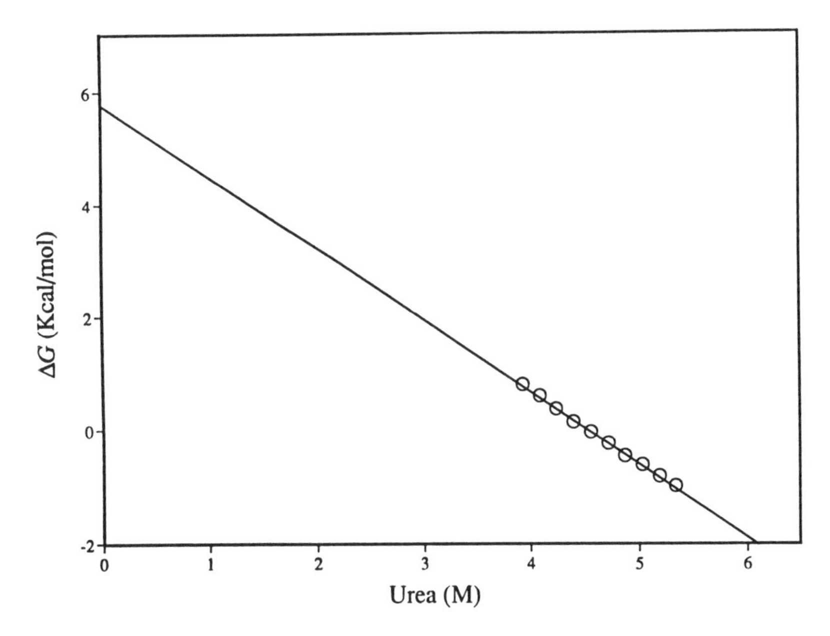

Fig. 3. The variation of ΔG_U with urea concentration for ribonuclease T1. The data points are from the transition region of Fig. 2 as analyzed in Table 4.

The value of *m* in this equation is a measure of the dependence of free energy on denaturant concentration, and depends on the amount and composition of the polypeptide chain that is freshly exposed to solvent on unfolding *(14–17)*. The determination of $\Delta G(H_2O)$ for ribonuclease T1 using linear extrapolation can be seen in Table 4 and Fig. 3.

It should be noted that there are other methods available for analyzing urea and guanidine hydrochloride denaturation curves. These methods are more complicated than the linear extrapolation model and at present there do not seem to be sufficient reasons to justify using them for estimating $\Delta G(H_2O)$ from equilibrium studies. These methods will not be discussed here but have been reviewed in detail elsewhere *(11,19)*.

3.5. Validity of the Two-State Approximation

There are several experimental characteristics expected for a two-state process. Two of these characteristics are observed when determining urea or guanidine hydrochloride denaturation curves. To avoid the incorrect interpretation of experimental data, it is best to apply these tests before attempting a thermodynamic analysis.

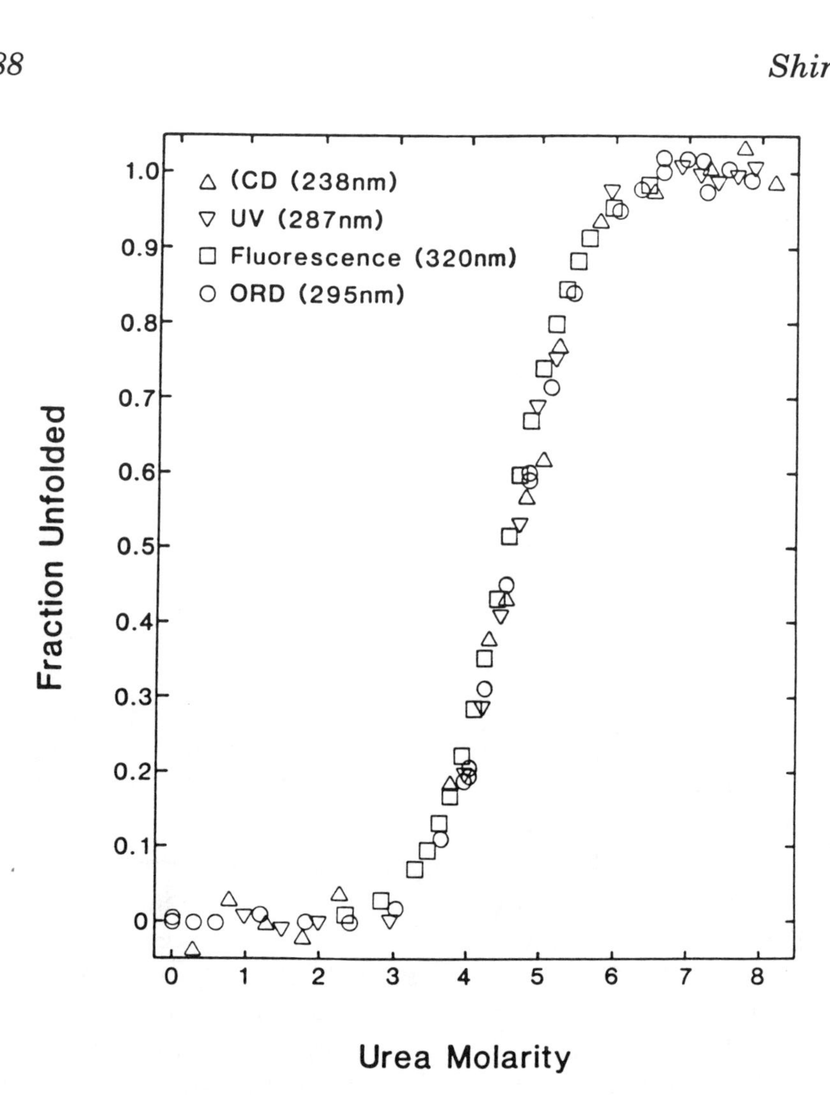

Fig. 4. Fraction unfolded ribonuclease T1 as a function of urea molarity from data obtained using the techniques indicated on the figure to follow unfolding. The fraction unfolded ribonuclease T1 was calculated using Eq. (4). Reprinted with permission from ref. *18.*

1. The transition from native to denatured protein should be characterized by an abrupt, single step and should not contain a plateau or even a shoulder.
2. When unfolding is followed with several different techniques, f_{obs}, the observed fraction of unfolded protein, should be independent of the observable parameter. This is sometimes referred to as the multiple-variable test *(20)* and is illustrated in Fig. 4 for ribonuclease T1 where four different

observable parameters were used to follow unfolding. Two of these techniques reflect the disruption of secondary structure, while two reflect disruption of tertiary structure. The presence of significant concentrations of intermediates at equilibrium will affect the value of f_{obs}. The effect of intermediates will generally not be the same for all observable parameters, and plots of f_{obs} vs denaturant concentration will be noncoincident. If the process is two-state then $f_{obs} = f_U$ and is thus clearly independent of the method of measurement.

Even if these characteristics are observed, the folding mechanism under a given set of conditions is not necessarily two-state. However, if either of these characteristics is not observed, the unfolding mechanism cannot be described as two-state. Other experimental characteristics expected for a protein with a two-state folding mechanism are discussed in Chapters 9 and 14.

4. Notes

1. Preparing a denaturation curve requires a large number of pipeting steps. It is important to use the highest quality pipets and check their accuracy frequently. The use of motorized pipets (e.g., Rainin Instruments, Woburn, MA) can make the task easier, and can increase precision by reducing the error that can result from repeated pipeting using manual pipets.
2. The pre- and posttransition baselines may either be flat or contain a significant slope. For ribonuclease T1, the example used in this chapter, the baselines are flat. However, for many proteins this is not the case and a least squares analysis of the baselines will be necessary to determine the value of y_F and y_U in the transition region.
3. Always use the highest quality glass for preparing all solutions. Experimental solutions can be most conveniently prepared in disposable borosilcate glass test tubes and sealed with parafilm for incubation.
4. Error in experimental measurements can be reduced by *never* removing the cuvet from the spectrophotometer during the experiment. Use a vacuum system to remove each solution before the next is added. Using this technique will result in much less error than removing, cleaning, and returning the cuvet to the instrument after each point on the curve. This technique also has the advantage of being much faster.
5. When using fluorescence it is important that any contribution from the denaturant be subtracted out of the final observed intensities before calculations are made. This can be achieved by preparing several solutions of intermediate denaturant concentration. The fluorescence intensity of each of these solutions can be measured and a least squares analysis used to

determine the fluorescence intensity contribution of the urea or buffer at any concentration. This correction can then be applied to the experimental solutions prior to analysis.

References

1. Alber, T. (1989) *Annu. Rev. Biochem.* **58**, 765–798.
2. Pace, C. N. (1990) *Trends Bichem. Sci.* **15**, 14–17.
3. Pace, C. N. (1975) *Crit. Rev. Biochem.* **3**, 1–43.
4. Hagel, P., Gerding, J. J. T., Feiggen, W., and Bloemendal, H. (1971) *Biochim. Biophys. Acta.* **243**, 366–370.
5. Stark, G. R. (1965) *Biochemistry* **4**, 1030–1036.
6. Kawahara, K. and Tanford, C. (1966) *J. Biol. Chem.* **241**, 3228–3232.
7. Warren, J. R. and Gordon, J. A. (1966) *J. Phys. Chem.* **67**, 1524–1527.
8. Nozaki, Y. (1972) in *Methods in Enzymology*, vol. 26 (Hirs, C. H. W. and Timasheff, S. N., eds.), Academic, New York, pp. 43–50.
9. Canter, C. R. and Timasheff, S. N. (1982) in *The Proteins*, vol. 5 (Neurath, H. and Hill, R. L., eds.), Academic, Orlando, FL, pp. 145–306.
10. Creighton, T. E. (1978) *Prog. Biophys. Mol. Biol.* **33**, 231–236.
11. Shirley, B. A. (1992) in *Stability of Protein Pharmaceuticals, Part A: Chemical and Physical Pathways of Protein Degradation* (Ahern, T. J. and Manning, M. C., eds.), Plenum, New York, pp. 167–194.
12. Shirley, B. A., Stanssens, P., Steyaert, J., and Pace, C. N. (1989) *J. Biol. Chem.* **264**, 11,621–11,625.
13. Pace, C. N., Shirley, B. A., and Thomson, J. A. (1989) in *Protein Structure: A Practical Approach* (Creighton, T. E., ed.), IRL, Oxford, pp. 311–330.
14. Tanford, C. (1964) *J. Am. Chem. Soc.* **86**, 2050–2090.
15. Greene, R. F., Jr. and Pace, C. N. (1974) *J. Biol. Chem.* **249**, 5388–5393.
16. Schellman, J. A. (1978) *Biopolymers* **17**, 1305–1322.
17. Schellman, J. A. (1987) *Biopolymers* **26**, 549–559.
18. Thomson, J. A., Shirley, B. A., Grimsley, G. R., and Pace, C. N. (1989) *J. Biol. Chem.* **264**, 11,614–11,620.
19. Pace, C. N. (1986) in *Methods in Enzymology*, vol. 131 (Hirs, C. H. W. and Timashef, S. N., eds.), Academic, New York, pp. 266–280.
20. Brandts, J. F. (1969) in *Structure and Stability of Biological Macromolecules* (Timasheff, S. N. and Fasman, G. D., eds.), Marcel Dekker, New York, pp. 213–290.

CHAPTER 9

Differential Scanning Calorimetry

Ernesto Freire

1. Introduction

Differential scanning calorimetry (DSC) is now generally accepted as the technique of choice to determine the energetics of protein folding/unfolding transitions and the thermodynamic mechanisms underlying those reactions. Despite this fact, the theoretical framework necessary for the analysis of calorimetric data is not available to the general scientist in a comprehensive way. The purpose of this chapter is to present in a succinct form the fundamental theory of differential scanning calorimetry, to summarize the information that can be obtained with this technique, and to alert the reader of some potential problems in data interpretation.

Differential scanning calorimetry measures the apparent molar heat capacity of a protein or other macromolecule as a function of temperature. Subsequent manipulation of this quantity yields a complete thermodynamic characterization of a transition. In general, three different types of information can be obtained from DSC:

1. The absolute partial heat capacity of a molecule;
2. The overall thermodynamic parameters (enthalpy change [ΔH], entropy change [ΔS], and heat capacity change [ΔC_p]) associated with a temperature induced transition; and
3. The partition function and concomitantly the population of intermediate states and their thermodynamic parameters.

In this chapter we will restrict our discussion to monomeric protein systems undergoing reversible folding/unfolding transitions under equi-

From: *Methods in Molecular Biology, Vol. 40: Protein Stability and Folding: Theory and Practice*
Edited by: B. A. Shirley

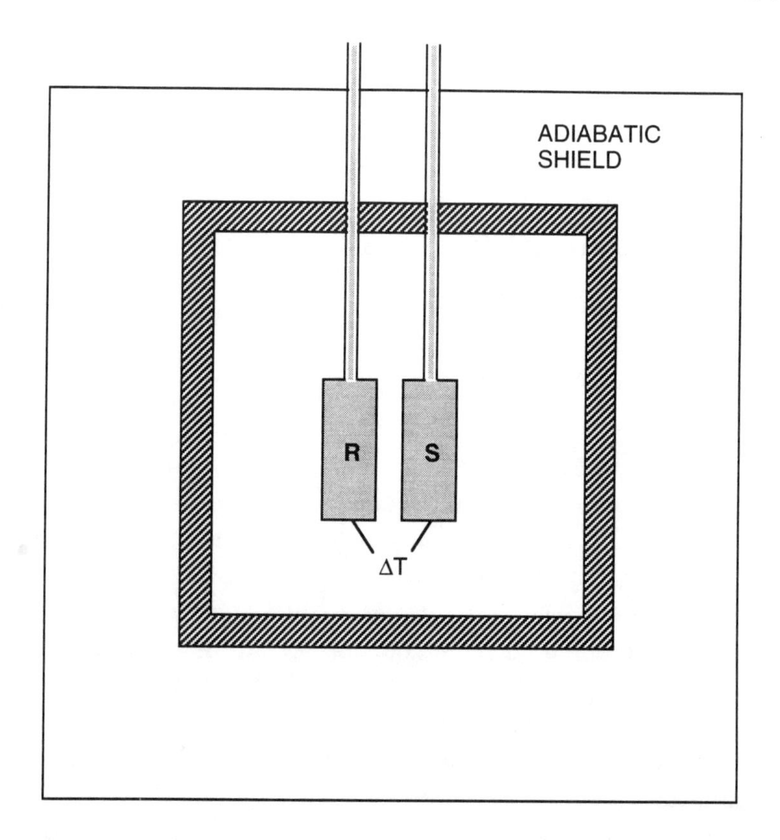

Fig. 1. Schematic illustration of the cell assembly of a differential scanning calorimeter. The temperature difference (ΔT) between the sample and reference cells (S and R) is maintained equal to zero by a feedback mechanism as the temperature is increased or decreased at a constant scanning rate. The differential power that needs to be applied to the sample cell in order to maintain $\Delta T = 0$ is monitored continuously as a function of temperature. This differential power is directly proportional to the heat capacity difference between the cells and constitutes the basic quantity measured by the instrument.

librium conditions; however, the equations and the general treatment are applicable to other macromolecular systems undergoing equilibrium monomolecular isomerization reactions.

2. Measurement

Figure 1 provides a schematic illustration of the cell compartment of a differential scanning calorimeter. It consists of two identical cells, one of which is used to hold the sample under investigation and the other the reference solution (usually the same buffer used in the preparation of the sample solution). The two cells operate in a differential mode such that the difference in heat capacity between them is the quantity measured. This is accomplished by continuously monitoring the differential electrical power required to maintain the temperature difference between the two cells equal to zero as the temperature of the entire assembly is scanned at a constant rate. This differential electrical power (usually given in μW = μJ/s or in μcal/s) after normalization by the scanning rate yields the difference in heat capacity between the two cells (units of μJ/deg or μcal/deg). Ideally, if the two cells were identical, a single scan would be enough to determine the difference in heat capacity between the sample and the reference solution. In practice, however, the two cells are never perfectly matched and two separate scans are required in order to subtract and eliminate instrumental effects.

A properly performed calorimetric experiment consists of a minimum of three runs: a buffer scan and two scans of the sample under study. The second scan of the sample is normally used to assess the reversibility of the transition. Usually, the reversibility is expressed in terms of the percentage of the original signal that is recovered in the second scan. This is an important check since an analysis based on equilibrium thermodynamics *cannot* be applied to a sample that undergoes an irreversible transition. Protein samples for calorimetric measurements need to be prepared following very rigorous and exact procedures that guarantee the purity and integrity of the protein under study. Protein concentrations must be determined with high accuracy since any error in concentration will be reflected directly in the measured thermodynamic parameters. For example, a 5% error in concentration will introduce a 5% error in the measured enthalpy change. Samples for calorimetric analysis must be dialyzed against the desired buffer, centrifuged to remove aggregates and dust particles, and degassed. The protein concentration must be measured after all sample preparation steps have been performed and the sample is ready to load in the calorimeter cell. The degassed buffer used in the final dialysis during sample preparation must be loaded into the reference cell of the instrument in order to insure that the solvents in the two cells are identical. This is especially critical if absolute values of the apparent heat capacities are needed.

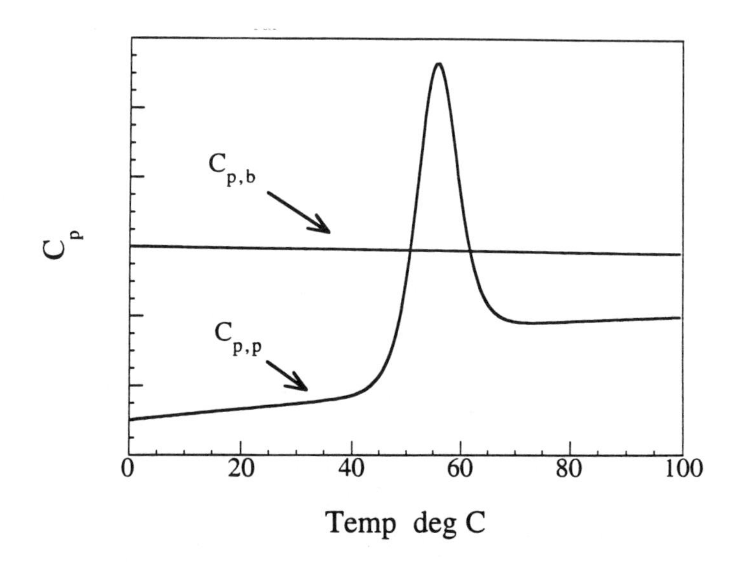

Fig. 2. Schematic representation of a typical calorimetric scan of a globular protein. The heat capacity of the protein solution ($C_{p,p}$) displays the characteristic peak associated with the unfolding transition. The heat capacity of the buffer solution ($C_{p,b}$) is higher than that of the protein solution. The partial heat capacity of globular proteins in the native state is on the order of 0.3 cal/K/g at 20°C *(1)*.

Figure 2 illustrates a typical set of results for a globular protein. The quantity obtained from a calorimetric scan is the heat capacity of the solution present in the calorimeter cell. In order to determine the heat capacity of the protein, the data from the protein solution scan and the buffer scan are needed. For the buffer (solvent) scan the measured heat capacity can be written as:

$$C_{p,b} = m_b \cdot C_{p,b}{}^o \quad (1)$$

where m_b is the mass of solvent in the cell and $C_{p,b}{}^o$ is the specific heat capacity of the buffer solution. Similarly, the heat capacity of the protein solution can be written as:

$$C_{p,p} = m_p \cdot C_{p,p}{}^o + m_b' \cdot C_{p,b}{}^o \quad (2)$$

where $C_{p,p}{}^o$ is the heat capacity of the protein per unit mass, m_p is the mass of protein in the calorimetric cell, and m_b' is the mass of solvent. $C_{p,p}{}^o$ can be obtained as follows:

$$C_{p,p}{}^o = [(C_{p,p} - C_{p,b}) + (m_b - m_b') \cdot C_{p,b}{}^o] / m_p \quad (3)$$

The quantity $(m_b - m_b')$ is equal to the mass of solvent displaced by the protein and can be written in terms of the partial specific volume of the protein as Eq. (1):

$$C_{p,p}{}^o = (C_{p,p} - C_{p,b})/m_p + C_{p,b}{}^o \cdot (V_p{}^o/V_b{}^o) \quad (4)$$

where $V_p{}^o$ and $V_b{}^o$ are the partial specific volumes of the protein and solvent, respectively. The molar heat capacity function (C_p) is simply equal to $C_{p,p}{}^o$ multiplied by the molecular weight of the protein. C_p is the main quantity measured by DSC and constitutes the center of our discussion.

3. Absolute Heat Capacity of Proteins

In the past, most analyses of DSC data were aimed at obtaining transition-related thermodynamic parameters. It was realized recently, however, that the absolute values of $C_{p,p}{}^o$ and C_p also contain fundamental information regarding the structural state of a protein. For example, it has been shown that the heat capacity of a protein in the unfolded state $(C_{p,u})$ can be calculated with high accuracy from the amino acid sequence *(2)*. If all the constituent groups are exposed to water, then $C_{p,u}$ obeys simple additivity rules and can be expressed in terms of the individual contributions from the amino acid side chains and the peptide backbone:

$$C_{p,u} = \left(\sum_{i=1}^{20} n_i \cdot C_{p,i}\right) + (N_{AA} - 1) \cdot C_{p,\,bb} + C_{p,\mathrm{NH2}} + C_{p,\mathrm{COOH}} \quad (5)$$

where the index in the summation refers to the 20 amino acids, n_i is the number of amino acids of type i and $C_{p,i}$ the molar heat capacity of its side chain; N_{AA} is the number of amino acids in the protein; $C_{p,bb}$ is the heat capacity of a peptide backbone unit (-CHCONH-), $C_{p,\mathrm{NH2}}$ the heat capacity of the amino terminus, and $C_{p,\mathrm{COOH}}$ the heat capacity of the carboxy terminus. Table 1 summarizes the parameters necessary to estimate $C_{p,u}$ from the amino acid sequence of a protein. The parameters in Table 1 were obtained by a polynomial fit of the data published by Privalov and Makhatadze *(2)* (Freire and Muphy, unpublished results).

The heat capacity of a protein in the native state $(C_{p,n})$ has been shown to be a linear function of temperature *(2)* within the range in which it can be measured (i.e., the temperature range in which the native state constitutes over 99% of the total population). Within that range the heat capacity of the native state is very similar for all proteins when normalized on a weight basis. Therefore, the molar heat capacity can be written as:

Table 1
Heat Capacity of Proteins in Unfolded State[a]

Amino acid	A	B	C	D
A. Side chain contributions				
ALA	177.957	–0.450	0.000153	
ARG	299.558	0.627	–0.00267	
ASN	68.395	0.884	–0.00160	
ASP	68.731	0.8206	–0.00107	
CYS	222.58	0.636	–0.0017	
GLN	165.183	0.6360	–0.0017	
GLU	164.742	0.636	–0.00168	
GLY	83.505	–0.226	-9.54×10^{-05}	
HIS	212.367	–1.699	0.0230	-8.36×10^{-05}
ILE	408.045	–0.233	0.000115	
LEU	386.787	–0.198	0.000225	
LYS	316.945	–1.365	0.0188	-7.434×10^{-05}
MET	200.523	–1.0456	0.00487	
PHE	398.560	–0.6340	0.00130	
PRO	218.452	–1.623	0.0075	
SER	74.736	0.233	–0.000108	
THR	196.165	–0.571	0.00589	
TRP	474.060	–0.634	0.00130	
TYR	314.312	–0.716	0.00785	-2.2×10^{-05}
VAL	325.226	–0.361	–0.000586	
B. Peptide backbone unit contribution (-CHCONH-)				
	1.273	0.613	–0.00286	
C. Amino terminal contribution (NH2)				
	–192.630	6.178	–0.0666	0.000157
D. Carboxy terminal contributions (COOH)				
	1.273	0.613	–0.00286	

[a]The coefficients were obtained by a polynomial least squares fit of the data published by Privalov and Makhatadze *(2)* (Freire and Murphy, unpublished results). $C_{p,u}(T) = \text{A} + \text{Bx}T + \text{Cx}T^2 + \text{Dx}T^3$ where C_p is in J/K /mol and T in °C.

$$C_{p,n} = (a_N + b_N \cdot \text{T}) \cdot M_r \quad (6)$$

where $a_N = 0.3161 \pm 0.013$ cal/K · g, $b_N = 0.0016 \pm 0.0003$ cal/K^2 · g, and M_r is the molecular weight of the protein.

Recently, it has been shown that the heat capacity difference (ΔC_p) between the unfolded and native states is directly proportional to the change in solvent accessible polar and apolar surfaces between those states *(3,4)*:

$$\Delta C_p = \Delta C^\circ_{p,ap} \cdot \Delta A_{ap} + \Delta C^\circ_{p,pol} \cdot \Delta A_{pol} \quad (7)$$

where $C^{\circ}_{p,ap}$ (0.45 ± 0.02 cal/K/[mol-Å^2]) and $C^{\circ}_{p,pol}$ (–0.26 ± 0.03 cal/K/[mol-Å^2]) are the elementary apolar and polar contributions to the total heat capacity increment; and ΔA_{ap} and ΔA_{pol} are the differences between the solvent accessible apolar and polar surface areas of the two states. The above approach accurately predicts the heat capacity values obtained for the native state of globular proteins and the magnitude of the heat capacity change associated with the complete unfolding of the protein. The heat capacity values provide a means to evaluate the degree of unfolding of a protein by comparing experimentally determined values to those calculated with Eqs. (5), (6), and (7). In conjunction with the measured changes in ΔH and ΔS on unfolding, the heat capacity values provide a rather complete assessment of the degree of unfolding of a protein.

Because of the curvature of $C_{p,u}$ and the linearity of $C_{p,n}$, ΔC_p decreases at higher temperatures. This decrease is, however, small within the temperature range of interest. For example, in most cases in which the temperature dependence of the enthalpy change has been measured over a wide temperature interval, no evidence of a diminished slope at high temperatures has been observed *(7)*. For practical purposes, the enthalpy change within the temperature range 0–80°C can be accounted for with a constant ΔC_p.

4. Excess Heat Capacity Function

DSC has been utilized primarily to study thermally induced structural transitions *(8)*. If a protein undergoes a transition, the heat capacity function will exhibit an anomaly at some characteristic temperature, usually called the transition temperature *(T_m)*. Under these conditions, the heat capacity function can no longer be ascribed to a single structural state since it contains contributions from all the states that become populated during the transition as well as the excess contributions arising from the existence of enhanced enthalpy fluctuations within the temperature transition region. These excess contributions give rise to the characteristic peak or peaks associated with thermally induced transitions *(9,10)*. The most important quantity for the thermodynamic analysis of the thermal unfolding of a protein is the excess heat capacity function ($\langle \Delta C_p \rangle$) that is obtained by subtracting the heat capacity of the native state from the measured heat capacity function *(11)*:

$$\langle \Delta C_p \rangle = C_p - C_{p,n} \tag{8}$$

Figures 3 and 4 illustrate the procedure required to estimate $\langle \Delta C_p \rangle$ from the experimental data. As indicated in the figure, this procedure involves

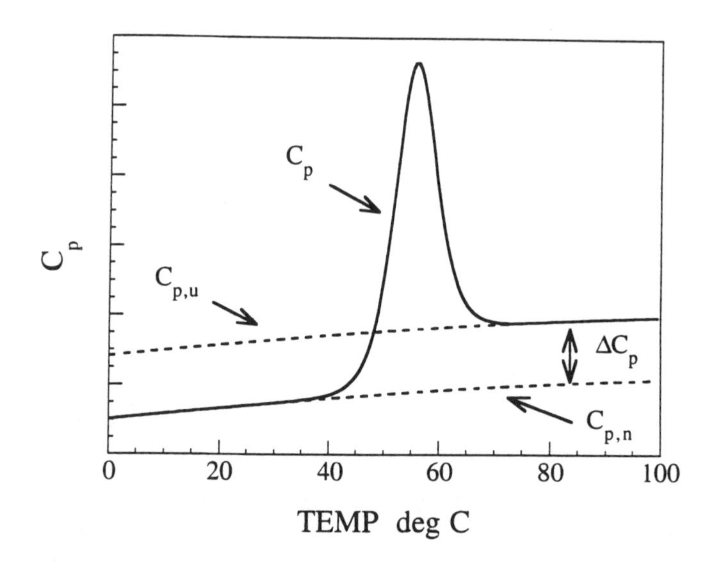

Fig. 3. The molar heat capacity function (C_p) obtained by Eq. (4). The molar heat capacities of the unfolded ($C_{p,u}$) and native ($C_{p,n}$) states are indicated by the dashed lines. The difference between these values define the heat capacity change for the transition ($\Delta C_p = C_{p,u} - C_{p,n}$).

extrapolation of $C_{p,n}$ to those temperature regions in which the native state is no longer the most significantly populated. This extrapolation might introduce systematic distortions in $<\Delta C_p>$ and therefore must be done with extreme care. Potential extrapolation errors can be minimized by:

1. Measuring a relatively long pretransition baseline so that its mathematical form can be determined; and
2. Performing several calorimetric scans under conditions in which the transition occurs at different temperatures and $C_{p,n}$ can be defined over a wider temperature range. The subtraction of $C_{p,n}$ from $<C_p>$ implicitly selects the native state as the reference state.

5. Statistical Thermodynamic Definition of the Excess Heat Capacity Function

In the analysis of DSC data, the most important quantity that needs to be defined using the tools of statistical thermodynamics is the average excess enthalpy function ($<\Delta H>$). This quantity is the sum of the enthalpy contributions of all the states that become populated during the transition *(11)*:

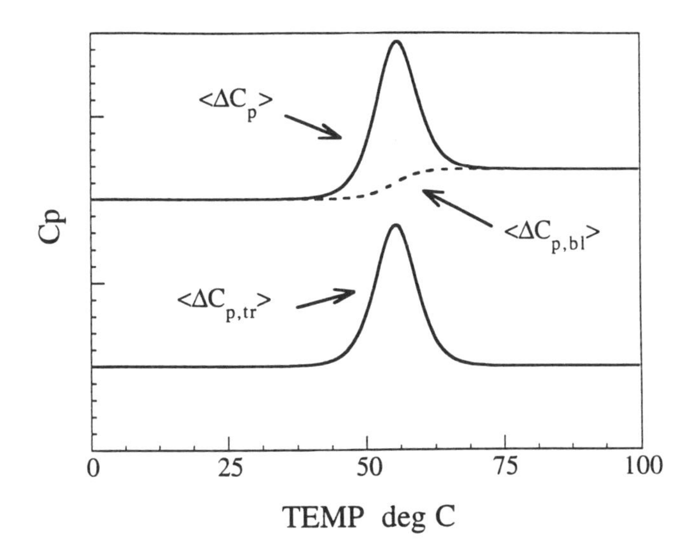

Fig. 4. The excess heat capacity function ($<\Delta C_p>$) is obtained by subtracting the molar heat capacity of the native state ($C_{p,n}$) from the molar heat capacity of the protein (C_p). $<\Delta C_p>$ is composed of two contributions: the transition excess heat capacity function ($<\Delta C_{p,tr}>$) and the sigmoidal function ($<\Delta C_{p,bl}>$) that gives rise to the baseline shift characteristic of transitions accompanied by a positive ΔC_p.

$$<\Delta H> = \sum_{i=0}^{N} P_i \cdot \Delta H_i \tag{9}$$

where P_i represents the population or probability of state *i*, and ΔH_i the enthalpy of the ith state relative to that of the native state that is taken as the reference state. $<\Delta C_p>$ is equal to the temperature derivative of $<\Delta H>$ at constant pressure:

$$<\Delta C_p> = \sum_{i=1}^{N} \Delta H_i \cdot (\partial P_i/\partial T) + \sum_{i=1}^{N} P_i \cdot \Delta C_{p,i} \tag{10a}$$

$$= <\Delta C_{p,tr}> + <\Delta_{Cp,bl}> \tag{10b}$$

The first term on the right hand side ($<\Delta C_{p,tr}>$) is the *transition excess heat capacity function* and defines the characteristic transition peak(s) in the heat capacity function (Fig. 4) *(12)*. The second term on the right hand side ($<\Delta C_{p,bl}>$) defines the "S-shape" shift in baseline usually associated with protein unfolding or other transitions characterized by positive changes in ΔC_p *(12)*.

$<\Delta H>$ is the cumulative integral of $<\Delta C_p>$:

$$<\Delta H> = \int_{T_o}^{T} <\Delta C_p> dT \qquad (11)$$

where T_o is a temperature in which the protein is in the native state (Fig. 4). Eqs. (9) and (10) are functions of the population of molecules in the different states accessible to the protein molecule. The population of molecules in state j (P_j) is equal to:

$$P_j = \exp(-\Delta G_j/RT)/[1 + \sum_{i=1}^{N} \exp(-\Delta G_i/RT)] \qquad (12)$$

where $\Delta G_j = \Delta H_j - T \cdot \Delta S_j$ is the Gibbs free energy difference between state j and the native state (state *0*) and R the gas constant. The quantity in the denominator in Eq. (12) is the folding/unfolding partition function Q defined as:

$$Q = 1 + \sum_{i=1}^{N} \exp(-\Delta G_i/RT) \qquad (13)$$

The partition function contains all the thermodynamic information of the system and is therefore sufficient to describe the thermal unfolding of a protein. It must noted that the above statistical thermodynamic treatment requires no *a priori* assumptions regarding the number of states or the magnitude of the thermodynamic parameters associated with the folding/unfolding equilibrium.

For convenience, Eq. (13) can be written as:

$$Q = 1 + \sum_{i=1}^{N-1} \exp(-\Delta G_i/RT) + \exp(-\Delta G_N/RT) \qquad (14)$$

where the terms under the summation include all intermediates that become populated during the transition. The first and last terms are the statistical weights of the native and unfolded states respectively. For a two-state transition the partition function reduces to $Q = 1 + \exp(-\Delta G_N/RT)$.

6. Overall Thermodynamic Parameters

The most important overall thermodynamic parameters associated with the thermal unfolding of proteins are the enthalpy (ΔH), entropy

(ΔS), and heat capacity (ΔC_p) changes between the unfolded and native states. The free energy change is defined in terms of ΔH, ΔS, and ΔC_p using the standard equation:

$$\Delta G(T) = \Delta H(T_R) + \int \Delta C_p dT - T[\Delta S(T_R) + \int \Delta C_p d\ln T] \quad (15a)$$

$$\Delta G(T) = \Delta H(T_R) + \Delta C_p \cdot (T - T_R) - T \cdot [\Delta S(T_R) + \Delta C_p \cdot \ln(T/T_R)] \quad (15b)$$

where Eq. (15b) is the classical equation for the case in which ΔC_p is temperature independent.

All these parameters are state functions, i.e., their value only depends on the nature of the unfolded and the native states and not on the specific transition pathway or the presence of partly folded intermediates. From a practical point of view, these parameters are independent of the shape of the measured heat capacity function. The heat capacity change is defined by the value of $<\Delta C_p>$ after completion of the transition, as indicated in Fig. 4. The enthalpy change is the area under the transition excess heat capacity function ($<\Delta C_{p,tr}>$):

$$\Delta H = \int_{T_o}^{T_f} <\Delta C_{p,tr}> dT \quad (16)$$

where the limits of integration are defined by the onset and completion temperatures of the transition (i.e., the temperatures at which essentially all molecules are in the initial and final states, respectively). The entropy change is simply evaluated by means of:

$$\Delta S = \int_{T_o}^{T_f} <\Delta C_{p,tr}> d\ln T \quad (17)$$

Both ΔH and ΔS, as defined by Eqs. (16) and (17), are referred to the transition temperature, T_m, i.e., $\Delta H = \Delta H(T_m)$ and $\Delta S = \Delta S(T_m)$. It must be noted that Eqs. (16) and (17) are defined in terms of the *transition* excess heat capacity curve, implying that $<\Delta C_{p,bl}>$ needs to be subtracted from the excess heat capacity function. Throughout the years, different subtraction methods have been utilized. In the past, $<\Delta C_{p,bl}>$ used to be approximated by a step function defined by the intersection of a vertical line centered at T_m with the extrapolated initial and final values of the heat capacity function (*see*, for example refs. *6,8*). Since the baseline is proportional to the degree of unfolding (*see* Eq. [10]) a more accurate way of estimating it can be achieved by defining its shape in terms of the normalized integral of the heat capacity function *(13,14)*. This alternative is easy to implement with computer digitized data and is mathematically exact for a two-state transition. In general, the overall thermodynamic parameters ΔH and ΔS are relatively insensitive to the exact method used to subtract $<\Delta C_{p,bl}>$ *(15)*. The situation is different, however, for the analysis of the shape of the heat capacity function.

7. Experimental Evaluation of the Partition Function

As indicated above, the excess enthalpy function plays a central role in the statistical thermodynamic analysis of DSC data because it provides a direct link between the experiment and the folding/unfolding partition function. $<\Delta H>$ can be calculated in terms of the cumulative integral of the measured $<\Delta C_p>$ and is also related to the partition function by the equation:

$$<\Delta H> = RT^2(\partial \ln Q/\partial T) \tag{18}$$

Freire and Biltonen *(11)* first realized that, by rewriting Eq. (18) in integral form, DSC could provide a direct numerical access to the folding/unfolding partition function:

$$\ln Q = \int_{T_o}^{T} <\Delta H>/RT^2 dT \tag{19a}$$

$$\ln Q = \int_{T_o}^{T} 1/RT^2 \left(\int_{T_o}^{T} <\Delta C_p> dT\right) dT \tag{19b}$$

The above equations provide a rigorous foundation to the deconvolution theory of the excess heat capacity function, since they establish a mathematical linkage between the experimental data and the most fundamental function in statistical thermodynamics. The uniqueness of the enthalpy function as a physical observable can be illustrated by comparing it with the observables measured by other techniques. The measured value of any arbitrary physical observable used to monitor a thermal transition can be written in terms of an equation similar to the one derived for the excess enthalpy function (Eq. [9]):

$$<\alpha> = \sum_{i=0}^{N} \alpha_i \cdot P_i \tag{20}$$

where $<\alpha>$ is the measured value of the observable and α_i is the characteristic value of the observable for the ith state. The key feature that distinguishes the excess enthalpy function from any other physical observable is that the population of each state (P_i) is a function of its enthalpy (ΔH_i) as seen in Eq. (12). This unique feature of DSC data has made possible the development of rigorous deconvolution algorithms aimed at obtaining a complete thermodynamic characterization of a folding/unfolding transition. For other physical observables the characteristic α_i values are not mathematically related to the P_i values, i.e., the amplitude of the melting curves are not related to a thermodynamic function, and the experimental data cannot be used to obtain a complete thermodynamic description of a transition.

8. Deconvolution of the Excess Heat Capacity Function

The main goal of the deconvolution analysis of the heat capacity function is the determination of the number of states that become populated during thermal unfolding and the thermodynamic parameters for each of those states. Throughout the years the deconvolution algorithms have been perfected in different ways (*see*, for example, *11,15–17*). Nowadays, the most effective algorithms involve a recursive deconvolution procedure that includes multiple cycling through each individual transition step combined with nonlinear least squares optimization, and conclude with a global nonlinear least squares optimization. A recent example can be found in the analysis of the multistate transition of the molecular chaperone DnaK *(17)*. Figure 5 illustrates some of the results obtained for this protein.

In performing a deconvolution analysis, special attention must be paid to the following points:

1. *Calculation of excess heat capacity function.* $<\Delta C_p>$ is obtained by subtracting the heat capacity of the initial state from the molar heat capacity function (C_p). It is important that at the initial temperature for integration essentially all molecules are in the native state, otherwise a substantial error will be introduced in all thermodynamic parameters *(12)*.
2. *General strategy.* The nonlinear least squares fit of $<\Delta C_p>$ must be per-

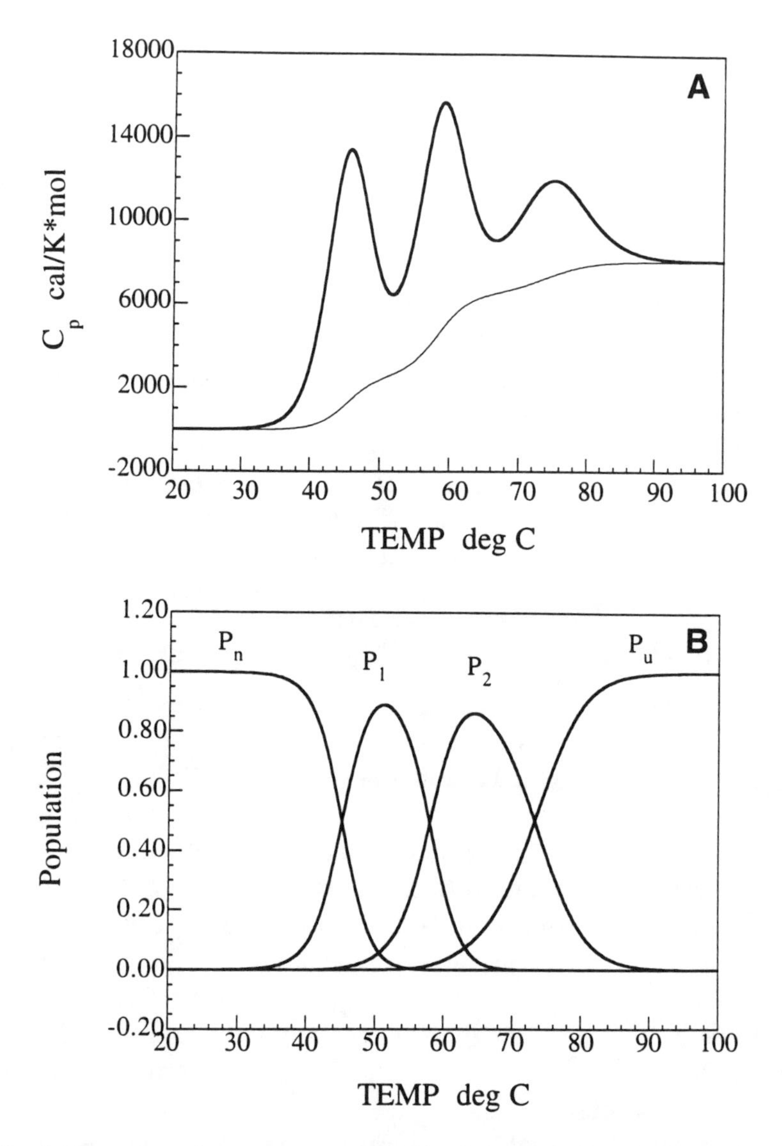

Fig. 5. The excess heat capacity function for the molecular chaperone DnaK obtained at pH 7.6 25 m*M* phosphate buffer **(A)** and the population of states resulting from the deconvolution of the heat capacity function **(B)**. The thermal unfolding of this protein involves four states, each of which is characterized by enthalpy, entropy, and heat capacity changes. All the thermodynamic parameters could be obtained by deconvoluting the heat capacity function as described by ref. *17*.

formed using Eq. 10, which is the exact equation for a folding/unfolding transition exhibiting an arbitrary number of states. After convergence, the nonlinear least squares procedure returns the best set of parameters that minimizes the sum of squared residuals (SSR) between the calculated and experimental values (SSR = Σ ($<\Delta C_p>_{calculated} - <\Delta C_p>_{experimental})^2$. The analysis must begin by assuming the simplest situation, i.e., the two-state case, and progressively increase the number of states. After completion of the nonlinear least squares optimization the quality of the fit must be evaluated in terms of the standard deviation of the fit and by performing an analysis of residuals *(18)*. Since there are three extra fitting parameters (ΔH, ΔS, ΔC_p) for each additional state included in the analysis, it is expected that the quality of the fit will increase with the number of states. It is necessary, then, to evaluate whether the increase in the quality of the fit actually reflects the existence of an additional state or is merely owing to the larger number of parameters used in the analysis. This can be done by performing an F-test statistics as described by ref. *18*.

3. *Estimation of initial parameter values.* In principle, initial parameter estimation can be performed in different ways: by visual inspection of the data; automatically with the deconvolution algorithm; or interactively by a combination of visual inspection of the data and a simulation program. In general, the most accurate methods are those that involve the recursive deconvolution algorithms since they yield parameter estimates that are already close to the final convergence values. This is especially true for strongly overlapped transitions. Also, it is important to check that the minimum in SSR is produced by a unique set of parameter values. Otherwise the system might exhibit multiple minima and not a unique solution.
4. *Variational analysis of baseline estimate.* In theory, the subtraction of the heat capacity of the initial state, $C_{p,n}$, can be incorporated into the nonlinear least squares algorithm. However, this practice is dangerous, especially if the variation in $C_{p,n}$ is left unrestrained. A better alternative is to perform a variational analysis in which $C_{p,n}$ is varied, within experimental confidence limits, after each cycle of nonlinear least squares optimization of the thermodynamic parameters. This procedure allows selection of the best baseline-parameter set combination.
5. *Error estimation of fitting parameters.* A convenient method to estimate joint confidence intervals for the fitted parameters involves the use of a nonlinear support plane method *(12,19,20)*. To estimate the confidence interval for a fitted parameter, that parameter is fixed at selected points around the convergence value and all other parameters are allowed to float. Then, the resulting variance of the fit is plotted as a function of the values

of the fixed parameter in order to evaluate the minimum of this function. A sharp minimum is indicative of a well defined and robust parameter estimation, whereas a shallow minimum or multiple minima are indicative of considerable uncertainty in the estimated parameter value.

It must be noted that the above analysis only includes the errors owing to the optimization procedure. Errors owing to protein concentration uncertainties, baseline subtraction, and so on, must be considered separately and added to the above error estimates. In general, it is a good idea to repeat the nonlinear least squares optimization for several baseline subtractions and for concentration values that bracket the limits of the experimental uncertainty.

9. van't Hoff Analysis

The deconvolution approach for characterizing folding/unfolding transitions of proteins is far superior than the classical approach based on the evaluation of the $\Delta H_{vh}/\Delta H$ ratio. In the past, the two state character of a folding/unfolding transition was inferred from the equality of the van't Hoff and calorimetric enthalpies. Usually, the van't Hoff enthalpy is calculated by means of the formula *(1)*:

$$\Delta H_{vh} = 4 \cdot R \cdot T_m^2 \cdot \langle \Delta C_{p,tr} \rangle_{\max} / \Delta H \qquad (21)$$

which utilizes a single temperature point from the entire heat capacity function to perform the calculation. For this reason the resulting value is not very sensitive to deviations that occur away from the temperature location of $\langle \Delta C_{p,tr} \rangle_{\max}$. In many cases the population of partly folded intermediates is not very large and as such it can be easily missed by a simple van't Hoff analysis. This situation can be illustrated with the example given in Fig. 6. In that figure, the excess heat capacity function and the population of states associated with a hypothetical three-state transition have been plotted as a function of temperature. As indicated in the figure, the population of the partly folded intermediate reaches a maximum of 25% within the transition region even though the $\Delta H_{vh}/\Delta H$ ratio is 0.9. This $\Delta H_{vh}/\Delta H$ ratio is within the range obtained by Privalov *(1,8)* for several globular proteins. If experimental uncertainties are included, this transition would have been mistakenly characterized as a two-state transition.

Figure 6 also show the result obtained by applying the deconvolution equation:

$$d \ln(Q - 1)/dT = \Delta H_1 + \langle \Delta H_1 \rangle \quad (22)$$

to the hypothetical data. In that equation ΔH_1 is the enthalpy difference between the first intermediate and the native state, and $\langle \Delta H_1 \rangle$ is the average excess enthalpy function that would exist if the first intermediate were the lowest enthalpy state and the native state did not exist *(11)*. As seen in the figure, application of the first recursive deconvolution equation clearly indicates the existence of an intermediate state.

The above example highlights the need for a more exhaustive analysis of the heat capacity function than that provided by a simple van't Hoff analysis. This is especially true now that it is known that not all small globular proteins undergo two-state folding/unfolding transitions and that, under certain solvent conditions, many globular proteins exhibit a significant population of intermediates *(21–24)*.

Another common situation that leads to anomalous $\Delta H_{vh}/\Delta H$ ratios occurs when the transition temperature of the protein under study is low and the onset of cold denaturation occurs before the native state is fully populated. Under those conditions ΔH is underestimated, ΔH_{vh} is overestimated, and therefore a $\Delta H_{vh}/\Delta H$ ratio >1 is observed even for two-state transitions. This situation was observed by Xie et al. *(22)* in their analysis of apo α-lactalbumin and has been discussed in detail by Straume and Freire *(12)*. In those cases, a straightforward interpretation of the $\Delta H_{vh}/\Delta H$ ratio yields a wrong assessment of the transition. In the case of apo α-lactalbumin, for example, a $\Delta H_{vh}/\Delta H$ ratio >1 is observed in the presence of moderate GuHCl concentrations even though the transition is characterized by the presence of a partly folded intermediate. In general, a single calorimetric curve obtained under such conditions cannot be analyzed correctly unless the heat capacity of the native state is known accurately. If this information is not available, a rigorous analysis can still be performed by introducing a second independent variable (e.g., pH, ligand concentration, and so on) and implementing a two-dimensional analysis of the data (*see* Section 11).

10. Heat and Cold Denaturation

As mentioned above, the unfolding of a protein is accompanied by a positive ΔC_p. As a result, both the enthalpy and entropy changes are

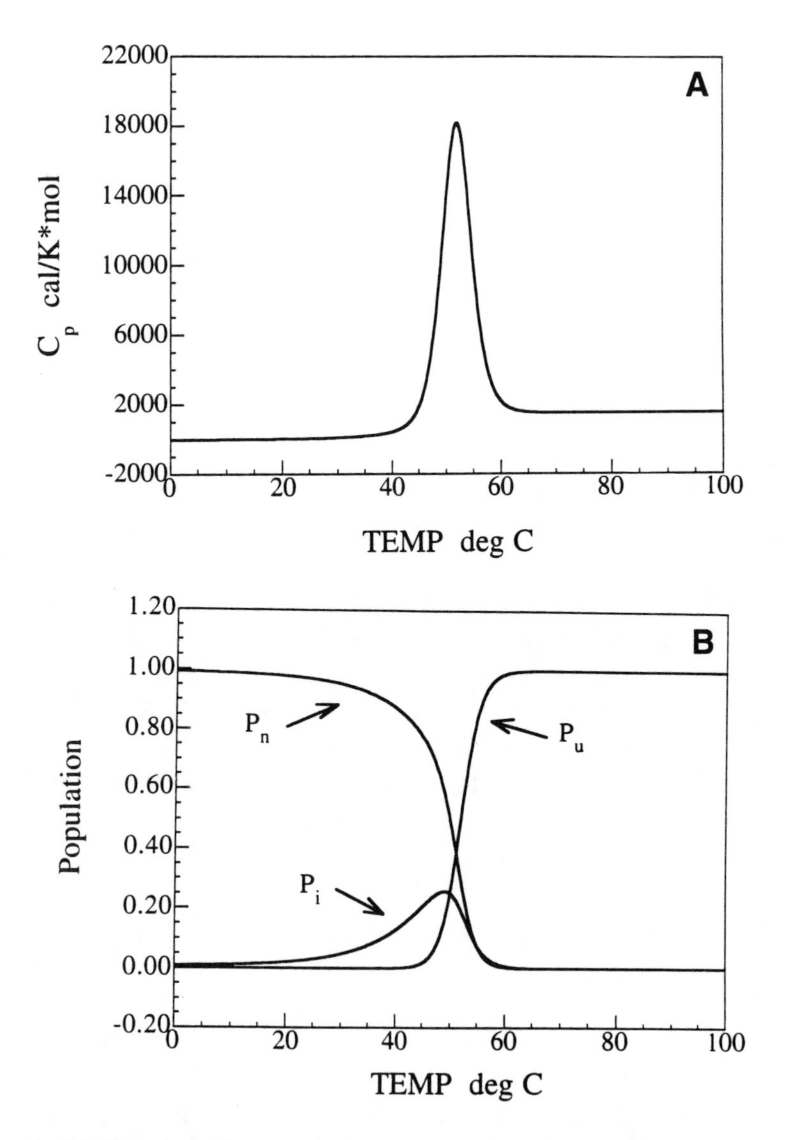

Fig. 6. (**A**) Simulated heat capacity function associated with a hypothetical three state transition characterized by the following parameters: $\Delta H_1(T_{m1}) = 30$ kcal/mol; $T_{m1} = 55°C$; $\Delta C_{p2} = 500$ cal/K/mol; $\Delta H_2(T_{m2}) = 100$ kcal/mol; $T_{m2} = 50°C$; $\Delta C_{p2} = 1000$ cal/K/mol. (**B**) As shown in this figure, the maximal population of the intermediate is close to 25% within the transition region. As explained in the text, the $\Delta H_{vh}/\Delta H$ is very close to one, implying that with that simplistic analysis this transition can be mistakenly characterized as a two-state transition, even though the intermediate population is significant.

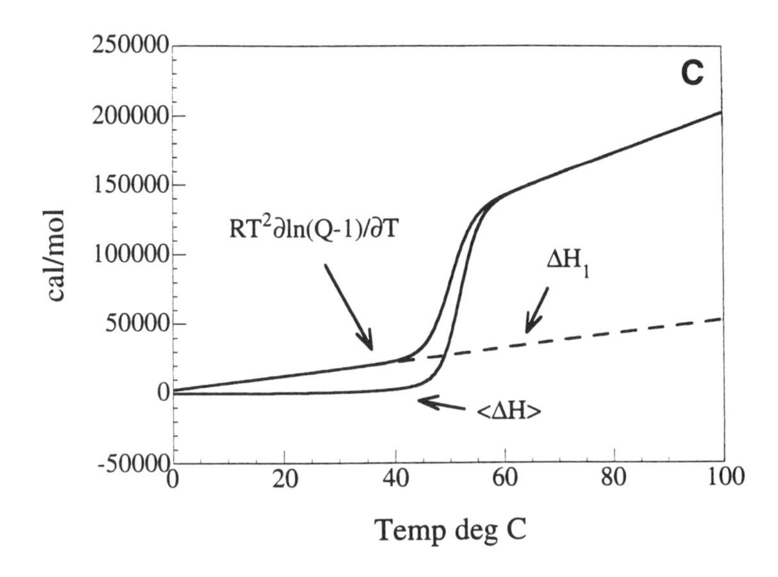

(C) A plot of the excess enthalpy function and the first recursive deconvolution function reveals a significant difference between them and illustrates the enthalpy change for the first intermediate.

increasing functions of temperature. The resulting Gibbs free energy function, on the other hand, exhibits a maximum at a characteristic temperature called the *temperature of maximal stability*. This temperature occurs at the point in which the entropy change is zero. Heating above this temperature or cooling below this temperature destabilizes the protein. This situation is illustrated in Fig. 7 for a typical globular protein. As shown in the figure, there are two temperatures at which the Gibbs free energy function is zero. These temperatures correspond to the temperature of cold and heat denaturation, respectively.

Most of the thermodynamic data for globular proteins have been obtained by performing DSC scans as a function of pH under conditions in which the enthalpic contributions owing to ionization are negligible *(1)*. Under those conditions the observed changes in protein stability are primarily entropic in origin. This situation is also illustrated in Fig. 7. It must be noted that, as the stability of the protein decreases, a point is reached in which ΔG is negative at all temperatures. Under those conditions, the system still exhibits a peak in a calorimetric scan, however the location of that peak cannot be interpreted as the temperature at which

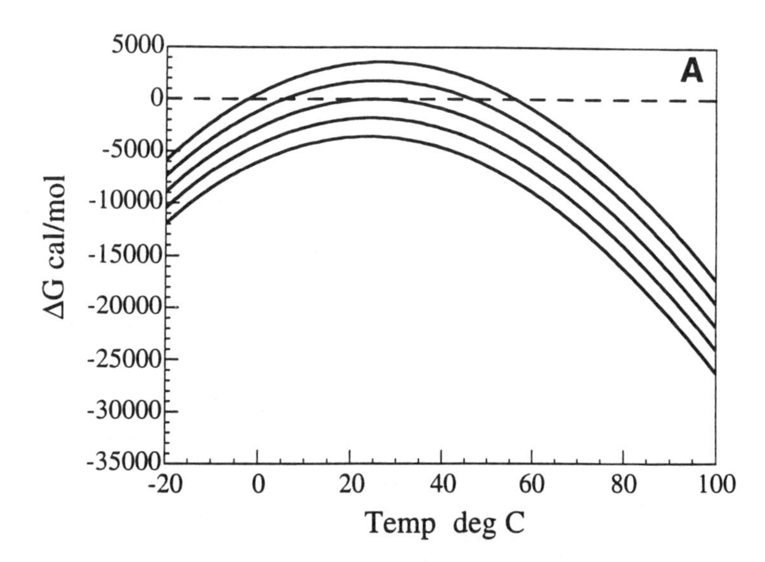

Fig. 7. (**A**) Gibbs' free energy (ΔG) vs temperature for a transition characterized by the following thermodynamic parameters: ΔC_p = 2500 cal/K/mol, $\Delta H(25) = 0$, and, from top to bottom $\Delta S(25) = -12, -6, 0, 6, 12$ cal/K/mol. It must be noted that as a protein is destabilized, a point is reached in which ΔG is never greater than zero. Formally, there is no transition temperature, even though transition peaks will still be observed in calorimetric scans. A conventional analysis of peaks obtained under destabilizing conditions will lead to substantial errors in the thermodynamic parameters.

$\Delta G = 0$ since that temperature does not exist. Of course, a conventional baseline subtraction yields an artificially low ΔH and an artificially high ΔH_{vh} as explained above. Further destabilization of the protein accentuates this trend and eventually leads to the absence of a transition, a situation that should not be interpreted as indicating that the native and denatured states are enthalpically equivalent. Because of the existence of a positive ΔC_p between the unfolded and native states there is a temperature at which $\Delta H = 0$ (sometimes called the *inversion* temperature), however, ΔH is different than zero at all other temperatures independently of whether a transition is observed or not.

The situation illustrated in Fig. 7 also highlights the need for a more rigorous analysis than that provided by conventional baseline subtraction followed by a van't Hoff analysis. In fact, we recommend that decisions about the two-state or multistate character of a transition not be

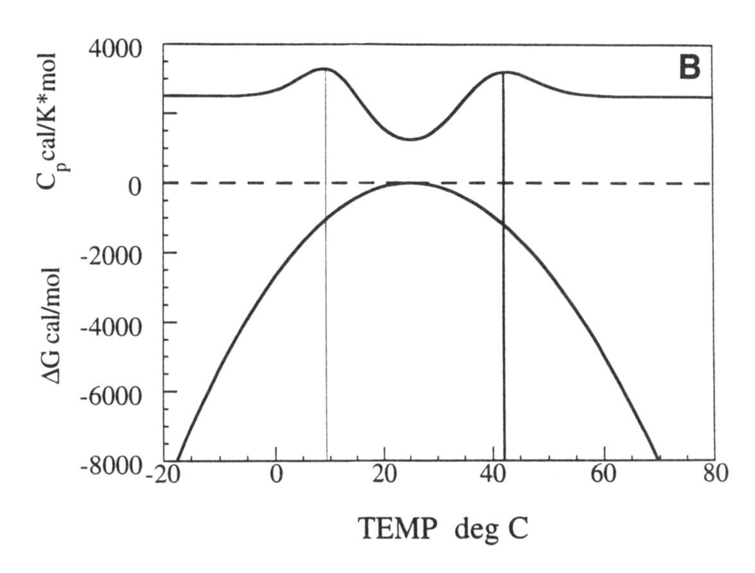

(B) Plot of ΔG and the heat capacity function for the case in which $\Delta S(25) = 0$. As seen in the figure, under those conditions the temperature location of the peaks in the heat capacity function do not correspond to the temperatures at which $\Delta G = 0$.

made on the basis of the value of the $\Delta H_{vh}/\Delta H$ ratio. Shrake and Ross *(25)* have also pointed out other situations in which the $\Delta H_{vh}/\Delta H$ ratio is different than one even for two-state transitions. Similarly, a $\Delta H_{vh}/\Delta H$ ratio of unity should not be interpreted as sufficient proof that a transition conforms to the two-state model.

11. Two-Dimensional Differential Scanning Calorimetry

The examples discussed in the previous section illustrate the dangers of performing a conventional van't Hoff analysis on protein folding/unfolding transitions characterized by very low transition temperatures. Under those conditions a rigorous alternative is to perform a two-dimensional analysis of the excess heat capacity function. A heat capacity surface is obtained by performing calorimetric scans as a function of a second variable. The resulting surface is then analyzed in terms of the linkage equations between the two variables. We will use pH as an example of a second variable but the equations can be easily generalized to other cases of ligand binding *(12)* or to the effects of denaturants *(13,22)*.

The folding/unfolding partition function including protonation effects can be written as:

$$Q = 1 + \sum_{i=1}^{N-1} [\exp(-\Delta G_i^o/RT) \cdot \prod_{j=1}^{m} (1 + K_{ij} \cdot \mathbf{a}_H)/(1 + K_{nj} \cdot \mathbf{a}_H)]$$

$$+ \exp(-\Delta G_N^o/RT) \cdot \prod_{j=1}^{m} (1 + K_{Nj} \cdot \mathbf{a}_H)/(1 + K_{nj} \cdot \mathbf{a}_H) \qquad (23)$$

which is similar in form to Eq. 14 except that each statistical weight is modified by the terms under the multiplication sign. For each state i, the term under the multiplication sign runs over all protonizable groups j and is a function of the protonation constants K_{nj} K_{ij}, and K_{Nj}, and the hydrogen ion activity $\mathbf{a}_H$. The population of each state is given by an equation similar to Eq. 12 and the average enthalpy function by Eq. 9. It must be noted that, in this case, the enthalpy of each state, ΔH_i, is given by:

$$\Delta H_i = \Delta H_i^o + \sum_{j}^{m} (F_{ij}\Delta H_{B,ij} - F_{nj}\,\Delta H_{B,nj}) \qquad (24)$$

where H_{Bij} is the protonation enthalpy for site j in state i, and F_{ij} is its fractional degree of saturation $[F_{ij} = K_{ij} \cdot \mathbf{a}_H/(1 + K_{ij} \cdot \mathbf{a}_H)]$.

The above equations permit a rigorous interpretation of the temperature and pH dependence of the heat capacity function. As an example, Fig. 8 illustrates the shape of the temperature-pH heat capacity surface calculated for apo-myoglobin. At pH values near neutral the thermal unfolding of apo-myoglobin conforms closely to a two-state transition. At lower pHs a partly folded intermediate becomes populated and the transition can be accounted for in terms of a three-state transition mechanism (Barrick and Baldwin, personal communication).

The experimental energetics for the overall folding/unfolding reaction of apo-myoglobin were measured calorimetrically by Griko et al. *(21)* and the energetic parameters for the intermediate were derived by Haynie and Freire *(24)*. The pKa values for the protonation of histidines in apomyoglobin were measured recently by NMR *(26)*. Four histidyl groups were found to have anomalously low pK_a values clustered around 5 (His 24, $pK_a < 4.8$; His 48, $pK_a = 5.2$; His 113, $pK_a < 5.5$; His 119, $pK_a = 5.3$). The titration of the histidines near to the heme binding site (His

64, His 93, and His 97) could not be observed. In the native state these histidyl residues are completely buried; in horse apo-Mb they do not titrate even at the lowest pH values that could be studied by NMR (pH 4.8). Since there are no buried carboxyl groups in the folded form of myoglobin and presumably also in apo-myoglobin, the low pH behavior appears to be attributable to the anomalous histidines.

As shown in Fig. 8, at pH values between 6 and 5 the heat capacity surface displays two well defined peaks corresponding to the heat and cold denaturation transitions. Under those conditions, the transition is experimentally close to a two-state transition. At pH values between 3 and 5, a significant intermediate population is found, especially at low temperatures, where it reaches about 100% of the molecules. The intermediate state denatures at constant temperature on lowering the pH below 3 or at constant pH on increasing the temperature. At pH values of 4 and lower, the heat capacity function does not exhibit a "measurable" peak *(see also* ref. *21).*

The case of ribonuclease A highlights another aspect of 2D-DSC. The stability of ribonuclease A as a function of temperature and the concentration of cytidine-2'-monophosphate (2'CMP) has been recently studied by two-dimensional differential scanning calorimetry *(12)*. It was shown that this approach is able to resolve simultaneously the folding/unfolding and the binding energetics. Although the partition function in this case is similar to that in Eq. (23), it must be noted that in a calorimetric experiment the total ligand concentration, and not the ligand activity or free ligand concentration, is the quantity experimentally known. Therefore the partition function and all other thermodynamic functions must be recast in terms of the total ligand concentration. This is done by finding the zero of the equation:

$$[X]_{\text{total}} - [X]_{\text{bound}} - [X]_{\text{free}} = 0 \tag{25}$$

as discussed by Straume and Freire *(12)*. Figure 9 depicts the temperature-[2'CMP] heat capacity surface for ribonuclease A. Under all ligand concentration conditions the transition conforms to the two-state mechanism. The protein concentration used in the experiments performed to generate the surface was 0.219 m*M*. At zero ligand concentration the $\Delta H_{vh}/\Delta H$ ratio is equal to one. On introduction of the ligand, the $\Delta H_{vh}/\Delta H$ ratio at first is reduced substantially at partially saturating ligand concentrations (e.g., at 0.16 m*M* 2'CMP the $\Delta H_{vh}/\Delta H$ ratio is only 0.746)

A

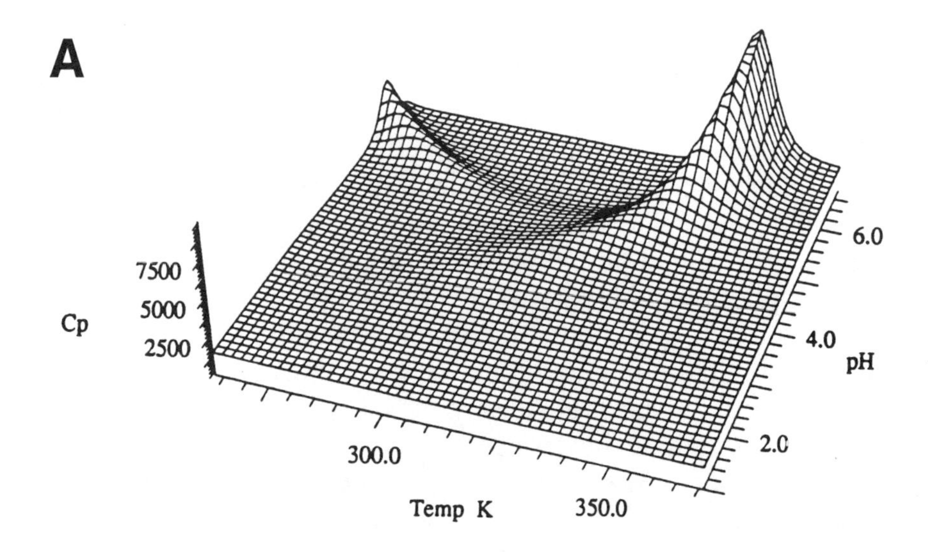

B

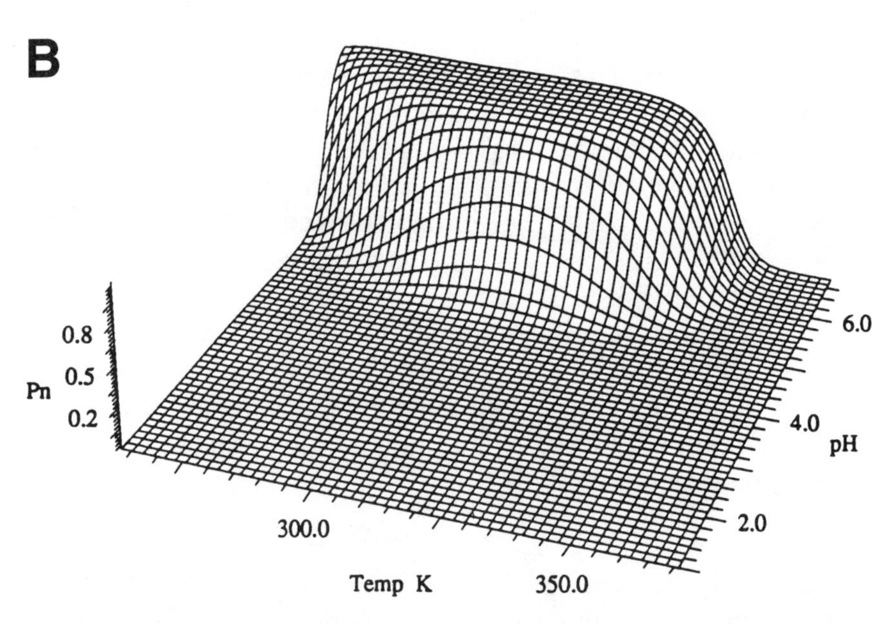

Fig. 8. Calculated heat capacity surface for apo-myoglobin as a function of temperature and pH (**A**) and population surfaces for the native (**B**), intermediate (**C**), and unfolded (**D**) states. The energetic parameter values used in the calculations were $\Delta H_I(25°C) = 8$ kcal/mol, $\Delta C_{p,I} = 850$ kcal/K/mol, $\Delta S_I(25°C) = 11.5$ cal/K/mol; $\Delta H_U(25°C) = 28$ kcal/mol, $\Delta C_{p,u} = 1550$ cal · cal/K/mol, $\Delta S_u(25°C) = 63.5$ cal/K/mol.

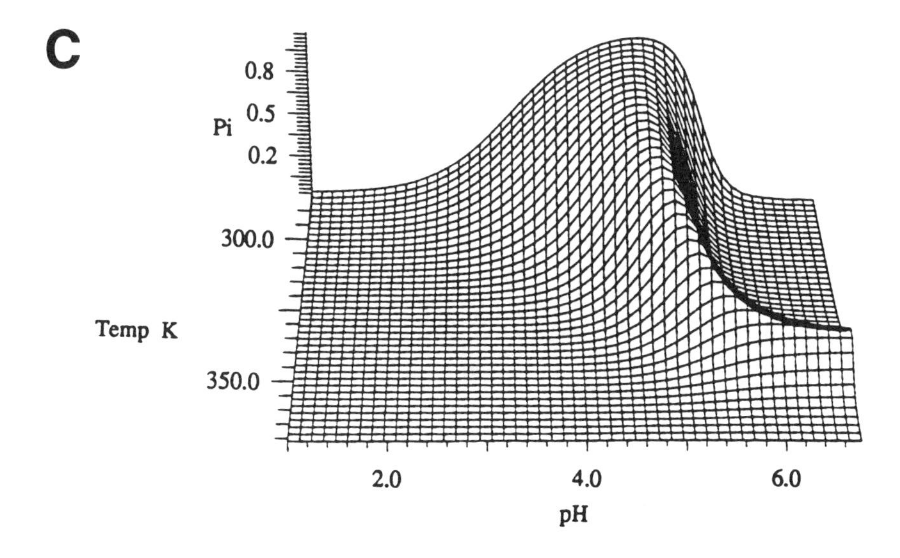

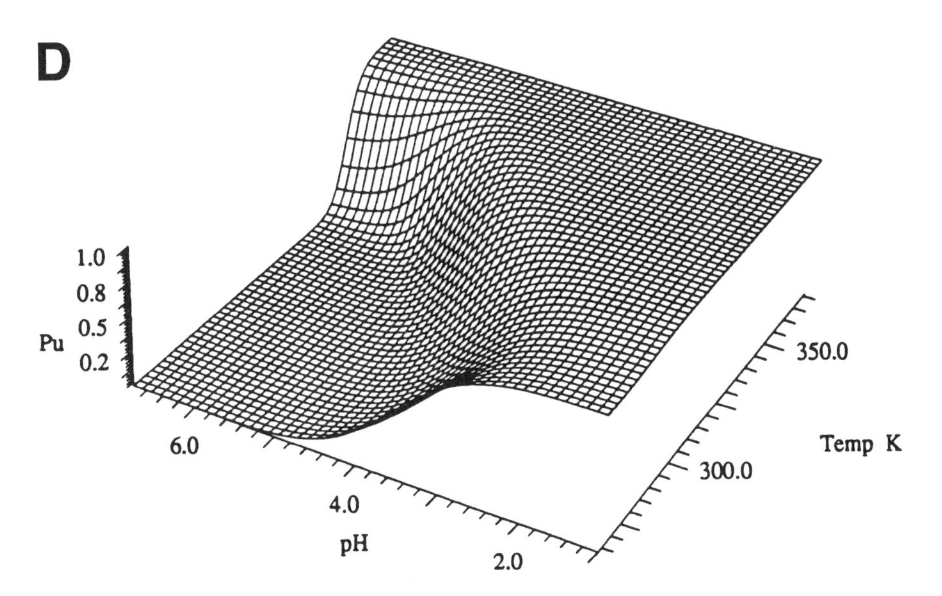

Fig. 8. *(continued)* For the simulation seven histidyl groups were assumed to have anomalous pK_as in the native state (4.5, 5.2, 5.2, 5.3, 2.5, 2.5, and 2.5). Two of the groups with pK_as around 2.5 were assumed to remain anomalous in the *I* state. The pK_as of the exposed histidyl groups were assumed to be 6 and the enthalpy of protonation of the histidyl residues –7 kcal/mol. Small variations in these values do not significantly affect the main features of the stability surfaces.

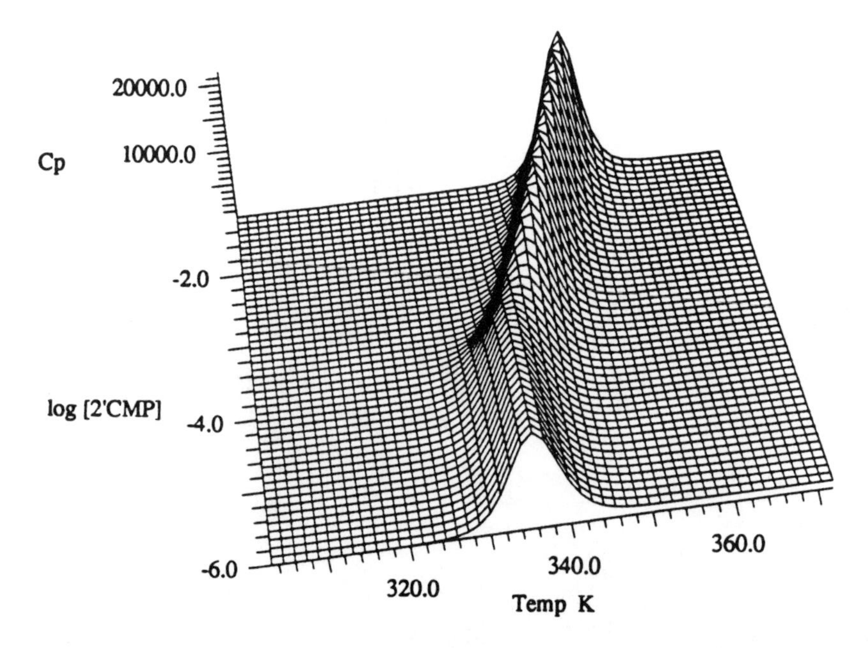

Fig. 9. Heat capacity surface as a function of temperature and 2'CMP for ribonuclease A. The surface was solved by Straume and Freire *(12)*, who obtained the following parameters by global nonlinear optimization of the experimental surface: $\Delta H = 109$ kcal/mol, $T_m{}^o = 61.9$ °C, $\Delta C_p{}^o = 1.2$ kcal/K/mo, $\Delta H_B = -18$ kcal/mol, log K(25) = 6.19. It must be noted that at intermediate saturating conditions the $\Delta H_{vh}/\Delta H$ ratio for the individual peaks is significantly <1 even though the transition is a two-state transition under all conditions (*see text* for details).

after which it again returns to a value of one at saturating levels of 2'CMP. This behavior arises because, on increasing the temperature, the free 2'CMP concentration increases as a result of the reduced concentration of protein molecules in the native state. The remaining molecules in the native state are further stabilized by the additional free 2'CMP resulting in a broader transition profile. At fully saturating ligand concentrations, the $\Delta H_{vh}/\Delta H$ ratio again returns to one because the degree of ligand-induced structural stabilization no longer varies with the degree of completion of the reaction. The case of RNase A-2'CMP provides a vivid illustration that the $\Delta H_{vh}/\Delta H$ ratio can be different than one even for a two-state transition.

12. Structural Thermodynamics

The uniqueness of calorimetry is that it is the only technique that provides a direct experimental access to the overall thermodynamics and to the statistical thermodynamic partition function associated with a thermally induced transition. This experimental link to the most fundamental quantity in statistical thermodynamics has permitted the development of deconvolution procedures aimed at deducing the population and thermodynamics of the different states that become populated during the transition. Recently, a significant amount of research has been directed at developing a different type of experimental linkage: that between the partition function and the structure of a protein *(3,4,27–29)*.

In theory, the folding/unfolding partition function can be evaluated from structural information. Essentially, the strategy to calculate the folding/unfolding partition function starting from the molecular structure of the folded state involves generation of the set of most probable partly folded states using the crystallographic structure as a template and evaluation and assignment of the relative Gibbs free energies of those states. Once this task is accomplished, the evaluation of the partition function is straightforward by simple application of Eq. 14. Recently, an experimentally accurate structural parametrization of the folding/unfolding energetics of proteins has been developed (*see* refs. *3,4* and Chapter 1). This parametrization permits an evaluation of the relative probabilities of arbitrary partly folded states and especially those that can be generated using the crystallographic structure of the protein as a template *(24,27,29)*. This strategy allows determination of the set of most probable partly folded states of a protein. The initial results of this approach are encouraging.

It is expected that in the near future the combination of high resolution calorimetry, structural information, and the tools of structural thermodynamics should provide sufficient information for accurate predictions of protein stability and the effects caused by specific mutations, solvent composition, or other environmental variables.

Acknowledgments

This chapter was supported by grants from the National Institutes of Health (RR-04328, GM-37911, and NS-24520) and the National Science Foundation (MCB-9118687).

References

1. Privalov, P. L. and Khechinashvili, N. N. (1974) *J. Mol. Biol.* **86,** 665–684.
2. Privalov, P. L. and Makhatadze, G. I. (1990) *J. Mol. Biol.* **213,** 385–391.
3. Murphy, K. P. and Freire, E. (1992) *Adv. Protein Chem.* **43,** 313–361.
4. Murphy, K. P., Bhakuni, V., Xie, D., and Freire, E. (1992) *J. Mol. Biol.* **227,** 293–306.
5. Privalov, P. L. and Makhatadze, G. I. (1992) *J. Mol. Biol.* **224,** 715–723.
6. Griko, Y. V., Privalov, P. L., Sturtevant, J. M., and Venyaninov, S. Y. (1988) *Proc. Natl. Acad. Sci. USA* **85,** 3343–3347.
7. Privalov, P. L., Griko, Y. V., Venyaminov, S. Y., and Kutyshenko, V. P. (1986) *J. Mol. Biol.* **190,** 487–498.
8. Privalov, P. L. (1979) *Adv. Protein Chem.* **33,** 167–239.
9. Freire, E. (1989) *Comments Mol. Cell. Biophys.* **6,** 123–140.
10. Freire, E., van Osdol, W. W., Mayorga, O. L., and Sanchez-Ruiz, J. M. (1990) *Annu. Rev. Biophys. Biophys. Chem.* **19,** 159–188.
11. Freire, E. and Biltonen, R. L. (1978) *Biopolymers* **17,** 463–479.
12. Straume, M. and Freire, E. (1992) *Anal. Biochem.* **203,** 259–268.
13. Ramsay, G. and Freire, E. (1990) *Biochemistry* **29,** 8677–8683.
14. Shriver, J. W. and Kamath, U. (1990) *Biochemistry* **29,** 2556–2564.
15. Privalov, P. L. and Potekhin, S. A. (1986) *Methods Enzymol.* **131,** 4–51.
16. Rigell, C., de Saussure, C., and Freire, E. (1985) *Biochemistry* **24,** 5638–5646.
17. Montgomery, D., Jordan, R., McMacken, R., and Freire, E. (1993) *J. Mol. Biol.* **232,** 680–692.
18. Draper, N. and Smith, H. (1981) *Applied Regression Analysis,* Wiley, New York.
19. Johnson, M. L. (1983) *Biophys. J.* **44,** 101–106.
20. Johnson, M. L. and Frasier, S. G. (1985) *Methods Enzymol.* **117,** 301–342.
21. Griko, Y. V., Privalov, P. L., Venyaminov, S. Y., and Kutyshenko, V. P. (1988) *J. Mol. Biol.* **202,** 127–138.
22. Xie, D., Bhakuni, V., and Freire, E. (1991) *Biochemistry* **30,** 10,673–10,678.
23. Kuroda, Y., Kidokoro, S.-I., and Wada, A. (1992) *J. Mol. Biol.* **223,** 1139–1153.
24. Haynie, D. T. and Freire, E. (1993) *Proteins: Struct. Func. Genet.* **16,** 115–140.
25. Shrake, A. and Ross, P. D. (1990) *J. Biol. Chem.* **265,** 5055–5059.
26. Cocco, M. J., Kao, Y.-H., Phillips, A. T., and Lecomte, J. T. J. (1992) *Biochemistry* **31,** 6481–6491.
27. Freire, E. and Murphy, K. P. (1991) *J. Mol. Biol.* **222,** 687–698.
28. Freire, E., Murphy, K. P., Sanchez-Ruiz, J. M., Galisteo, M. L., and Privalov, P. L. (1992) *Biochemistry* **31,** 250–256.
29. Freire, E., Haynie, D. T., and Xie, D. (1993) *Prot. Structure Func. Genet.* **17,** 111–123.

CHAPTER 10

Disulfide Bonds in Protein Folding and Stability

Nigel Darby and Thomas E. Creighton

1. Introduction

1.1. Disulfide Bond Formation and Protein Folding

Many proteins rely on disulfide bonds for the stability of their folded state. This implies that disulfide bond formation can be coupled to folding and assembly, and it is a unique tool in studying both phenomena *(1–5)*.

An important aspect of disulfide bonds is that their formation and breakage is an oxidation-reduction process, requiring the presence of electron acceptors and donors, respectively. Consequently, disulfide bonds are not made and broken in isolation, but only under certain conditions that are under experimental control. The rates and equilibrium of the process can be varied and measured, making it possible to measure experimentally the relative free energies of individual disulfide bonds. This provides unique information about protein stability and folding that is not obtainable directly with other interactions, such as hydrogen bonds or van der Waals interactions *(2,6–8)*. It also makes it possible to trap in a stable form the intrinsically unstable partly-folded intermediates that define pathways of protein folding and assembly. The techniques described here have been developed primarily using bovine pancreatic trypsin inhibitor (BPTI) *(9–17)*, but they have also been applied to homologues of BPTI *(18,19)*, ribonuclease A *(20–23)*, ribonuclease T_1 *(24)*, and α-lactalbumin *(25–27)*.

From: *Methods in Molecular Biology, Vol. 40: Protein Stability and Folding: Theory and Practice*
Edited by: B. A. Shirley

1.2. Thiol–Disulfide Exchange

The experimental approach described here *(28–30)* uses thiol–disulfide exchange with low-mol-wt thiol and disulfide reagents to make or reduce protein disulfide bonds. This reaction can be controlled by varying the concentrations of the reactants and can be stopped at any time, trapping the disulfide-bonded species present; they can then be quantified and characterized using a variety of analytical techniques.

Thiol–disulfide exchange involves the reaction between an ionized thiol group, the thiolate anion (the nucleophile, *nuc*), and a disulfide bond. The disulfide sulfur atom attacked is the central (*c*) atom of the linear transition state and is part of the new disulfide bond, whereas the other (the leaving group, *lg*) becomes an ionized thiol *(31–33)*:

$$-S^{-}_{nuc} + S_c{-}S_{lg} \xrightarrow{k_{S^-}} S_{nuc}{-}S_c + -S^{-}_{lg} \quad (1)$$

The rate of thiol–disulfide exchange depends on the extent of ionization of the nucleophilic thiol, and therefore generally increases as the reaction pH is increased, until the pK of the nucleophilic thiol group is exceeded. The thiols of proteins in the unfolded state generally have pK values similar to those reported for comparable monothiols, in the range 8.7–8.9, but their pK values in the folded state can vary enormously *(32)*.

The rate of thiol–disulfide exchange depends on the reactivities of all three sulfur atoms involved *(31,32)* (Eq. 1), which is reflected by their ionization when in the thiol form. If these pK values are known, the rate constant for thiol–disulfide exchange (k_{ex}) can be predicted using the empirical equation of Szajewski and Whitesides *(31,32)*, which also accounts for the ionization of the nucleophilic thiol group at the pH used:

$$\log k_{ex} = 5.2 + 0.5\ pK_{nuc} - 0.27\ pK_c - 0.73\ pK_{lg} - \log(1 + 10^{pK_{nuc} - pH}) \quad (2)$$

This equation gives the rate of reaction at 25°C, in units of $s^{-1}\ M^{-1}$, of one thiol group at one particular sulfur atom of a disulfide. To predict the observed rate of any given interchange reaction, all possible reactions between all the thiol and disulfide sulfur atoms present need to be included. The rate of the thiol–disulfide exchange reaction varies with temperature, having an activation energy of about 58 kJ/mol *(34)*.

1.2.1. Types of Thiol and Disulfide Reagents

Disulfide reagents have either linear, "intermolecular" or cyclic, "intramolecular" disulfide bonds. The reduction of so-called intermolecular disulfide reagents results in the formation of two molecules, each with a single thiol group; examples include the disulfide forms of glutathione (GSSG), β-mercaptoethanol, and cystamine. Intramolecular reagents are reduced to a single molecule with two thiol groups; they include the oxidized forms of dithiothreitol (DTT^{S}_{S}) and lipoic acid.

1.2.2. Disulfide Bond Formation

The process of forming and breaking a disulfide bond in an unfolded protein using an intermolecular disulfide (RSSR) and thiol (RSH) reagent is:

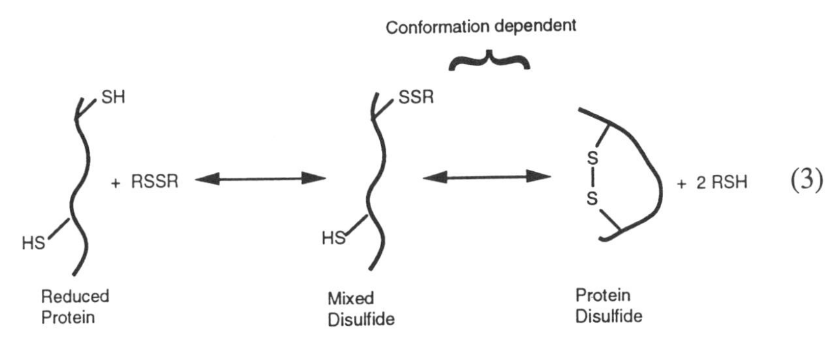

(3)

Formation of the mixed disulfide from the reduced state involves only the simple thiol–disulfide exchange reaction between the protein and the disulfide reagent (Eq. [1]). The subsequent step that leads to protein disulfide bond formation involves conformational changes in the protein molecule, and is the step most relevant to protein folding. Whether or not the mixed-disulfide accumulates to a substantial extent depends on the protein, the type of reagent, and the kinetic circumstances. If it does, the mixed disulfide can occur with either or both thiols.

A further possibility that must be considered in proteins with more than two cysteine residues is the intramolecular rearrangement of protein disulfide bonds or mixed-disulfides:

SH (A), S—S (B–C) ⟷ S—S (A–B), SH (C) (4)

This is simply an intramolecular version of the thiol–disulfide exchange reaction. The disulfide isomer with the lowest free energy predominates at equilibrium, which can be extremely useful in studying protein conformational stability *(17,25–27)*.

1.3. Trapping Protein Disulfide Bonds

To determine the rates and equilibria of formation, breakage, and rearrangement of specific disulfide bonds, it is usually necessary to "trap" the thiol and disulfide species present, by quenching the thiol–disulfide exchange reaction. Two approaches often used are acidification and alkylation of the free thiol groups.

1.3.1. Acidification

Reducing the pH of a thiol–disulfide exchange reaction from 8 to 2 should slow the reaction about 10^6-fold, by reducing the concentration of the reactive thiolate anion. Acidification works rapidly and is effective even when the thiol groups are relatively inaccessible. It only slows the reaction, however, so changes in the distribution of species may occur after trapping. On the other hand, the reversibility of acid trapping has the advantage that it is possible to isolate individual species that contain particular disulfides and free thiols and to reinitiate thiol–disulfide exchange by subsequently raising the pH *(17,35,36)*. Of course, the lower the pH of trapping the better, but it should be borne in mind that the protein may unfold or be modified at very acid pH. Acid trapping is most compatible with analysis methods that work at acid pH, such as reverse-phase HPLC.

1.3.2. Covalent Modification

Identification of the disulfide bonds in the trapped species ultimately requires that the protein thiol groups be modified irreversibly and completely; even traces of residual thiol may cause extensive disulfide bond rearrangements at alkaline pH. Iodoacetamide and iodoacetic acid are most frequently used to alkylate thiol groups. The two are similar, but the former introduces a negative charge, whereas the latter is neutral. They react with the thiolate anion, so trapping is pH-dependent. At pH 8.7 and 25°C, the half-time for reaction with a normal thiol–group is 1–3 s at a reagent concentration of 0.1*M*.

A potential problem is that protein species in rapid equilibrium in the reaction mixture might react with the alkylation reagent at different rates;

the effective removal of one species by covalent trapping can pull the equilibrium and distort the levels of species present, the more reactive species being trapped in increased quantity *(37)*. A final problem is that thiol groups buried within a protein may not be accessible at all to the trapping reagent *(37,38)*. Despite these potential problems, the alkylating agent usually reacts with all the thiol groups of a protein species that is not fully folded at nearly equal rates and adequately quenches the reaction. More reactive reagents are available and are being developed *(39)*.

Trapping with iodoacetic acid, which alters the net charge of the protein, can be used in conjunction with ion exchange chromatography (*see* Section 1.4.1.) or electrophoresis (*see* Section 1.4.2.) to separate the trapped species on the basis of the number of free cysteine residues present in the protein at the time of trapping. Trapped molecules with the same number of disulfide bonds frequently have similar properties, so disulfide intermediates can be separated into groups of one-disulfide, two-disulfide species, and so on. The members of each group are often in rapid equilibrium during folding by intramolecular disulfide rearrangements and can be treated as kinetically equivalent, simplifying the analysis of disulfide bond folding pathways that involve many intermediates (*see* Section 1.8.). Disulfide reagents with charged groups can also be used to separate molecules on the basis of the number of mixed-disulfides they contain *(35,40)*.

Using both the neutral iodoacetamide and the acidic iodoacetate, it is possible to count the integral number of cysteine residues *(14,41)* and disulfide bonds *(26,42)* present in a protein species.

After alkylation, the thiol groups are usually trapped irreversibly, and any disulfide bonds present will be stable in the absence of external reagents that react with disulfide bonds such as thiols, hydroxide ions, and cyanide *(43)*. The trapped proteins can be analyzed by a wide range of techniques, including those that work at alkaline pH, although alkaline hydrolysis of the disulfide bond may become a problem at pH values greater than 9 *(43)*.

1.4. Separating the Trapped Species

1.4.1. Chromatography

Ion exchange chromatography *(11,12,15)* and reverse phase HPLC *(25,26,36)* are suitable for both preparative and analytical separations of the trapped species. Both allow direct quantification of the levels of the various species by integration of the absorbance traces. Reverse-phase separations are based on a large number of parameters, such as the polarity of the protein, its state of folding, the numbers of free cysteine residues and disulfide bonds, and whether any cysteine thiol groups are alkylated or present as mixed-disulfides. The complexity of such separations makes them most useful when there are only a few species present that can be totally resolved. In simple systems, such as a peptide containing two cysteine residues, the reduced and oxidized states and the mixed-disulfides can all generally be resolved using conventional HPLC methods and aqueous solvent systems in acetonitrile-trifluoroacetic acid (Fig. 1). This technique can also separate and quantify the thiol and disulfide forms of the reagent *(44)*.

Ion exchange separations are suitable for separating species trapped with charged alkylating reagents or as charged mixed-disulfides into groups of species, simplifying the analysis of complex systems. The state of folding can also influence the net charge distribution on the molecule, so it is sometimes possible to separate a few species with the same apparent net charge *(12)*.

The principal disadvantage of HPLC methods is the requirement for sophisticated equipment. Simple ion-exchange separations can be run on conventional low-pressure columns, but use large amounts of protein. Chromatographic analysis also can be time consuming, because it is generally feasible to run only a single sample at a time.

1.4.2. Electrophoresis

Polyacrylamide gel electrophoresis (PAGE) allows the rapid simultaneous analysis of multiple samples. Nonreducing SDS-PAGE may distinguish proteins with and without disulfide bonds, but native PAGE is most useful *(10,45)*. Native gels separate protein species on the basis of their net charge and their degree of compactness (Fig. 2). In general, the more folded a protein, the more compact it will be and the faster it will migrate in the gel. With unfolded proteins, the size of the disulfide loop also matters; large disulfide loops increase the mobility more than do small loops. Conformational effects other than those of disulfide loop size can be minimized by including 8*M* urea in the gel, allowing separations to be carried out principally on the basis of net charge (i.e., on the number of charged blocking moieties or mixed disulfides).

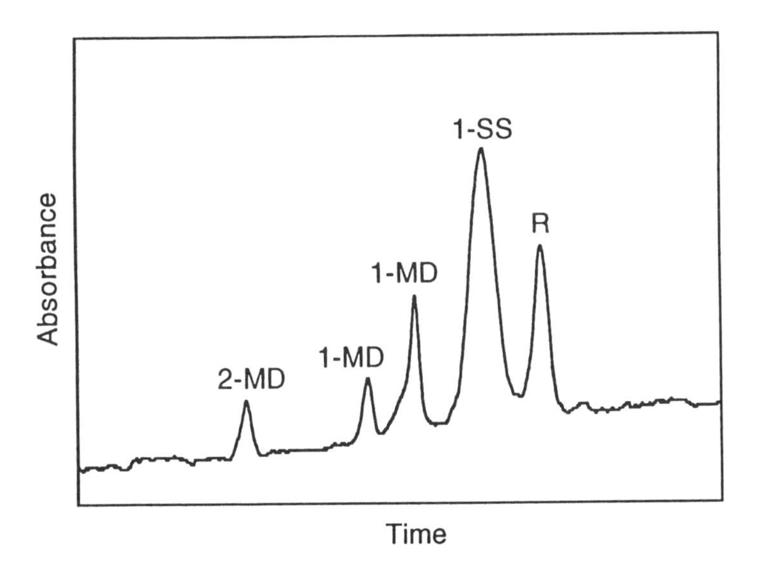

Fig. 1. HPLC analysis of disulfide bond formation in a 28-residue peptide containing two cysteine residues. It is based on the BPTI sequence 4–31, with cysteine residues at residues 5 and 30, but that normally at position 14 replaced by serine. Disulfide bond formation in the reduced peptide was carried out in 0.1*M* Tris-HCl, pH 7.4, using 0.5 m*M* GSSG-1.4 m*M* GSH. After 15 min the reaction was quenched by reducing the pH to 2. The trapped species were separated using a 0.46 × 25 cm Vydac TP518 column on a 27–34% (v/v) gradient of acetonitrile in 0.1% (v/v) aqueous TFA at a flowrate of 1 mL/min. Peaks were identified by mol mass determination using electrospray mass spectroscopy. The chromatographic procedure resolves all the possible species, which are reduced peptide (R) and its disulfide form (1-SS), the two possible single mixed-disulfides with glutathione (1-MD) and the double mixed-disulfide (2-MD). Note that in this case the differential accumulation of the single mixed-disulfides indicates that the two cysteine residues are not equally reactive.

Identifying the mixed disulfide is most straightforward if the disulfide reagent has a different charge from the trapping reagent, or is radioactive *(12,15)*. Analyzing mixtures generated with neutral and charged disulfide reagents, and trapped in parallel with iodoacetic acid and iodoacetamide can usually clarify the situation *(15,40)*; if the conformations of the trapped species do not depend on the trapping reagent, the two electrophoretic patterns can be compared to determine the number of disulfide bonds in each trapped species *(10)*.

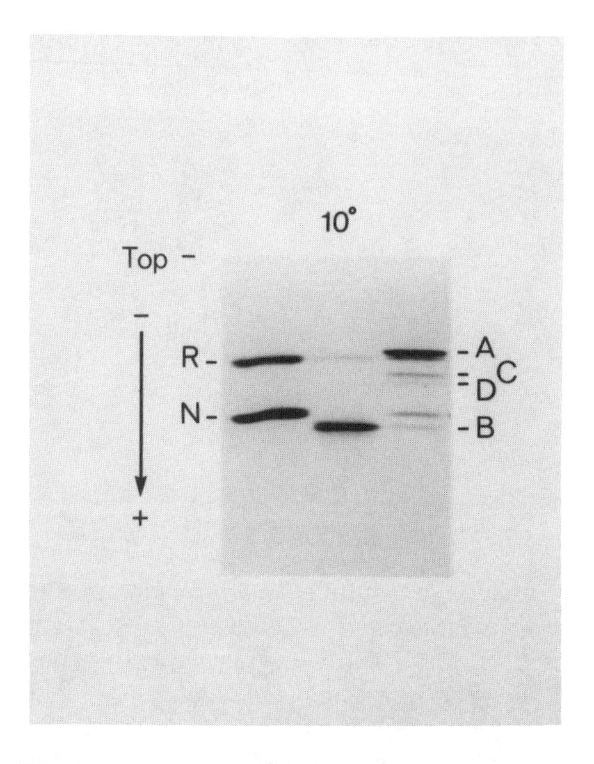

Fig. 2. Gel electrophoretic analysis at 10°C of species trapped by iodoacetic acid during the unfolding/refolding of RNase T_1. The left hand track contains the fully reduced (R) and the native (N) protein, which has disulfide bonds Cys2–Cys10 and Cys6–Cys103. The reduced protein carries four more negative charges, from blocking groups, than the native state, which increase the electrophoretic mobility, but its electrophoretic mobility is lower because it is much less compact, being very unfolded. The central track contains intermediate B, which has the native Cys6–Cys103 disulfide bond. It has a native-like conformation, but carries two more negative charges from blocking groups than the native protein, which increase its relative mobility. The right hand track shows a mixture of RNase T_1 trapped during refolding by 0.1 m*M* GSSG. Band A is a mixture of unfolded one-disulfide intermediates that have a lower mobility than the reduced protein because they each carry two less negatively charged blocking groups. Bands C and D are both proteins with two nonnative disulfide bonds. These disulfide bonds increase the compactness of these species relative to the reduced protein, but they are not as compact as the native protein, so they have an intermediate electrophoretic mobility. Taken from *(24)* with permission.

In very complex cases, differences between different disulfide isomers can cause the spectra of species with different numbers of disulfide bonds to overlap. Double trapping (*see* Section 3.1.6.) may then be required in order to generate interpretable electrophoretic patterns. The protein is first trapped with iodoacetamide, the disulfides in it are reduced, and the free thiol groups are blocked with iodoacetic acid. The resulting species are fully unfolded in 8*M* urea, where the electrophoretic mobility is determined only by the number of charged carboxymethyl groups introduced, not their positions in the protein. Consequently, the species are separated by the number of disulfide bonds or mixed disulfides originally present *(26,27,42)*.

1.4.3. Identifying and Characterizing the Trapped Species

Disulfide species in which all cysteine thiols are involved in disulfide bonds, either intramolecular or mixed-disulfides, or blocked irreversibly, may be characterized by any procedure that does not break disulfide bonds *(20,26,27,46–54)*. A first step is usually to determine which cysteine residues are paired in disulfide bonds, but there is no satisfactory general method for doing so. The two-dimensional diagonal technique of Brown and Hartley *(30,55)* is ideally suited *(11,12,15,21,24,56)*, except that it requires large amounts of protein and is not amenable to modern chromatographic techniques *(27)*. The most satisfactory approach is to cleave the protein between all the cysteine residues, to separate the resulting peptides, and to identify all those linked by disulfide bonds. It is easier to identify which cysteine residues are free, either on trapping or after reducing any disulfide bonds in species trapped irreversibly with another reagent, by blocking them irreversibly with a trapping reagent that introduces a suitable chromophore, such as *N*-iodoacetyl-*N'*-(5-sulfo-1-naphthyl)ethylenediamine *(36)*, but this does not identify the cysteine residues paired in disulfide bonds.

1.5. Measuring Rates of Protein Disulfide Bond Formation and Breakage

Although inter- and intramolecular reagents form and break disulfide bonds by the same basic mechanism, there are important practical differences in their application and in interpretation of the data.

1.5.1. Intermolecular Reagents

These reagents are particularly useful to ensure accumulation of intermediates in disulfide bond formation and when accumulation of the mixed-disulfide provides important information about slow intramolecular steps in forming a protein–disulfide bond.

1.5.1.1. Proteins with Two Cysteine Residues

A protein with two free thiol groups, P^{SH}_{SH} can react with an intermolecular disulfide reagent, RSSR, to form single and double mixed-disulfide species (P^{SSR}_{SH} and P^{SSR}_{SSR}, respectively), plus the intramolecular disulfide bond, P^{S}_{S}:

$$P^{SH}_{SH} \underset{k_{-1MD}\,[RSH]}{\overset{k_{1MD}\,[RSSR]}{\rightleftharpoons}} P^{SSR}_{SH} \underset{k_r\,[RSH]}{\overset{k_{intra}}{\rightleftharpoons}} P^{S}_{S}$$

$$P^{SSR}_{SH} \underset{[RSH]\,k_{-2MD}}{\overset{k_{2MD}\,[RSSR]}{\rightleftharpoons}} P^{SSR}_{SSR} \quad (5)$$

There are two possible P^{SSR}_{SH} species, with the mixed-disulfide on different cysteine residues; the present analysis will treat them as being indistinguishable, although this need not be the case (Fig. 1). The equilibrium constants for these reactions are defined as follows:

$$K^{RSSR}_{1MD} = \frac{k_{1MD}}{k_{-1MD}},\ K^{RSSR}_{2MD} = \frac{k_{2MD}}{k_{-2MD}},\ K^{RSSR}_{SS} = K^{RSSR}_{1MD}\frac{k_{intra}}{k_r} \quad (6)$$

In an ideal fully unfolded reduced protein, forming the mixed-disulfides should simply reflect the chemistry of thiol–disulfide exchange. Therefore, k_{1MD} and k_{-1MD} should have values of $2k_{ex}$ and $1/2\ k_{ex}$, respectively, where k_{ex} is the intrinsic rate constant for thiol–disulfide exchange between a normal thiol and either sulfur atom of a normal disulfide bond (Eq. [2]); it has a value at pH 8.7 at 25°C of about 10/s/*M*. The statistical factors reflect the possible number of thiol nucleophiles and sulfur leaving groups. The values of k_{2MD} and k_{-2MD} should ideally be k_{ex}. Consequently, the equilibrium constants for making the one (K_{1MD}^{RSSR}) and two (K_{2MD}^{RSSR}) mixed-disulfide species should be about 4 and 1, respectively. Deviations from these ideal rate and equilibrium constants may reflect differences in the pK values of the protein and reagent thiol groups, or an interaction between the protein and the reagent. These may be minimized experimentally by studying thiol–disulfide interchange at a pH of about 8.7, where the effects of different pK values on the ionization and the intrinsic reactivity cancel out *(40)*, and by a substantial ionic strength to minimize electrostatic interactions.

The rate constants k_{intra} and k_r are those of greatest importance to studying protein folding and stability. The former reflects the energetics of the protein conformational transitions involved in bringing two cysteine residues into suitable proximity for making a disulfide bond and the latter the role of the disulfide bond in the protein.

All the rate and equilibrium constants shown in Eq. (5) can be determined directly if all the species accumulate to measurable extents and can be resolved by the analytical technique used. It is almost always possible to make mixed-disulfides accumulate by altering the concentrations of RSSR and RSH.

To determine the rate constants, the change as a function of time of the amount of each species is measured with varying concentrations of RSSR and RSH, with the other conditions kept constant. The appropriate kinetic scheme can be modeled by numerical integration to simulate the changes in the levels of the various species, including RSSR and RSH, expected with any combination of rate constants (Fig. 3). The aim of the simulation is to match the experimental data under a variety of concentrations of RSH and RSSR to curves generated with a single set of rate constants for Eq. (5). Both forward and reverse rate constants can be determined simultaneously by studying disulfide bond formation or reduction in the presence of both RSSR and RSH. The rates of all the reactions involving RSSR and RSH should be demonstrated to depend linearly on the concentration of the appropriate reagent. The precision with which each rate constant is defined by the data can be established by systematically varying the values of all the rate constants and determining when the experimental data can no longer be fitted. If it is impossible to fit the data obtained across a range of concentrations to a single set of rate constants,

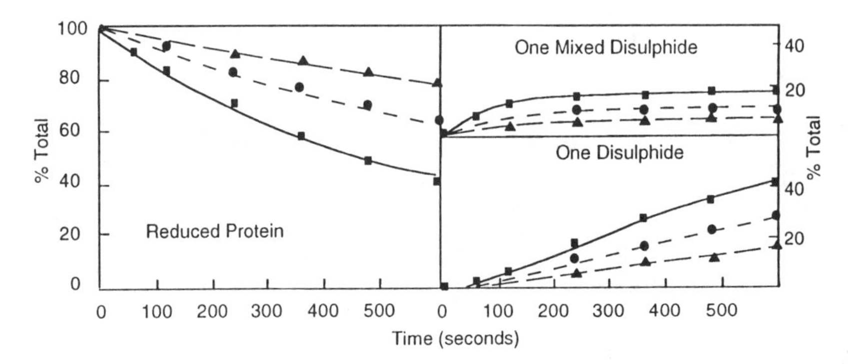

Fig. 3. The kinetics of disulfide bond formation in the peptide from Fig. 1. Disulfide bond formation in the reduced protein was carried out under three different redox conditions at pH 7.4. At various times aliquots were removed, acidified, and analyzed by HPLC as described in Fig. 1; the levels of the trapped species were quantified by integration of the elution profiles. The curves shown were generated from computer simulations of Eq. (5) using the folowing rate constants: $k_{1MD} = 3.5/s/M$; $k_{-1MD} = 0.8/s/M$; $k_{intra} = 0.007/s$; $k_r = 1/s/M$. The points shown are experimental data using 0.125 m*M* GSSG–0.7 m*M* GSH (▲), 0.25 m*M* GSSG–1 m*M* GSH (●), and 0.5 m*M* GSSG–1.4 m*M* GSH (■). The two single mixed-disulfide species were treated as being kinetically equivalent in the simulation.

the assumptions made may be invalid, steps other than thiol–disulfide exchange may be rate-determining, or all the species present are not being resolved by the analytical technique used.

The equilibrium constants can be most simply derived from measurements of the amounts of each of the species present at equilibrium. In many cases, the value of one rate constant can be estimated if the other rate constant and the equilibrium constant are known (Eq. [6]).

1.5.1.2. Three or More Cysteine Residues

In situations where multiple disulfides can form between *n* cysteine residues, avoiding the accumulation of mixed-disulfides is desirable to reduce the analytical complexity. Mixed-disulfides do not accumulate when their rate of formation is rate-limiting, i.e., $n\,k_{ex} < k_{intra}$. On the other hand, the accumulation of any particular mixed-disulfide is positive evidence that this cysteine residue has a relatively small value of k_{intra} in forming a protein disulfide bond with any other free cysteine residue *(15,57)*.

The overall rate constant for disulfide formation with intermolecular reagents when mixed disulfides do not accumulate is determined by the accessibility and reactivity of the cysteine thiol groups and is not relevant to the folding process. The observed overall rate of reducing the protein disulfide bond should be proportional to $[RSH]^2$, and the value of the third-order rate constant, k_{red}^{RSH}, is proportional to the value of k_{intra}:

$$k_{intra} = k_{ex}\, k_{red}^{RSH} / 2k_r \quad (7)$$

This equation holds only when the value of k_{intra} is greater than k_{1MD} [RSH], so that reduction of the mixed-disulfide and reformation of the protein disulfide bond are competing processes. It assumes that the mixed-disulfide is normal, so that $k_{-1MD} = k_{ex} / 2$ (Eq. [5]). The value of $k_r \approx k_{red}^{DTT} / 2$ (*see* Eq. [11], Sections 1.5.2. and 1.6.). The overall value of k_{intra} is obtained by multiplying this value by the number of disulfides that can be formed *(37)*.

1.5.1.3. Competition Between Mixed-Disulfide and Protein Disulfide Formation

When very high concentrations of a linear disulfide reagent are used to oxidize a reduced protein, in the absence of thiol reagent, formation of protein disulfide bonds and intermolecular mixed-disulfides are competing processes. The final distribution of the species depends on the relative values of k_{intra} and k_{ex} (Fig. 4A). It is assumed that each thiol of the protein reacts with the reagent with the expected rate constant, $k_{SR} = k_{ex}$ [RSSR]. Although individual microscopic intramolecular rate constants for forming the various protein disulfide bonds can be used, it is simpler to use an average rate constant, k_{SS}.

The expected spectrum of final species with different numbers of protein disulfide bonds, and as mixed-disulfides, can be calculated for different values of k_{SR} and $k_{SS.}$ For example, in a protein with four free thiols, the probability of *not* forming a protein disulfide at each stage with *i* mixed-disulfides, $P_{SRi,}$ will be

$$P_{SR1} = 3k_{SR}/(3k_{SR} + 3k_{SS4}) = 1/[1 + (k_{SS4}/k_{SR})] \quad (8)$$

$$P_{SR2} = 2k_{SR}/(2k_{SR} + 4k_{SS4}) = 1/[1 + (2k_{SS4}/k_{SR})] \quad (9)$$

$$P_{SR3} = k_{SR}/(k_{SR} + 3k_{SS4}) = 1/[1 + (3k_{SS4}/k_{SR})] \quad (10)$$

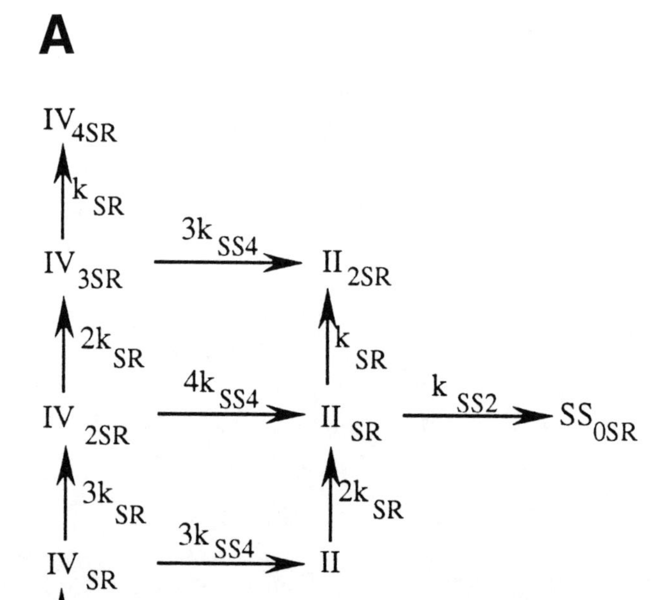
A
IV4SR
kSR
IV3SR
3kSS4
II2SR
2kSR
kSR
IV2SR
4kSS4
IISR
kSS2
SS0SR
3kSR
2kSR
IVSR
3kSS4
II
4kSR
IV

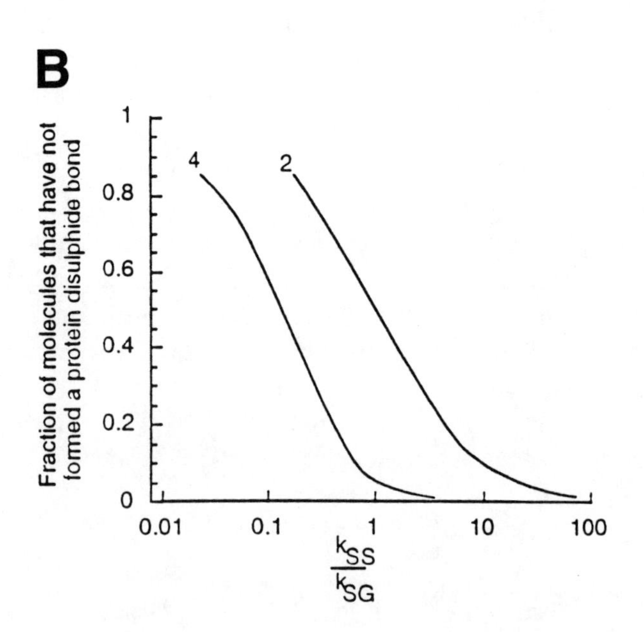
B
Fraction of molecules that have not formed a protein disulphide bond
1
0.8
0.6
0.4
0.2
0
4
2
0.01
0.1
1
10
100
kSS/kSG

Fig. 4. Kinetic competition between forming intramolecular protein disulfide bonds and intermolecular mixed-disulfides with a linear disulfide reagent, RSSR, in a fully unfolded protein with 2 or 4 cysteine thiol groups. **(A)** The kinetic scheme. The Roman numerals represent the number of cysteine residues that are not paired in intramolecular protein disulfide bonds; the subscript gives the number of mixed-disulfides on the cysteine residues. Species SS_{0SR} has all cysteine residues paired in disulfide bonds. Each species with disulfide bonds, either intra- or intermolecular, will be a mixture of various isomers, but these are not distinguished.

The protein is assumed to approximate under all circumstances a random coil, in which all cysteine thiols are equivalent. The intermolecular reaction with RSSR to generate each mixed-disulfide is depicted by a vertical arrow, but the reagent is omitted for clarity. The rate of this reaction will be the number of cysteine thiol groups multiplied by k_{SR}, the rate of reaction between one thiol and RSSR ($=k_{ex}$ [RSSR]). Intramolecular protein disulfide bond formation in the protein molecule with both cysteine thiols and mixed-disulfides occurs in the horizontal direction. The rate of this reaction is assumed to be given by the product of the number of thiols and the number of mixed-disulfides, multiplied by the average rate for forming an intramolecular protein disulfide bond k_{SS}. The value of k_{SS} is assumed to depend on how many protein disulfides are present, and these different values of k_{SS} can be measured. The rate of forming a disulfide bond at each stage is assumed to be a constant average over all the possible combinations of cysteine residues. The reverse reactions are assumed to be negligible: The concentration of RSH should be minimal, ideally no more than that generated by the reaction of RSSR with the protein thiols.

On completion of the reaction, all molecules end up as species at the top of one of the vertical lines (e.g., IV_{4SR}, II_{2SR}, SS_{0SR}). These are distinguished only by the number of mixed-disulfides and protein disulfides they contain, not by which cysteine residues are involved.

The above scheme can be extended with additional pairs of cysteine residues, and a similar scheme can be generated for odd numbers of cysteine residues. **(B)** The probabilities of unfolded protein molecules with 2 or 4 thiol groups forming only mixed-disulfides with reagent, and no further intramolecular protein disulfide bonds, as a function of the relative values of the two processes (k_{SS}/k_{SG}). The graphs were calculated for the model in (A) for an ideal totally unfolded protein molecule with 4 or 2 free thiol groups. In each case, the fraction of the molecules trapped with 4 or 2 mixed-disulfides does not consider molecules that have greater numbers of intramolecular protein disulfide bonds.

The probability of this protein forming no protein disulfides, but four mixed-disulfides, will be the product of these three probabilities. The probabilities of the other final species with intramolecular disulfide bonds being generated are calculated similarly.

The reduced protein is rapidly and efficiently mixed with various high concentrations of the intermolecular disulfide reagent, chosen so that a significant fraction of molecules end up with and without the disulfide bonds of interest. The time-course of the reaction is not followed, only the final distribution of products as a function of [RSSR]. The final products are trapped, since all cysteine residues are disulfide bonded. The thiol form of the reagent that is generated during the reaction may catalyze exchange of the disulfides, however, so a trapping reagent should be used if the products are not examined immediately.

The proportions of the species trapped in the various forms are determined by a method such as electrophoresis (*see* Section 1.4.2.), which separates the species by the number of mixed-disulfides present. The observed proportions of the various species are used to determine the value of k_{SS}/k_{SR} from plots like those shown in Fig. 4B. k_{intra} is the total rate constant for forming a protein disulfide bond in all the species with one mixed-disulfide, so it is the value of k_{SS} multiplied by the number of different disulfide bonds that can be generated (1,6,15,28, when there are, respectively, 2,4,6, and 8 cysteine thiols present initially).

The values of k_{intra} determined in this way for forming the first intramolecular disulfide bonds in a reduced protein should agree with those measured in other ways, after taking into account slightly different reactivities of the different reagents *(40)*. Second, and subsequent, protein disulfide bonds may be formed at different rates when measured in this way, however, if intramolecular rearrangements of the initial disulfide bonds are normally important to generate nonrandom mixtures of disulfide intermediates. Such rearrangements are unlikely to occur to the normal extent under conditions of rapid disulfide bond formation.

1.5.2. Intramolecular Disulfide Reagents

Disulfide bond formation with an intramolecular reagent, such as DTT^{S}_{S}, involves the same reactions as for intermolecular reagents (Eq. [3]), but differs significantly in that reversal of mixed-disulfide formation is now an intramolecular process:

$$P^{SH}_{SH} + DTT^{S}_{S} \underset{}{\overset{2K^{DTT}_{MD}}{\rightleftharpoons}} P^{S\text{-}S}_{SH}\ DTT_{SH} \underset{k^{DTT}_{r}}{\overset{k_{intra}}{\rightleftharpoons}} P^{S}_{S} + DTT^{SH}_{SH} \quad (11)$$

The formation of the mixed-disulfide is rapidly reversed at a rate of 5×10^2/s at pH 8.7 and 25°C, so it barely accumulates. So long as k_{intra} is lower than this rate, the overall rate at which a protein disulfide is formed, k^{DTT}_{SS}, is proportional to the value of k_{intra},

$$k^{DTT}_{SS} = 2\ K^{DTT}_{MD}\ k_{intra} \quad (12)$$

Deriving k_{intra} from the observed rate constant requires that the value of K^{DTT}_{MD}, the equilibrium constant for mixed-disulfide formation, be known. This parameter has been estimated from the equilibrium between DTT^{S}_{S} and *GSH (40,44)*. The value is dependent on the pH and on the pK values of the relevant protein thiols, but in the range of pK values normally encountered (8.5–9.1), it varies no more than twofold. With protein thiol pK values of 8.8, it has a value of $0.01M^{-1}$ at pH 8.7.

This is the easiest method of determining k_{intra} in complex systems where many disulfide bonds and intermediate species accumulate. The absence of the mixed-disulfide of each species considerably simplifies the kinetic analysis.

The rate of reduction of a protein disulfide by DTT^{SH}_{SH} provides an estimate of the value of k_r for Eq. (7), since the subsequent reduction of the mixed-disulfide to the fully reduced protein is generally not rate-limiting. The observed rate constant, k^{DTT}_{red}, is predicted from the intrinsic reactivities of the two thiol groups of DTT^{SH}_{SH}, and observed to be about 2 k_r of Eq. (7) *(40)*.

1.6. Interpreting Rates of Disulfide Bond Formation and Reduction

Intramolecular rate constants for disulfide bond formation reflect the conformational transitions required to bring the two thiol groups into the correct orientation to form a disulfide bond. The free energy of the transition state for this process, $\Delta G^{\ddagger}$, is given by

$$\Delta G^{\ddagger} = -RT \ln k_{intra}\, h / kT \quad (13)$$

where h is Planck's constant, k Boltzmann's constant, R the gas constant, and T the absolute temperature.

Large values of k_{intra} are encountered when a protein structure holds the two participating thiol groups in the correct orientation for disulfide formation. Smaller values are observed in unfolded proteins, and very low values may be encountered if formation of the disulfide bond requires some particularly unfavorable conformational change, perhaps in the transition state for disulfide bond formation.

In some circumstances, a protein disulfide bond may be formed only slowly because both of the thiols involved in forming the disulfide bond are inaccessible to the disulfide reagent. In this case, mixed-disulfides should not accumulate on these thiol groups using an intermolecular reagent. If they do accumulate, the low rate of forming the disulfide bond is likely to be caused by conformational factors *(57)*.

The disulfide-forming tendency of a pair of cysteine residues is indicated by their effective concentration, which reflects the tendency of the polypeptide chain to keep the thiol groups in proximity in the absence of the disulfide bond. It is given by the ratio of the intramolecular rate constant k_{intra} to the comparable intermolecular rate constant for reduction of their mixed-disulfides by a monothiol, k_{-1MD} in Eq. (5) *(40)*. In unfolded polypeptides, the effective concentrations of pairs of cysteines have been found to be in the low millimolar range, the exact value depending on their separation in the chain. In folded proteins, values of zero occur when the thiol groups are kept apart by the conformation and values as high as $10^5 M$ have been measured when the thiol groups are kept in proximity *(37)*.

The rate at which a protein disulfide bond is reduced to the mixed-disulfide, k_r in Eq. (3), provides information about the role of the disulfide bond in the protein. The value of k_r should be approx k_{ex}, unless structural strain increases this rate or structural stabilization decreases it. Low rates of reduction are also observed if the disulfide bond is buried; this is owing, in part, to inaccessibility to the thiol reagent, but buried sulfur atoms are also likely to have elevated pK values and consequently lowered reactivities (Eq. [2]). It should be noted that if the transition state for forming such a buried disulfide bond is the same as that in reducing the bond, forming that disulfide bond will also be slow *(1,52)*.

Detailed comparison of k_{intra} and k_r values obtained with different reagents suggests that there should be slight differences in the values obtained with the two types of reagent, as a result of differences in the pK values of their thiol groups (Eq. [2]); for example, $k_r^{DTT} \approx 2\ k_r^{GSH}$ and

$k_{intra}^{GSS} \approx 2\, k_{intra}^{DTT}$ have been found experimentally at pH 8.7 and predicted by considering the relevant pK values in Eq. (2) *(40)*.

1.7. Equilibrium Constants

Equilibrium constants for various steps in making or breaking protein disulfide bonds may be derived from the rates of the forward and reverse reactions, or from measuring the levels of the various species present at equilibrium. If the mixed-disulfide does not accumulate, only the overall equilibrium constant can be determined:

$$K_{SS} = \frac{[P_{S}^{S}]\,[RSH]^2}{[P_{SH}^{SH}]\,[RSSR]} \text{ or } \frac{[P_{S}^{S}]\,[DTT_{SH}^{SH}]}{[P_{SH}^{SH}]\,[DTT_{S}^{S}]} \quad (14)$$

If it is assumed that the simple thiol–disulfide equilibrium (K_1 in Eq. [5]) behaves in a predictable manner, the overall equilibrium constant can be interpreted solely in terms of the more important conformation-related equilibrium (k_{intra}/k_r) (Eq. [5]). This is satisfactory in many instances, but there may be occasions where some unusual factor affects the value of K_1. The overall equilibrium constant for forming a protein–disulfide bond with intermolecular disulfide reagents is often equated with the value of the effective concentration. Note that this is not appropriate in circumstances where the protein with the disulfide bond adopts a folded conformation that decreases the rate of disulfide reduction. In these circumstances, the kinetic definition of effective concentration (*see* Section 1.6.) is more appropriate.

The difference in free energies of the disulfide bonds that are interconverted, ΔG_{SS}, is given by the equilibrium constant K_{SS} and the redox potential:

$$\Delta G_{SS} = -RT \ln K_{SS} \frac{[P_{SH}^{SH}]}{[RSH]^2} \text{ or } -RT \ln K_{SS} \frac{[P_{SH}^{SH}]}{[DTT_{SH}^{SH}]} \quad (15)$$

Using this value and the value of the transition state free energy (Eq. [13]) for each step, it is possible to construct a complete thermodynamic profile for disulfide bond formation and reduction at a given redox potential *(16,24,26,37)*.

There is thermodynamic linkage between the stabilities of protein conformations and disulfide bonds, so equilibrium constants for either can be used to measure the effect of a disulfide bond on protein stability and vice versa:

$$\begin{array}{ccc} U^{SH}_{SH} & \overset{K^{U}_{SS}}{\rightleftharpoons} & U^{S}_{S} \\ K^{SH}_{F}\updownarrow & & \updownarrow K^{SS}_{F} \\ N^{SH}_{SH} & \overset{K^{N}_{SS}}{\rightleftharpoons} & N^{S}_{S} \end{array} \qquad (16)$$

Because this is a closed cycle,

$$K^{N}_{SS}/K^{U}_{SS} = K^{SS}_{F}/K^{SH}_{F} \qquad (17)$$

The contribution of the disulfide bond to the free energy of the folded state, $\Delta\Delta G_{fold}$, is given by:

$$\Delta\Delta G_{fold} = -RT \ln K^{SS}_{F}/K^{SH}_{F} \qquad (18)$$

and the contribution of the folded conformation to the stability of the disulfide bond, $\Delta\Delta G_{SS}$, is given by:

$$\Delta\Delta G_{SS} = -RT \ln K^{N}_{SS}/K^{U}_{SS} \qquad (19)$$

According to Eq. (16), the two must be the same.

$$\Delta\Delta G_{SS} = \Delta\Delta G_{fold} \qquad (20)$$

By determining the values of equilibrium constants for making the disulfide bonds in the presence (K^{U}_{SS}) and absence of denaturant (K^{N}_{SS}), $\Delta\Delta G_{fold}$ can be measured indirectly. Estimates made in this way are similar to those made by more direct means *(29,58)*.

If a protein disulfide bond is stabilized by the folded conformation, as is usually the case, the stability of the disulfide bond is greater in the folded state than in the unfolded state, $\Delta\Delta G_{SS} < 0$, and the disulfide bond stabilizes the native state, $\Delta\Delta G_{fold} < 0$. There are cases, however, where the opposite occurs and the protein disulfide bond is less stable in the folded state and destabilizes the native folded conformation *(58)*.

1.8. Reconstructing the Pathway of Disulfide Bond Formation and Breakage

Pathways of disulfide bond formation and breakage are reconstructed by identifying the intermediates involved and determining their relationship by kinetic analysis. Two intermediates that differ only in one disulfide bond *cannot* be assumed to be interconverted directly by simply making or breaking that disulfide bond *(15)*.

The rates of interconversions that involve the formation or reduction of disulfide bonds depend on the concentrations of the disulfide and thiol reagents, respectively, whereas there is no such dependence for intramolecular disulfide rearrangements. Systematic alterations in the concentrations of the reagents thus reveal what type of process is rate-limiting. The ability to simulate the experimental data with a particular kinetic scheme does not prove its validity, but such analysis can be used to exclude alternative pathways. Thermodynamic restrictions must be observed, so the net free energy change around any closed cycle in the pathway must be zero, and the value of each equilibrium constant must be determined by the values of the others in the cycle. It is usually assumed that all protein molecules with the same disulfide bonds are kinetically equivalent, but this may not be the case if they differ by some slow conformational transition, for example by *cis-trans* isomerization of peptide bonds.

Kinetic analysis can be assisted by using purified acid trapped intermediates *(17,26,36)*, but not if the rates of intramolecular disulfide rearrangements are significant compared to the rates of disulfide bond formation or reduction; in this case the kinetic behavior of all the species in rapid equilibrium are the same, and it is impossible to determine which of these species is the precursor of any other. It is always possible that important kinetic intermediates in the folding pathway will not accumulate during folding and may be missed. Indeed, the accumulation of a particular species does not indicate its kinetic importance and may indicate the opposite. With particularly stable proteins, formation of the native state may only require the formation of a subset of native disulfide bonds (*see* Section 4.2.). The accumulation of such quasinative states during protein folding is a reflection of their stability and not of their kinetic importance. To overcome these problems, kinetic studies must be performed on proteins in which particular cysteine residues are either blocked irreversibly or replaced by site-directed mutagenesis, so that they cannot take part in disulfide bond formation or rearrangement *(15,26,40,59,60)*. Steps involving these residues in disulfide bonds should be absent in the modified protein, allowing the contributions of individual cysteine residues and intermediates to be dissected *(61,62)*.

2. Materials

2.1. Gel Electrophoresis

Any type of electrophoresis apparatus may be used, but best results are obtained in systems that are cooled efficiently. Gels 3 mm thick have proved most satisfactory in running samples of iodoacetate-trapped BPTI species; thinner gels exhibit poorer resolution of species and distortion across the tracks. With iodoacetamide-trapped samples, the advantages of the thicker gels are less clearcut, but the gels still need to be at least 1 mm thick. The other dimensions of the gel are less important, and the resolution seen on 10 × 10 cm gels is often equivalent to that obtained on larger gels.

Gels are often photopolymerized to reduce the generation of free radicals that often accompany polymerization with ammonium persulfate. Sunlight can be used to do this, but light from a bank of fluorescent tubes or even tungsten light bulbs placed close to the gels can be used.

2.1.1. Solutions for Gel Electrophoresis

2.1.1.1. Stock Solutions for Low-pH Gels *(63)*

1. Solution A: 48 mL 1*M* KOH and 1 mL *N,N,N',N'*-tetramethylethylenediamine (TEMED) diluted to 70 mL in water. Adjust the pH to 4.3 with acetic acid and make up the vol to 100 mL. Recheck the pH and adjust as necessary. Store at 4°C.
2. Solution B: 30 g acrylamide and 0.2 g bisacrylamide in a final vol of 100 mL.
3. Solution C: 40 g acrylamide and 0.3 g bisacrylamide in a final vol of 100 mL.
4. Solution D: 4 mg riboflavin in 100 mL water. Store in the dark at 4°C.
5. Solution E: 48 mL 1*M* KOH and 0.46 mL TEMED diluted to 70 mL with water. Adjust the pH to 6.7 with acetic acid and make up the volume to 100 mL. Recheck the pH and adjust as necessary. Store at 4°C.
6. Solution F: 10 g acrylamide and 2.5 g bisacrylamide in a final vol of 100 mL.
7. Overlay solution: 50% (v/v) ethanol.
8. Running buffer: (4X stock) 31.2 g β-alanine in 1 L water with the pH adjusted to 4.5 with glacial acetic acid.
9. Sample buffer: 100 mg methyl green dissolved in 50 mL of 50% (w/v) glycerol. **Caution:** TEMED, acrylamide, and bisacrylamide are all highly toxic.

2.1.1.2. Stock Solutions for High pH Gels *(64)*

1. Solution G: 30 g acrylamide and 0.8 g bisacrylamide in a final vol of 100 mL.
2. Solution H: 36.3 g Tris base in 48 mL of 1*M* HCl. Adjust the pH to 8.8

with 5*M* HCl and make up the vol to 100 mL. Recheck the pH and adjust as necessary.

3. Solution I: 6 g Tris base in 40 mL water. Adjust the pH to 6.8 with 1*M* HCl and make up the vol to 100 mL. Recheck the pH and adjust as necessary.
4. Solution J: 0.5 g ammonium persulfate in 4.5 mL of water, freshly prepared.
5. Overlay solution: 50% (v/v) ethanol.
6. Running buffer: 3 g Tris base and 14.4 g glycine dissolved in water and made up to a final vol of 1000 mL.
7. Sample buffer: 5 g glycerol mixed with 5 mL of 0.02% (w/v) bromophenol blue.

2.1.2. Stain and Destain Solutions

1. Stain: 0.25 g Coomassie blue R-250 is dissolved in 200 mL of water with stirring for 1 h. 25 g Trichloroacetic acid and 25 g sulfosalicylic acid are then dissolved in this solution. **Caution:** Trichloroacetic acid and sulfosalicylic acid are very caustic.
2. Wash: 500 mL methanol and 100 mL acetic acid made up to a vol of 1000 mL with water.
3. Destain: 100 mL methanol and 100 mL acetic acid made up to a final vol of 1000 mL with water.

2.2. Trapping Reagents

Iodoacetic acid and iodoacetamide are available in highly purified forms and should be white crystalline solids; yellow crystals should not be used. They can be easily recrystallized from carbon tetrachloride and water, respectively, but care should be taken to avoid contact with the skin. Trapping reagents are prepared immediately before use, because they gradually decompose. Solutions of iodoacetic acid require neutralization before use, and all solutions should be buffered to provide control over the pH of the trapping reaction. The practical limits of solubility at neutral pH and 25°C for iodoacetic acid and iodoacetamide are about 1 and 0.5*M*, respectively. Typical stock solutions of trapping reagents are:

1. 0.5*M* iodoacetic acid: 92 mg iodoacetic acid dissolved in 0.5 mL of 1*M* KOH and 0.5 mL of 1*M* Tris-HCl, pH 7.3.
2. 0.5*M* iodoacetamide: 92 mg iodoacetamide dissolved in 1 mL 0.5*M* Tris-HCl, pH 7.3.

Table 1
Determining Thiol and Disulfide Concentrations

	Extinction coefficients of disulfide compounds			
	pH	Wavelength, nm	$\varepsilon(M^{-1}/cm^{-1})$	Ref.
DTT^S_S	8	283	273	*69*
	8	310	110	*69*
GSSG	7	248	382	*44*

2.3. Thiol and Disulfide Reagents

2.3.1. Purification and Preparation

The most commonly used reagents are the oxidized and reduced forms of glutathione, β-mercaptoethanol, cystamine, and dithiothreitol. The commercially available reagents may contain impurities that interfere with their use. In particular, DTT^S_S usually contains traces of more reactive linear disulfide species, so it is purified as follows.

1. Prepare a 40 × 1.5 cm column of Amberlite IRA-410 in the OH^- form. The column is prewashed with 1*M* HCl, regenerated with 1*M* NaOH, and equilibrated with 0.1*M* NH_3.
2. Dissolve 2 g of DTT^S_S in 100 mL 0.1*M* NH_3 under N_2. Add 0.2 g DTT^{SH}_{SH} and incubate at room temperature for at least 1 h.
3. Filter the solution and apply to the column, followed by washing with H_2O.
4. Monitor the elution of DTT^S_S by its absorbance at 283 nm (Table 1). The DTT^S_S should be eluted directly from the column, but it is usually eluted only by a large vol of H_2O.
5. Measure the elution of the thiol-containing compounds using the Ellman method (*see* Section 2.3.2.). Thiol groups should be ionized at the alkaline pH and should absorb to the resin, so none should be eluted from the column.
6. Pool the eluate containing DTT^S_S, but free from thiol groups and lyophilize.
7. Store the lyophilized DTT^S_S at –20°C in the presence of dessicant.

Traces of disulfides in reduced thiol reagents can be quantified by HPLC and taken into account in the kinetic analysis. The separation of the oxidized and reduced forms of glutathione is shown in Fig. 5, and similar separations may be devised for other reagents.

Solutions of thiol and disulfide reagents are prepared immediately before use, neutralizing acidic and basic groups where necessary with NaOH and HCl solutions, respectively. Reduced forms of the reagents

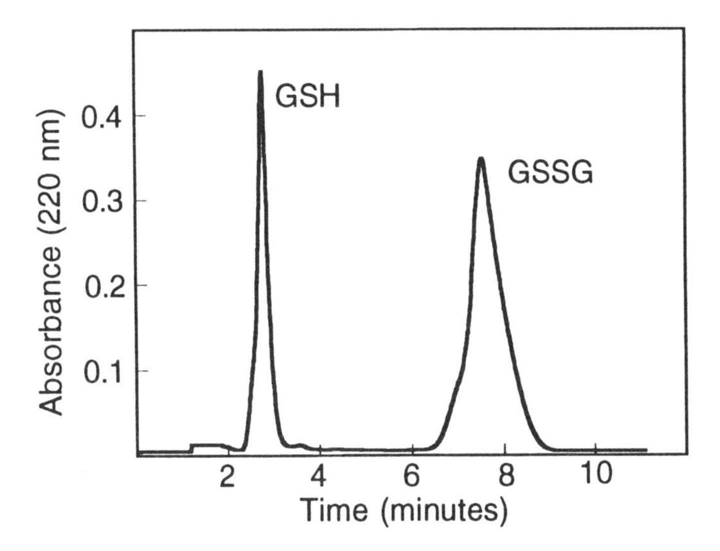

Fig. 5. Separation of 80 nmol of each of the oxidized and reduced forms of glutathione on a 0.39 × 15 cm Waters C-18 Delta-Pak column. The column was eluted at a flow rate of 1 mL/min for the first 5 min after injection with 0.1% (v/v) TFA and then with a 10-min linear gradient to 3% (v/v) acetonitrile in 0.1% (v/v) TFA.

should not be kept at neutral pH or above for long, since they will progressively air-oxidize. Note that it is difficult to prepare solutions of oxidized DTT more concentrated than 0.1*M*. Extinction coefficients are available for a number of the disulfide reagents to check their concentrations (Table 1), whereas the concentration of thiol reagents can be determined using the Ellman assay for free thiol groups.

2.3.2. Ellman's Assay for Free Thiol Groups (28)

1. At least 2 nmol of thiol compound are mixed with 1 mL 6*M* guanidine hydrochloride, 0.1*M* sodium phosphate, pH 7.3, and 1 m*M* EDTA.
2. Add 50 µL 3 m*M* 5,5'-dithio-bis(2-nitrobenzoic acid) (DTNB) in 0.1*M* sodium phosphate, pH 7.3. Measure the change in absorbance of the sample at 412 nm, compared to the appropriate blank, until it stabilizes.
3. Nitrothiobenzoate is liberated quantitatively by reaction of DTNB and the thiol, and its extinction coefficient is 13,700/*M*/cm (14,150/*M*/cm in the absence of guanidine hydrochloride) *(28)*.

2.4. Other Solutions

2.4.1. Protein Reduction and Gel Filtration

1. 6*M* Guanidine hydrochloride, 0.1*M* Tris-HCl, pH 8.7, 0.1*M* DTT^{SH}_{SH}: Solid DTT^{SH}_{SH} is dissolved in the 6*M* guanidine hydrochloride, 0.1*M* Tris-HCl, pH 8.7, immediately before use.
2. 10 m*M* HCl.

2.4.2. Disulfide Bond Formation and Reduction

1. 1*M* Tris-HCl, pH 8.7.
2. 10 m*M* EDTA, pH 7.
3. 2*M* KCl.

2.4.3. Double Trapping

1. 0.1*M* Tris-HCl, pH 8.7, 0.2*M* KCl, 1 m*M* EDTA, 8*M* urea: Prepared immediately before use.
2. 0.1*M* DTT^{SH}_{SH}: Prepared immediately before use.

3. Methods

3.1. Kinetics of Disulfide Bond Formation and Reduction

The methods given here have been used to study a number of small model proteins, but they may need to be modified if the solubility or other properties of another protein differ (*see* Notes 1 and 2).

3.1.1. Preparation of Gels

Stock solutions are described in Section 2.1.1.

3.1.1.1. Preparation of Low pH Gels

1. For a 15% (w/v) polyacrylamide separating gel, mix 5 mL solution A + 20 mL solution B + 5 mL solution D + 10 mL water.
2. Degas the mixture under vacuum in an aluminum foil-covered vessel and then pour into the gel apparatus.
3. Overlay the acrylamide solution with 50% (v/v) ethanol and illuminate it with sunlight or artificial light for 30–60 min to polymerize it. If problems are encountered with the polymerization, 10 µL of a 10% (w/v) ammonium persulfate solution can be added to increase the rate.
4. Once the gel has set, pour off the overlay solution and rinse with water.
5. For the stacking gel, mix 2 mL solution E + 4 mL solution F + 2 mL solution D + 8 mL water.
6. Degas as above and then rinse the top of the separating gel with a few milliliters of the stacking gel solution. Pour the solution into the gel appa-

ratus, insert the sample well-forming comb, and photopolymerize as above. If necessary, polymerization may be improved by adding 5 µL of 10% (w/v) ammonium persulfate.

3.1.1.2. Low pH Gels in the Presence of 8*M* Urea

1. For a 15% (w/v) polyacrylamide separating gel, mix 22.5 mL of solution C + 5 mL of solution A + 3.75 mL solution D + 28.8 g urea. Adjust the vol to 60 mL with water and cast the gel as described in Section 3.1.1.1.
2. For the stacking gel, mix 2 mL solution E + 4 mL solution F + 2 mL solution D + 7.7 g urea. Adjust the vol to 16 mL with water and cast the gel as described in Section 3.1.1.1.

3.1.1.3. Preparation of High pH Gels

1. For a 15% (w/v) separating gel mix 45 mL solution G + 11.25 mL solution H + 33 mL water + 45 µL TEMED.
2. Degas the mixture and then add 0.675 mL solution J. Pour into the gel apparatus and overlay with 50% (v/v) ethanol.
3. Once the gel has set, rinse away the overlay solution with water.
4. For the stacking gel, mix 5 mL solution G + 10 mL solution I + 20 mL water + 5 mL solution D + 30 µL TEMED.
5. Degas the mixture in darkness (*see* Section 3.1.1.1.), then rinse the top of the separating gel with a few milliliters of stacking gel solution. Pour into the gel apparatus and insert the sample well-forming comb. Photopolymerize as described in Section 3.1.1.1.

3.1.1.4. High-pH Gels in the Presence of 8*M* Urea

1. For a 12% (w/v) polyacrylamide separating gel mix 34.5 mL of solution G + 11.25 mL solution H + 10.45 mL water, 45 µL TEMED, and 43.2 g of urea. Allow the urea to dissolve and adjust the vol to 90 mL.
2. Degas the mixture under vacuum and then add 0.675 mL of solution J. Pour into the gel apparatus and overlay it with 50% (v/v) ethanol.
3. Urea is not normally added to the stacking gel, which is prepared as described in Section 3.1.1.3.

3.1.2. Preparation of the Reduced Protein

1. Dissolve the protein (about 10 mg) in 0.5 mL of 6*M* guanidine hydrochloride, 0.1*M* Tris-HCl, pH 8.7, 0.1*M* DTT^{SH}_{SH}, and leave for 30 min to 1 h to reduce all the disulfide bonds.
2. Desalt the protein on a 1.5 × 25 cm column of G-25 Sephadex run in 10 m*M* HCl (*see* Notes 2 and 3), and determine its concentration using its extinction coefficient. Adjust the protein concentration to about 0.5 mg/mL.

3.1.3. Disulfide Bond Formation and Reduction

1. Saturate all the reagent solutions with N_2 or Ar prior to use (*see* Note 4).
2. Prepare stock solutions of the thiol and disulfide reagents (*see* Section 2.3.). The concentrations should be sufficient for the desired final concentration of reagent to be reached when they are diluted three- to tenfold into the reaction mixture.
3. Mix 0.4 mL of protein solution (0.2 mg protein) with 0.1 mL Tris-HCl (pH 8.7), 0.1 mL 10 m*M* EDTA, and 0.1 mL 2*M* KCl in a Pierce Reacti-Vial™ sealed with a septum stopper. Add sufficient water so that the volume of the reaction after addition of thiol and disulfide reagents (*see* step 4) is 1 mL.
4. Hypodermic needles are inserted through the septum so that a constant flow of water-saturated N_2 or Ar can be passed over the surface (*see* Note 4). Immerse the vials in a water bath at 25°C and leave to equilibrate for 5 min.
5. Initiate the reaction by addition of the disulfide and thiol reagents in sufficient volume to bring the final reaction vol to 1 mL.

3.1.4. Trapping the Species Present Irreversibly

1. At the time intervals required, withdraw a portion of the reaction mixture containing 5–10 µg protein and add 1/4 vol 0.5*M* iodoacetamide or iodoacetic acid solution buffered to pH 7.3 (*see* Section 2.2.). Using these solutions, the final pH of the trapping reaction is about 8.
2. After 2 min at 25°C, place the trapped mixture on ice until it is analyzed (*see* Section 3.1.5.).

If quasinative intermediates are suspected (*see* Sections 1.8. and Note 5), the trapping reactions are doubled in scale. One half of the reaction mixture is placed on ice after the 2 min at 25°C. The other half is saturated with urea and left for a further 20 min at 25°C before being placed on ice.

3.1.5. Separation of the Trapped Species by Electrophoresis

1. To the trapped protein solution, add 1/4 vol of the sample buffer (*see* Note 6) appropriate for the gel type that is being used (*see* Sections 2.1.1.1. and 2.1.1.2.)
2. Load the samples onto gels precooled to 4°C.
3. Initially run the gels at 10 mA per gel constant current. Once all the tracking dye has entered the gel, the current can be increased to 25 mA /gel (*see* Note 6).
4. When the dye front has run 5–10 cm, remove the gels from the apparatus and immerse them in staining solution (*see* Section 2.1.2.).
5. Stain for at least 2 h with constant agitation to ensure that the gels do not

stick to each other or to the vessel. The high concentration of acid in the stain fixes most proteins rapidly, and the low solubility of the dye under these conditions produces low background staining.

6. Briefly rinse the gels in wash solution (*see* Section 2.1.2.) to remove stain precipitated on the gel surface and then place them in destain solution (*see* Section 2.1.2.).

3.1.6. Double Trapping (26)

This is an alternative procedure to that described in Section 3.1.4. and is used to obtain a separation that is solely dependent on the number of disulfide bonds or mixed disulfides that were originally present in the molecule (*see* Section 1.4.2.).

1. Prepare gels containing 8*M* urea for resolving acidic or basic proteins (*see* Sections 3.1.1.2. and 3.1.1.4.).
2. Set up disulfide bond formation and reduction reactions (*see* Section 3.1.3.).
3. At the desired timepoints withdraw an aliquot of the reaction mixture and add 1/4 vol 0.5*M* iodoacetamide (*see* Section 2.2.). Leave for 2 min at 25°C and then place the mixture on ice.
4. Desalt the trapped reaction mixtures on a small (about 2 mL) G-25 gel filtration column run in 10 m*M* HCl or desalt by reverse-phase HPLC (*see* Note 3). Whichever method is used, no residual alkylating reagent must be left in the sample.
5. Lyophilize the desalted protein.
6. Resuspend the protein in 0.1*M* Tris-HCl, pH 8.7, 0.2*M* KCl, 1 m*M* EDTA, and 8*M* urea, and add 1/8 vol of 0.1*M* DTT^{SH}_{SH} to reduce the disulfide bonds.
7. After 15 min add 1/4 vol 0.5*M* iodoacetic acid (*see* Section 2.2.), leave for 2 min at 25°C, and then place on ice.
8. Add 1/4 vol of the appropriate sample buffer (*see* Sections 2.1.1.1. and 2.1.1.2.) and analyze by electrophoresis (*see* Section 3.1.5.).
9. Stain and destain the gels as in Section 3.1.5.

Since the gels contain 8*M* urea the mobility of the species is dependent solely on the number of charged carboxymethyl groups present, and hence the number of protein or mixed disulfides originally present. The occurrence of species with odd numbers of carboxymethyl cysteine residues is indicative of the occurrence of mixed-disulfides with the disulfide reagent. The system can be calibrated by reacting the fully reduced protein with varying ratios of iodoacetamide to iodoacetic acid, which produces a “ladder” of molecules differing by the charge of a single carboxymethyl group *(26,27,41)*.

3.1.7. Quantification

After the gels are completely destained, they can be scanned for quantification of the various species. We currently analyze video images of gels using Image 1.43 software (Wayne Rasband, NIH, Bethesda, MD). The density profile of each track can be determined, and the area under each peak normalized. A number of commercial gel scanners that provide extensive data manipulation options are also available. Older designs of gel scanner that simply produce a paper trace output of the density profile along each track have also been used successfully.

3.2. Equilibrium Measurements

Equilibrium measurements are made using the techniques described above, but instead of following the changing levels of the species with time, the reactions are left for sufficient time to come to equilibrium before sampling and analysis. A range of thiol:disulfide reagent concentrations is used, and the mixture is sampled at at least three well-separated time points to check that equilibrium is reached. The concentrations of the reagents chosen ideally should be sufficient to ensure that equilibrium is reached rapidly and that the changes in the concentrations of redox reagents during the reaction are insignificant compared to their overall concentrations. In circumstances where a large excess of the reagent cannot be used, the actual levels of the thiol and disulfide forms at equilibrium can be quantified by reverse-phase HPLC *(65,66)*. The ultimate demonstration that equilibrium has been attained is to obtain the same result by the reverse reaction.

3.3. Analysis by HPLC

Proteins that are to be analyzed by HPLC are normally acid trapped by addition of an appropriate volume of HCl to reduce the pH to about 2. No general HPLC conditions can be given, because these depend on the nature of the protein used, although most investigations have used acetonitrile with trifluoroacetic acid as the aqueous modifier and 300 Å reverse-phase column packings *(26,27,36,58,65,66)*. Illustrative separation conditions are in Fig. 1. Acid trapping only slows the reaction, so careful controls are required to check that the length of time before the analysis and the residency time on the column have no influence on the levels or types of species seen. Snap freezing the samples and storage at

–20°C prior to analysis is possible with some proteins, but this should be checked in individual cases.

4. Notes

1. The major considerations in adapting the method are the isoelectric point (pI) and the size of the protein. The low-pH electrophoresis system has been used to study the basic proteins BPTI *(9–17)*, its homologs *(18,19)* and ribonuclease A *(20–23),* whereas the high-pH system has been used for α-lactalbumin *(25–27)* and ribonuclease T1 *(24)*. Compendiums of appropriate systems for native gel electrophoresis at various pH values are available *(67,68)*, so it should be possible to find a system suitable for any protein. In deciding on such a system it is important to take account of the changes in the net charge of the molecule that can occur in the trapping reactions.

   The 12–15% polyacrylamide separating gels described have proved suitable for running proteins with mol wt of 6000–12,000 Daltons. Larger proteins may require that a lower percentage acrylamide separating gel is used.
2. The pI of the protein influences the way in which the protein is handled. Desalting and manipulating the protein under acid conditions is desirable to reduce the rate of air oxidation, but some proteins are insoluble at low pH. The protein being studied must be reasonably soluble in all forms under the conditions necessary for thiol–disulfide interchange, because experiments must be carried out at a protein concentration that is high enough to permit analysis on gels.
3. If only small amounts of protein are available, the scale of the gel filtration column used for desalting the reduced protein can be decreased. With very small amounts of protein, the reduced protein can be absorbed onto a small reverse-phase HPLC cartridge, such as a 0.46 × 3 cm Brownlee Aquapore RP-300 *(48)* and eluted with a steep acetonitrile gradient in 0.1% (v/v) trifluoroacetic acid, at a low flowrate, so it is obtained in a minimal volume. The solvent can be removed by centrifugation under a vacuum, and the dry reduced protein dissolved in 10 m*M* HCl or another appropriate solution in which it is soluble.
4. Irreproducibility of results often arises from problems related to control of the reaction pH, since even small differences can have significant kinetic effects. The pH of the complete reaction mixture should be routinely checked. Oxidation of protein and reagent thiol groups can also be a problem, especially if long time courses are followed. To check for this, examine whether protein disulfide formation is occurring in the absence of redox reagents and how the levels of thiols measured by the Ellman method (*see* Section 2.3.2.) change with time.

5. Quasinative states are problematic in the analysis of disulfide-linked folding because they can occur in substantial levels and can be stable. In some circumstances, the quasinative conformation can bury cysteine thiol groups, rendering them unreactive to both thiol–disulfide reagents and alkylating reagents *(37,38)*. Comparing the electrophoretic patterns of samples trapped in the presence and absence of urea should reveal the presence of such quasinative states. One general approach to minimize the occurrence of these quasinative states is to carry out the folding reactions at as high a pH as possible. This increases the ionization of the thiol groups, which resists any tendency for them to be buried.
6. A very common technical problem is a smeared electrophoretic pattern; usually it can be alleviated to some extent by using thicker gels. It also arises from poorly polymerized gels, or electrophoresis at too high a current or without cooling. The problem is exacerbated if the density of the sample is too low. Smearing caused by the protein aggregating during folding, trapping, or electrophoresis may be cured by reducing the protein concentration or by inclusion of urea in the gel. High protein concentrations during the reactions may also lead to intermolecular disulfide-bonded aggregates.

References

1. Creighton, T. E. (1978) *Prog. Biophys. Mol. Biol.* **33,** 231–297.
2. Creighton, T. E. (1985) *J. Phys. Chem.* **89,** 2452–2459.
3. Creighton, T. E. (1988) *BioEssays* **8,** 57–64.
4. Creighton, T. E. (1990) *Biochem. J.* **270,** 1–16.
5. Creighton, T. E. (1992) in *Protein Folding* (Creighton, T. E., ed.), Freeman, New York, pp. 301–351.
6. Creighton, T. E. (1983) *Biopolymers* **22,** 49–58.
7. Creighton, T. E. (1983) in *Functions of Glutathione: Biochemical, Physiological, Toxicological and Clinical Aspects* (Larsson, A., Orrenius, S., Holmgren, A., and Mannervik, B., eds.), Raven, New York, pp. 205–213.
8. Goldenberg, D. P. and Creighton, T. E. (1985) *Biopolymers* **24,** 167–182.
9. Creighton, T. E. (1974) *J. Mol. Biol.* **87,** 563–577.
10. Creighton, T. E. (1974) *J. Mol. Biol.* **87,** 579–602.
11. Creighton, T. E. (1974) *J. Mol. Biol.* **87,** 603–624.
12. Creighton, T. E. (1975) *J. Mol. Biol.* **95,** 167–199.
13. Creighton, T. E. (1975) *J. Mol. Biol.* **96,** 767–776.
14. Creighton, T. E. (1975) *J. Mol. Biol.* **96,** 777–782.
15. Creighton, T. E. (1977) *J. Mol. Biol.* **113,** 275–293.
16. Creighton, T. E. (1977) *J. Mol. Biol.* **113,** 295–312.
17. Creighton, T. E. (1977) *J. Mol. Biol.* **113,** 313–328.
18. Hollecker, M., Creighton, T. E., and Gabriel, M. (1981) *Biochimie* **63,** 835–839.
19. Hollecker, M. and Creighton, T. E. (1983) *J. Mol. Biol.* **168,** 409–437.

20. Creighton, T. E. (1977) *J. Mol. Biol.* **113,** 329–341.
21. Creighton, T. E. (1979) *J. Mol. Biol.* **129,** 411–431.
22. Galat, A., Creighton, T. E., Lord, R. C., and Blout, E. R. (1981) *Biochemistry*, **20,** 594–601.
23. Wearne, S. J. and Creighton, T. E. (1988) *Proteins: Struct. Funct. Genet.* **4,** 251–261.
24. Pace, C. N. and Creighton, T. E. (1986) *J. Mol. Biol.* **188,** 477–486.
25. Ewbank, J. J. and Creighton, T. E. (1991) *Nature* **350,** 518–520.
26. Ewbank, J. J. and Creighton, T. E. (1993) *Biochemistry* **32,** 3677–3693.
27. Ewbank, J. J. and Creighton, T. E. (1993) *Biochemistry* **32,** 3694–3707.
28. Creighton, T. E. (1984) *Methods Enzymol.* **107,** 305–329.
29. Creighton, T. E. (1986) *Methods Enzymol.* **131,** 83–106.
30. Creighton, T. E. (1989) in *Protein Structure: A Practical Approach* (Creighton, T. E., ed.), IRL, Oxford, pp. 155–167.
31. Szajewski, R. P. and Whitesides, G. M. (1980) *J. Am. Chem. Soc.* **102,** 2011–2026.
32. Houk, J., Singh, R., and Whitesides, G. M. (1987) *Methods Enzymol.* **143,** 129–140.
33. Lees, W. J. and Whitesides, G. M. (1993) *J. Org. Chem.* **58,** 642–647.
34. Creighton, T. E., Hillson, D. A., and Freedman, R. B. (1980) *J. Mol. Biol.* **142,** 43–62.
35. Konishi, Y., Ooi, T., and Scheraga, H. A. (1982) *Biochemistry* **20,** 3945–3955.
36. Weissman, J. S. and Kim, P. S. (1991) *Science* **253,** 1386–1393.
37. Creighton, T. E. and Goldenberg, D. P. (1984) *J. Mol. Biol.* **179,** 497–526.
38. States, D. M., Dobson, C. M., Karplus, M., and Creighton, T. E. (1984) *J. Mol. Biol.* **174,** 411–418.
39. Houk, J., Singh, R., and Whitesides, G. M. (1987) *Methods Enzymol.* **143,** 129–140.
40. Darby, N. J. and Creighton, T. E. (1993) *J. Mol. Biol.* **232,** 873–896.
41. Creighton, T. E. (1980) *Nature* **284,** 487–489.
42. Hirose, M., Takahashi, N., Oe, H., and Doe, E. (1988) *Anal. Biochem.* **168,** 193–201.
43. Jocelyn, P. C. (1972) *Biochemistry of the SH Group*, Academic, London.
44. Chau, M.-H. and Nelson, J. W. (1991) *FEBS Lett.* **291,** 296–298.
45. Goldenberg, D. P. and Creighton, T. E. (1984). *Anal. Biochem.* **138,** 1–18.
46. Creighton, T. E., Kalef, E., and Arnon, R. (1978) *J. Mol. Biol.* **123,** 129–147.
47. Kosen, P. A., Creighton, T. E., and Blout, E. R. (1980) *Biochemistry* **19,** 4936–4944.
48. Kosen, P. A., Creighton, T. E., and Blout, E. R. (1981) *Biochemistry* **20,** 5744–5754.
49. Kosen, P. A., Creighton, T. E., and Blout, E. R. (1983) *Biochemistry* **22,** 2433–2440.
50. States, D. J., Creighton, T. E., Dobson, C. M., and Karplus, M. (1987) *J. Mol. Biol.* **195,** 731–739.
51. van Mierlo, C. P. M., Darby, N. J., Keeler, J., Neuhaus, D., and Creighton, T. E. (1993) *J. Mol. Biol.* **229,** 1125–1146.
52. van Mierlo, C. P. M., Darby, N. J., Neuhaus, D., and Creighton, T. E. (1991) *J. Mol. Biol.* **222,** 353–371.
53. van Mierlo, C. P. M., Darby, N. J., Neuhaus, D., and Creighton, T. E. (1991) *J. Mol. Biol.* **222,** 373–390.

54. van Mierlo, C. P. M., Darby, N. J., and Creighton, T. E. (1992) *Proc. Natl. Acad. Sci. USA* **89,** 6775–6779.
55. Brown, J. R. and Hartley, B. S. (1966) *Biochem. J.* **101,** 214–228.
56. Creighton, T. E. (1980) *FEBS Lett.* **118,** 283–288.
57. Creighton, T. E. (1981) *J. Mol. Biol.* **151,** 211–213.
58. Zapun, A., Bardwell, J. C., and Creighton, T. E. (1993) *Biochemistry* **32,** 5083–5092.
59. Goldenberg, D. P. (1988) *Biochemistry* **27,** 2481–2489.
60. Kosen, P. A., Marks, C. B., Falick, A. M., Anderson, S., and Kuntz, I. D. (1992) *Biochemistry* **31,** 5705–5717.
61. Creighton, T. E. (1992) *BioEssays* **14,** 195–199.
62. Goldenberg, D. P. (1992) *Trends Biochem. Sci.* **17,** 257–261.
63. Reisfeld, R. A., Lewis, U. J., and Williams, D. E. (1962) *Nature* **195,** 281–283.
64. Davis, B. J. (1964) *Ann. NY Acad. Sci.* **121,** 404–427.
65. Huyghues-Despointes, B. M. P. and Nelson, J. W. (1992) *Biochemistry* **31,** 1476–1482.
66. Chau, M.-H. and Nelson, J. W. (1992) *Biochemistry* **31,** 4445–4450.
67. Jovin, T. M. (1976) *Biochemistry* **12,** 879–890.
68. Jovin, T. M. (1976) *Biochemistry* **12,** 890–898.
69. Iyer, K. S. and Klee, W. A. (1973) *J. Biol. Chem.* **248,** 707–710.

CHAPTER 11

Solvent Stabilization of Protein Structure

Serge N. Timasheff

1. Introduction

The stability of the native structure of globular proteins is known to be affected strongly by a variety of substances that act at high concentration (usually $\geq 1M$). We will refer to these as co-solvents, whether they are liquids or solids. Some of these substances are known to stabilize protein structure, others to destabilize it, whereas still others can act either as stabilizers or destabilizers, depending on their concentration and on the solution pH *(1)*. The first class encompasses sugars (sucrose, trehalose), glycerol, other polyols (sorbitol, mannitol), some amino acids (glycine, proline), methyl amines (sarcosine, trimethylamine-*N*-oxide), in fact most osmolytes used by nature *(2–4)*, and also some salting out salts, such as $MgSO_4$. The second class consists of the strong denaturants, such as urea, guanidine hydrochloride, and sodium dodecyl sulfate. The last class is composed of weakly acting agents, such as guanidine sulfate *(5)*, $MgCl_2$ *(6)*, the glutamate ion, and dimethyl sulfoxide (DMSO).

In analyzing the stabilizing/destabilizing capacity of any co-solvent relative to any given protein it is necessary to determine three functions *(7)*:

1. A precise transition curve for the equilibrium $N \rightleftharpoons D$ (where N is native protein and D is denatured protein) as a function of the co-solvent concentration, ascertaining that the effect is reversible;
2. The extent of thermodynamic (dialysis equilibrium) binding of the co-solvent to the protein as a function of co-solvent concentration; and
3. The transfer free energy of the protein from water to the co-solvent at each co-solvent concentration used.

From: *Methods in Molecular Biology, Vol. 40: Protein Stability and Folding: Theory and Practice*
Edited by: B. A. Shirley

The last two parameters must be determined for the protein both in the native and the denatured states. Knowledge of these parameters at any solvent conditions gives a complete thermodynamic description (characterization) of the stabilizing/destabilizing action of a co-solvent toward the given protein. Since the methodology resides principally in the proper analysis of the experimental measurements and the calculation of the pertinent parameters, Section 3. will be devoted mostly to these parameters.

2. Materials

Any proteins used in such studies must be of a high degree of purity, e.g., a single band in SDS-PAGE analysis, as well as be free of aggregates, i.e., they must elute as a single symmetrical peak from a gel filtration column, such as Sephadex, in the absence of co-solvent, i.e., in the dilute buffer selected for the study. All co-solvents used must be known to be pure (do not rely on the manufacturer's label!), or be purified by distillation or recrystallization, depending on the material.

The instruments necessary are:

1. For the transition measurements, a good precision UV spectrophotometer and/or a spectrofluorimeter. No particular brands are listed, since most such laboratory instruments will be adequate.
2. The interaction measurements require the ability to measure very small changes in the concentration (10–100 μ*M* out of a total concentration of 1*M*) of the co-solvent during the course of the dialysis. Such determinations can be done routinely on a high precision densimeter (a Mettler DA-310 Density Meter instrument is quite adequate), or on a high precision photoelectric differential refractometer (the author is not aware of any commercial instrument; a description of the one built and used in his laboratory has been published *[8]*).

3. Methods

Since the methods consist mostly in the proper analysis and interpretation of the experimental measurements, these will be presented first, with a brief description of the experimental procedures at the end.

3.1. Methods of Analysis

Thermodynamically, the stabilizing/destabilizing action of a co-solvent on the native structure of a protein may be defined with respect to three reference states *(1,7)*:

1. The co-solvent solution of the given composition;
2. Pure water (in practice this is taken as the dilute buffer free of co-solvent); and
3. A co-solvent solution of a concentration different from that being examined.

We will take each case in turn. In the discussion we will define component 1 as water, component 2 as protein, and component 3 as co-solvent *(9,10)*.

3.1.1. The Reference State Is the Solvent of the Given Composition

In the first reference state the effect of a co-solvent on protein stability is expressed by the Wyman linkage relation *(11,12)*, which states that if an equilibrium, such as $N \rightleftharpoons D$, is affected by a ligand (a co-solvent in the present context), the effective binding of the co-solvent to the protein must change during the course of the reaction. Therefore, if the equilibrium constant, *K*, of the denaturation reaction is affected by the co-solvent, then, at constant temperature and pressure:

$$- d\Delta G^\circ / (RT \text{ d} \log a_3) = \text{d} \log K / \text{d} \log a_3 = \nu_3^{Prod} - \nu_3^{React} = \Delta\nu_3 \qquad (1)$$

where a_3 is the activity of the co-solvent, i.e., its concentration, c_3, multiplied by its activity coefficient, f_3 ($a_3 = c_3 f_3$), ν_3 is the extent of binding of co-solvent to the protein. In what follows we will use the molal concentration scale, m_i, i.e., moles of substance i /1000 g water. The extent of binding, ν_3, is identical to the value of the thermodynamic binding measured by a direct technique, such as dialysis equilibrium. Thermodynamically, then:

$$\nu_3 \equiv (\partial m_3/\partial m_2)\mu_3 = - [\partial\mu_2/\partial m_3)_{m_2}] / [(\partial\mu_3/\partial m_3)_{m_2}] \qquad (2)$$

The statement of this equation is that dialysis equilibrium binding is the change of concentration of free co-solvent caused by addition of the protein at chemical equilibrium, i.e., at constant chemical potential of the co-solvent, μ_3. This, in turn, is equal to the perturbation of the chemical potential of the protein by the co-solvent at constant protein concentration. The denominator of the last term of Eq. (2) accounts for the thermodynamic nonideality of the co-solvent solution. Now, taking the definition of the chemical potential of any component i:

$$\mu_i = \mu_i^0 + RT \ln m_i + RT \ln \gamma_i \qquad (3)$$

Table 1
Dialysis Equilibrium Measurements of Co-solvent Binding to Proteins

Protein	System	ν, moles bound/mole protein	Effect
β-Lactoglobulin	20% ClETOH	+108	Denat
β-Lactoglobulin	80% ClETOH	–145	Denat
RNase A	1*M* Sucrose	–7.6	Stabil
CTGen, native	20% PEG-1000	–1.0	Nat
			⥮
CTGen, denat	20% PEG-1000	+1.3	Denat
BSA, native	1*M* GuHCl	+17.5	Destabil
BSA, native	0.5 Gua_2SO_4	+2.6	Destabil
BSA, native	1*M* Gua_2SO_4	–16.4	Stabil
Lysozyme, denat	8*M* Urea	+12.0	Denat
β-Lactoglobulin	2*M* Glycine	–23.8	Stabil

where μ_i^0 is the standard chemical potential, and γ_i is the activity coefficient of component *i,* gives for the denominator:

$$(\partial\mu_3/\partial m_3)_{m_2} = RT\,[(1/m_3) + (\partial \ln \gamma_3/\partial m_3)] \tag{4}$$

The activity coefficients of a number of commonly used co-solvents have been measured and are listed in the reference literature.

3.1.2. Preferential Interactions

Thermodynamic, dialysis equilibrium, binding, as seen in Eq. (2), specifies only that the concentration of ligand (co-solvent) in the free state is changed by the introduction of the protein into the system. It does not state whether this change is positive or negative, nor does it state whether the mechanism of this change is transport of co-solvent molecules or water molecules across the dialysis membrane. As such, this parameter can assume negative, as well as positive values *(13).* Some typical examples are given in Table 1. The reason for this situation resides in the nature of the binding (interaction) process. This is illustrated in Fig. 1, which shows that the binding of a ligand molecule to the surface of a protein requires that water molecules be displaced. In other words, the binding process is an exchange reaction *(14–17)*:

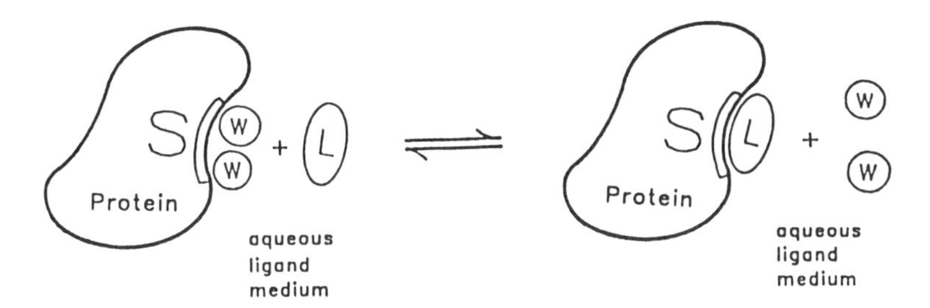

Fig. 1. Schematic representation of the competition between a ligand molecule, L, and water, W, at site, S, on a protein, as defined by Eqs. (5) and (6). (Timasheff [1992] *Biochemistry* **31,** 9857–9864; reprinted with permission from the American Chemical Society.)

$$P \cdot H_2O_m + L \overset{K_b}{\rightleftharpoons} PL + mH_2O$$
$$K_b = ([PL][H_2O]^m) / ([P \cdot H_2O_m][L]) \tag{5}$$

where K_b is the binding (exchange) constant. Depending on the relative affinities of ligand (co-solvent) and water molecules for a locus on the surface of the protein molecule, the site will have a preference for occupancy by the co-solvent or by water. Hence, this interaction is called *preferential interaction* and the binding of the co-solvent is called *preferential binding*. When the affinity for water is greater than that for the ligand, the result is occupancy of the locus (site) by water and the situation is that of *preferential hydration*. This is frequently referred to as *preferential exclusion* of the ligand. Therefore, the negative values of the binding measured in Table 1 mean that the protein has a higher affinity for water than for the co-solvent. Preferential binding and preferential hydration are related by a reciprocity equation *(18):*

$$m_3(\partial m_1/\partial m_2)_{\mu_1,\mu_3} = m_1(\partial m_3/\partial m_2)_{\mu_1,\mu_3} \tag{6}$$

In conformity, the term $(\partial\mu_2/\partial m_3)_{m_2}$ in Eq. (2) is known as the *preferential interaction* parameter.

Typical results of such measurements are presented in Fig. 2, which shows the dialysis equilibrium measurements for two co-solvent systems, 2-chloroethanol *(19)* and $MgCl_2$ *(6)*. We notice that for 2-chloroethanol the preferential binding is positive in the lower concentration range, then after attaining a maximum, it decreases, crosses zero at 65% chloro-

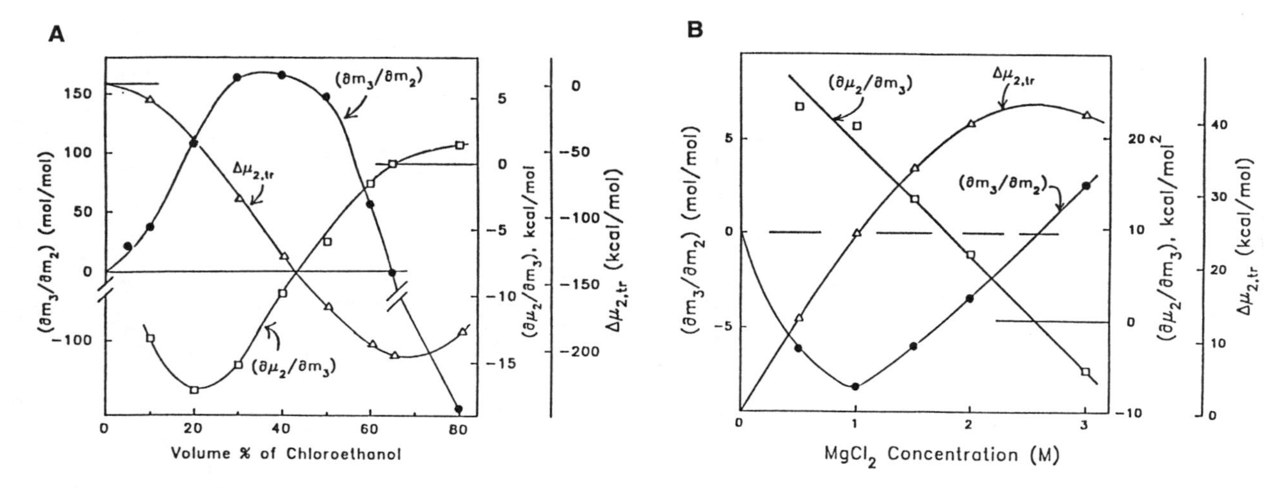

Fig. 2. Variation of the thermodynamic interaction parameters with cosolvent concentration: preferential binding $(\partial m_3/\partial m_2)_{T,\mu_1,\mu_3}$ (●), preferential interaction parameter $(\partial \mu_2/\partial m_3)_{m_2}$ (□), and transfer free energy, $\Delta\mu_{2,tr}$ (Δ). **(A)** β-lactoglobulin in aqueous 2-chloroethanol; **(B)** β-lactoglobulin in aqueous $MgCl_2$ solution at pH 3.0 (plotted from the data of Inoue and Timasheff [1968] *J. Am. Chem. Soc.* **90,** 1890–1897 and Arakawa et al. [1990] *Biochemistry* **29,** 1914–1923). Note that $\Delta\mu_{2,tr}$ is at its peak (minimal for the favorably interacting 2-chloroethanol and maximal for the unfavorably interacting $MgCl_2$) at the point where the binding measured by dialysis equilibrium is zero. (The drawn points at 65% 2-chloroethanol are not experimental data; they were interpolated from the preferential binding measurements.) (Timasheff [1993] *Annu. Rev. Biophys. Biomol. Struct.* **22,** 67–97; reprinted with permission from Annual Reviews Inc.)

ethanol, and finally assumes negative values. In the case of $MgCl_2$, just the opposite is observed. At low concentrations its preferential binding to the protein is negative. Then, after attaining a minimum value, it crosses zero and becomes positive at high salt concentrations. For both systems the preferential interaction parameter, $(\partial\mu_2/\partial m_3)_{m_2}$, was calculated from the dialysis equilibrium results by Eq. (2). The values plotted in Fig. 2 follow concentration dependences quite distinct from those of the dialysis equilibrium results. This reflects the contribution of the co-solvent nonideality term in Eq. (2), as defined by Eq. (4). For 2-chloroethanol the concentration dependence is complex: It starts by being strongly negative, passes through a minimum, reverses its course, and, after passing through zero, becomes positive. This means that, *at each point,* the interaction between protein and co-solvent is first highly favorable with reference to the given solvent composition (one may view this as with reference to a concentration infinitesimally smaller), but becomes unfavorable at high co-solvent concentration. In the case of $MgCl_2$, $(\partial\mu_2/\partial m_3)_{m_2}$ follows a linear dependence on concentration, from strongly positive (unfavorable interaction) at low concentration to negative (favorable interaction) at high salt concentration. These two examples show that the variations with co-solvent concentration of the interaction parameters, expressed either as binding or as thermodynamic interaction, can be very complex and, at first sight, might appear as unrelated.

3.1.3. The Reference State Is Water

In the second reference state, that of pure water, the effect of a co-solvent on protein stability is expressed by the difference in the transfer free energies of the protein from water to the co-solvent solution of the given composition in the denatured and native states, i.e., $\Delta\,\mu_{2,tr}^{D} - \Delta\,\mu_{2,t}^{N}$. The transfer free energy, $\Delta\,\mu_{2,tr}$, is the total free energy of interaction (binding) of the protein with the co-solvent. It is related to experimentally measurable quantities by:

$$\partial\Delta\mu_{2,tr} = \int_{o}^{m3} (\partial\mu_2/\partial m_3)dm_3 \tag{7}$$

Referring to Fig. 2, the transfer free energy is the integral under the $(\partial\mu_2/\partial_{m_3})$ curves from zero to the given concentration of the co-solvent. As such, it is experimentally measurable by the determination of dialysis equilibrium binding at a series of co-solvent concentrations and the numerical integration under the plot of:

$$-\,RT\ (\partial m_3/\partial m_2)_{\mu_3}[(1/m_3) + (\partial \ln \gamma_3/\partial m_3)]\ \text{vs}\ m_3$$

Typical results of such calculations are presented in Fig. 2. If we examine this figure, we find that, for both systems, $\Delta\mu_{2,tr}$ starts at zero and then varies monotonely with co-solvent concentration until it reaches an apex and reverses its course. For 2-chloroethanol, the value of $\Delta\mu_{2,tr}$ is negative at all concentrations. This means that the interaction of the protein with the co-solvent is favorable relative to water over the entire range. Exactly the opposite is true for $MgCl_2$. This is reflected by the preferential binding pattern, which for 2-chloroethanol is positive up to a point at which it crosses zero, the converse situation being true for $MgCl_2$. Further comparison of the three curves for each of the two systems reveals that, in both, $\Delta\mu_{2,tr}$ reaches its apex at the co-solvent concentration at which preferential binding and, hence $(\partial\mu_2/\partial m_3)_{m_2}$ have zero values and reverse sign. If this is viewed in terms of Eqs. (2) and (7), we find that the preferential interaction parameter is the derivative of the transfer free energy at any point:

$$(\partial\mu_2/\partial m_3)_{m_2} = (\partial\Delta\mu_{2,tr}/\partial m_3)_{m_2} \tag{8}$$

This means that the preferential interaction parameter is the rate of change of the transfer free energy with co-solvent concentration. This is reflected in the preferential binding. Therefore, the point at which $(\partial\mu_2/\partial m_3)_{m_2}$ and $(\partial m_3/\partial m_2)_{\mu_3}$ reverse signs must be the maximum or minimum in the transfer free energy curve. In other words, taking the 2-chloroethanol system, the interaction of the protein with the co-solvent is first increasingly favorable with reference to water. Then, at high concentration, it becomes less favorable and the dialysis equilibrium result passes from preferential binding to preferential exclusion of the co-solvent.

The third reference state, that of any given co-solvent concentration, is exactly the same as the last one, except that now the lower limit in Eq. (7) is the reference solvent concentration and the transfer free energy simply refers to the difference in the free energy of the interaction of the protein with co-solvent at two concentrations of the latter.

3.1.4. Stabilization/Destabilization by Co-solvents

In order to find out the stabilizing/destabilizing capacity of a co-solvent, both the preferential binding and the transfer free energy must be measured for each of the two end states of the unfolding (denaturation) equilibrium. Let us take each in turn.

Relative to water, the stabilizing/destabilizing capacity of a co-solvent is expressed by the difference between the transfer free energies of the protein in the denatured and native states from water to the co-solvent solution:

$$\partial\Delta\mu_{2,tr} = \Delta\,\mu_{2,tr}^{D} - \Delta\,\mu_{2,tr}^{N} \tag{9}$$

Since Eq. (9) represents the total contribution of the co-solvent to the denaturation reaction, it must also represent the difference between the standard free energies of denaturation in water and the co-solvent system. Therefore:

$$\delta\Delta\mu_{2,tr} = \Delta\,G^{o}_{m3} - \Delta\,G^{o}_{W} = \delta\,\Delta\,G^{o} \tag{10}$$

This is represented in Fig. 3 by the "Thermodynamic Boxes" for two systems, the denaturation of chymotrypsinogen by polyethylene glycol 1000 (PEG) *(20)* and the stabilization of ribonuclease by sorbitol *(21).* The numbers in the figure are values of the transfer free energies calculated from preferential binding measurements, as described above (*see* Eq. [7]), and standard free energies of denaturation obtained from transition curves *(see* Section 4.5*)*. In both cases the boxes close in conformity to Eqs. (9) and (10).

Experimentally, the efficacy of the co-solvent can be determined by comparing either the free energies of denaturation in water and co-solvent or the transfer free energies of the protein in the native and unfolded states from water to the co-solvent system. In practical terms, frequently it is not possible to measure the denaturation equilibrium constant in water because of the stability of the protein. Under these circumstances the co-solvent efficacy must be obtained from measurements of the transfer free energies of the protein in the two states. This requires the performance of dialysis equilibrium experiments as a function of co-solvent concentration on both the native and unfolded states of the protein and integration of the data according to Eq. (7) for each. In such measurements, the unfolded state of the protein may be obtained by its chemi-

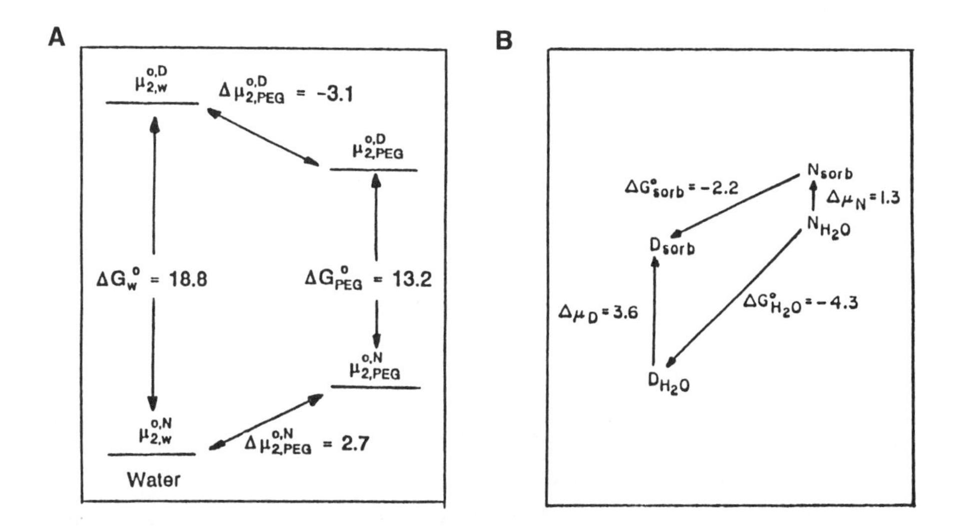

Fig. 3. Thermodynamic analysis of the effect of co-solvents on the protein denaturation equilibrium. **(A)** Unfolding of chymotrypsinogen by 20% polyethylene glycol 1000 (PEG) at 20°C; contact between the protein and PEG is unfavorable ($\Delta\mu_{2,tr}$ positive) in the native state, but favorable ($\Delta\mu_{2,tr}$ negative) in the denatured state. Hence $\delta\Delta\mu_{2,tr} = -5.8$ kcal/mol reduces $\Delta G^{o,N-D}$ by an equal amount, making PEG into a denaturant [calculated from the data of Lee and Lee [1987] *Biochemistry* **26,** 7813–7819). **(B)** Stabilization of ribonuclease by 30% sorbitol at 48°C; the contact between protein and sorbitol is unfavorable ($\Delta\mu_{2,tr}$ positive) for both the native and denatured protein, but more so for the latter. Hence $\delta\Delta\mu_{2,tr} = +2.3$ kcal/mol raises $\Delta G^{o,N-D}$ by an equal amount, which makes sorbitol into a stabilizer (G. Xie and S. N. Timasheff, manuscript in preparation). (Timasheff [1993] *Annu. Rev. Biophys. Biomol. Struct.* **22,** 67–97; reprinted with permission from Annual Reviews Inc.).

cal modification, such as the reduction and carboxymethylation of disulfide bridges, or simply by keeping a reducing agent (mercaptoethanol, dethiothreitol, and so on) in the medium *(22).*

At the *given constant concentration of the co-solvent,* its efficacy is defined by Eq. (1). In other words, whether a co-solvent is a stabilizer or destabilizer at the given solvent composition is defined by the difference between the preferential binding of the co-solvent to the protein in the denatured and native states. This can be measured in two ways: from the slope of the double logarithmic plot of the denaturation equilibrium constant vs the thermodynamic activity of the co-solvent; or by deter-

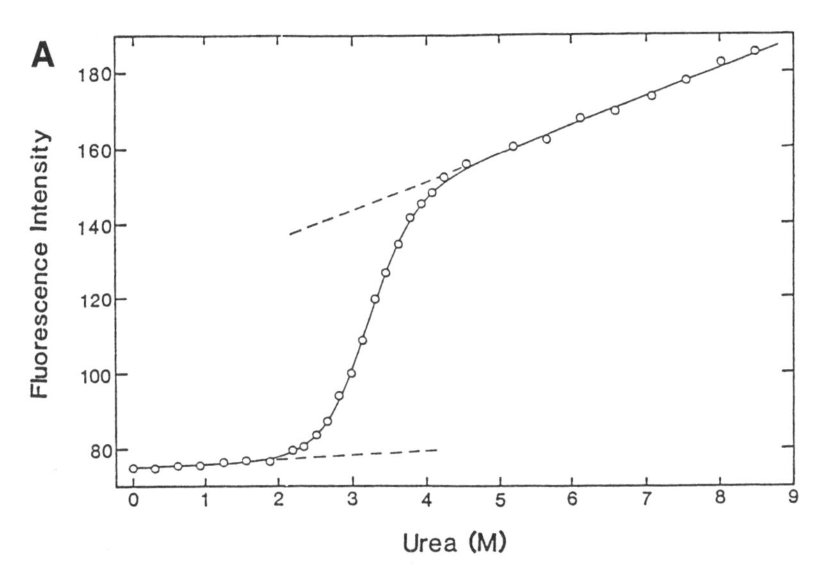

Fig. 4. (**A**) Urea unfolding curve for RNase A at pH 3.55 measured by fluorescence. The dashed lines are based on a least-squares analysis of the pre- and posttransition base lines. (Pace et al. [1990] *Biochemistry* **29,** 2565–2572; reprinted with permission from C. Nick Pace and the American Chemical Society.)

mining in dialysis equilibrium experiments the bindings of the co-solvent to the protein in the denatured and native states and taking the difference. In practical terms, again denaturation may not be proceeding at the conditions used, which would preclude the use of the first approach. In the second approach, once again the unfolded form of the protein can be obtained by its chemical destabilization (e.g., use of disulfide reducing agents). These results can be used then in the calculation of the transfer free energies.

4.5. Transition Curve

The effectiveness of a co-solvent on protein stability as a function of co-solvent concentration, and hence the change in preferential binding during the denaturation reaction, can be evaluated from measurements of a transition curve and its analysis in terms of Eq. (1). A transition curve can be constructed by measuring the change of some property of the protein as it undergoes unfolding. Most frequently this is done by following a spectral property. After carefully correcting for the baselines, the data are transformed into the form of Eq. (1). A typical transition curve carefully measured by fluorescence is shown in Fig. 4A for the unfolding of

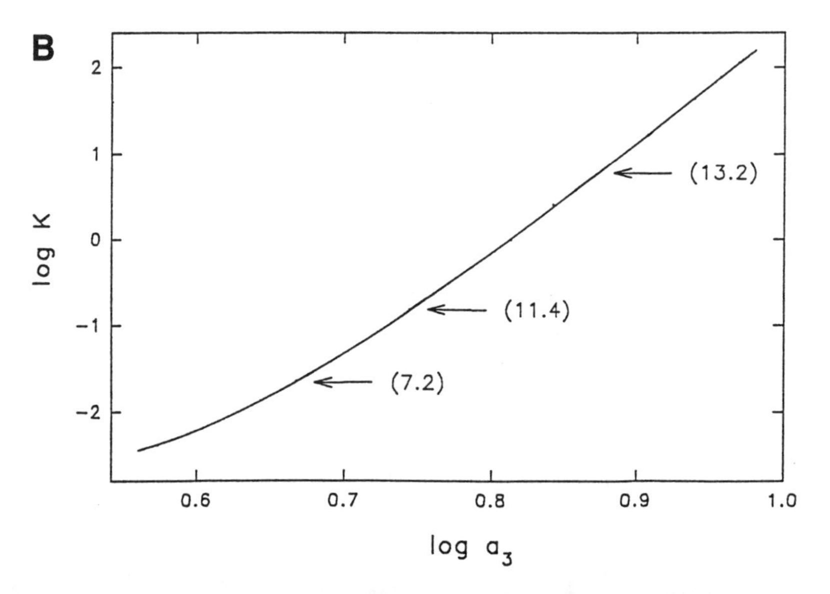

(B) Unfolding transition of RNase A at pH 7.0 presented as a Wyman plot according to Eq. (1). The numbers next to the curve represent the slope and, as such, are equal to $\Delta\nu_3$, i.e., the preferential binding of urea to RNase A increases by these amounts at the corresponding urea concentrations when the protein undergoes its transition from the native to the denatured (unfolded) state. (Calculated from the data of Greene and Pace [1974] *J. Biol. Chem.* **249,** 5388–5393).

ribonuclease A by urea at pH 3.55 *(23)*. The dashed lines represent the baselines to be used in calculating the extent of the transition. At any concentration of co-solvent (urea in Fig. 4A), the $N \rightleftharpoons D$ equilibrium constant is given by:

$$K = (y_N - y)/(y - y_D) \tag{11}$$

where y is the measured value of the parameter and y_N and y_D are the values extrapolated linearly (baselines) from the native (low urea concentration) and denatured (high urea concentration) states. This is then plotted in logarithmic form as a function of the logarithm of the co-solvent activity. Such a plot is shown in Fig. 4B for the urea unfolding of ribonuclease A at pH 7.0 *(24)*. The slope at any point gives $\Delta\nu = (\nu^D - \nu^N)$, i.e., the change in preferential binding of the co-solvent to the protein during its unfolding at the given protein concentration. The numbers marked next to the curve are values of $\Delta\nu$. From these one can calculate $\Delta(\partial\mu_2/\partial m_3)_{m_2}$ and by integration obtain directly $\delta\Delta\mu_{2,tr}$. A note of caution is required: This type of analysis is based on the assumption that the protein property (fluorescence, UV absorption, optical rotation) varies linearly with the extent of the unfolding reaction. Comparison of the results obtained from analysis of transition curves with those derived from dialysis equilibrium gives a rigorous test of the reliability of the measured parameters, since these must be identical when deduced from the two types of experiments.

5. Experimental Methods

The measurement of preferential binding involves essentially the performance of very careful dialysis equilibrium experiments. Since in these experiments the ligand (co-solvent) concentration is very high (≥1*M*) and its changes on interaction with proteins are extremely small (10–100 µ*M*), one must use high precision detection techniques. Such techniques are high precision densimetry and differential refractometry. Since no adequate differential refractometers are commercially available to the best knowledge of this author, we will refer to the densimetric technique. In fact, no description of the density measurements is needed, since, once the sample is introduced into the instrument, the actual measurement is routine and well described in manufacturers' instruction manuals. Three cautions are recommended:

1. Check the instrument constant with distilled water several times during the course of the day;
2. Keep the instrument in a room with the temperature controlled close to that of the experiment; and
3. Carefully wash the cell after using it with protein. This is done by injecting into it several doses of 8*M* urea and then washing thoroughly with distilled water.

The experimental techniques have been described in detail *(25,26)*. Success of such measurements depends on careful dialysis equilibrium. The first requirement is that the dialysis bags be well boiled out to extract all diffusible material from them. This is usually accomplished by boiling in distilled water, then dilute salt (sodium bicarbonate can be used), then again distilled water. The second requirement is that solutions be transferred to the densimeter cell without any loss of water by evaporation.

This can be accomplished by gentle withdrawing of the solution from the dialysis bag with a syringe, being careful that no bubbles are formed. The needle is then removed, the syringe is fitted directly to the densimeter by the luer lock, and the protein solution is introduced gently into the cell. Protein concentration must be known accurately. This is normally done spectrophotometrically after careful determination of the extinction coefficient with a protein sample that has been dried (vacuum oven at 40°C, over P_2O_5, for example). An in-depth discussion of protein concentration measurements is available in the literature *(27)*.

In the determination of the preferential interaction, density measurements must be performed on two sets of protein solutions, one of which has been dialyzed (constant chemical potential of solvent), and the other not dialyzed (constant composition of solvent). The second type of measurement requires that the protein be dried carefully. It is then weighed out rapidly and dissolved in the experimental medium. Any extraneous water molecules will change the solvent composition sufficiently to affect the density measurement. Remember: You are looking for change in the fifth decimal place of density! For each solvent composition, one must do measurements as a function of protein concentration and extrapolate to zero. Normally 5–10 protein concentrations between 2 and 25 mg/mL are sufficient.

The preferential binding is then calculated from the results of the density measurements. The working equation is *(28,29)*:

$$(\partial m_3/\partial m_2)_{T,\mu_1,\mu_3} = (M_2/M_3)\,(\partial g_3/\partial g_2)_{T,\mu_1,\mu_3} \qquad (12A)$$

$$(\partial g_3/\partial g_2)_{T,\mu_1,\mu_3} = [(1 - \phi_2'^{\circ}\rho_o) - (1 - \phi_2^{\circ}\rho_o)] \,/\, (1 - \bar{v}_3\rho_o) \qquad (12B)$$

where g_i is concentration of component *i* in units of grams of *i* per gram of water, ρ_o is the density of the solution in g/mL, $\bar{v}_3$ is the partial specific volume of the co-solvent at the experimental concentration, ϕ_2 is the apparent partial specific volume of the protein measured under conditions at which the molal concentration of the diffusible component (co-solvent), m_3, is kept identical in the solvent and the protein solution, ϕ_2' is the apparent partial specific volume of the protein measured under conditions at which the chemical potential of the co-solvent, μ_3, is kept constant between the protein solution and the reference solvent. Operationally this is attained within a close approximation by bringing the system to dialysis equilibrium. The two parameters, ϕ_2 and ϕ_2', are known as the isomolal and isopotential apparent partial specific volumes of the protein. The superscript "o" indicates that this is the value extrapolated to zero protein concentration.

Note: For these relations (Eq. [12]) to be valid, the co-solvent concentration must be expressed in molal concentration units (moles of co-solvent/1000 g of water). Hence all solution preparation must be done gravimetrically, i.e., by weighing out the components, and *not* by adjusting volumes.

The parameters ϕ_2 and ϕ_2' at each protein concentration are obtained from density measurements *(30)*:

$$\phi = (1/\rho_o)[1 - (\rho - \rho_o)/C] \tag{13}$$

where ρ is the density of the protein solution in g/mL, ρ_o that of the solvent, and C the protein concentration in g/mL. These values are then plotted as a function of protein concentration. Their extrapolation to zero protein concentration gives the values to be used in Eq. (12) to obtain the preferential binding parameter, $(\partial m_3/\partial m_2)_{T,\mu_1,\mu_3}$. Typical results are given in Fig. 5 *(31)*.

The derived parameter, $(\partial \mu_2/\partial m_3)_{m_2}$ requires a knowledge of the activity coefficients of the co-solvent as a function of its concentration. This is needed to evaluate $(\partial \mu_3/\partial m_3)_{m_2}$ (*see* Eqs. [2] and [4]). For many of the co-solvents used, activity coefficients have been measured and are available in the literature. The pertinent references are: urea *(32–35);* alkyl ureas *(36)*; guanidine salts *(37,38)*; sucrose *(39–42)*; glycerol *(32)*; various salts *(43)*; amino acids and methylamines *(40,44–48)*; and polyethylene glycol *(49,50)*. For co-solvents for which no such data are available, activity coefficients can be measured easily by vapor pressure osmometry. An instrument that we have found to be suitable is the Knauer Vapour Pressure Osmometer. No experimental details need to be given, since the manufacturer's instructions are fully adequate. Two cautions should be added: It must be installed in a constant temperature room and it must be kept clear of air currents.

Finally, calculation of the transfer free energy, $\Delta\mu_{2,tr}$ is done by numerical integration under the $(\partial \mu_2/\partial m_3)_{m_2}$ vs m_3 curve.

Acknowledgment

This work was supported in part by NIH grants GM-14603 and CA-16707.

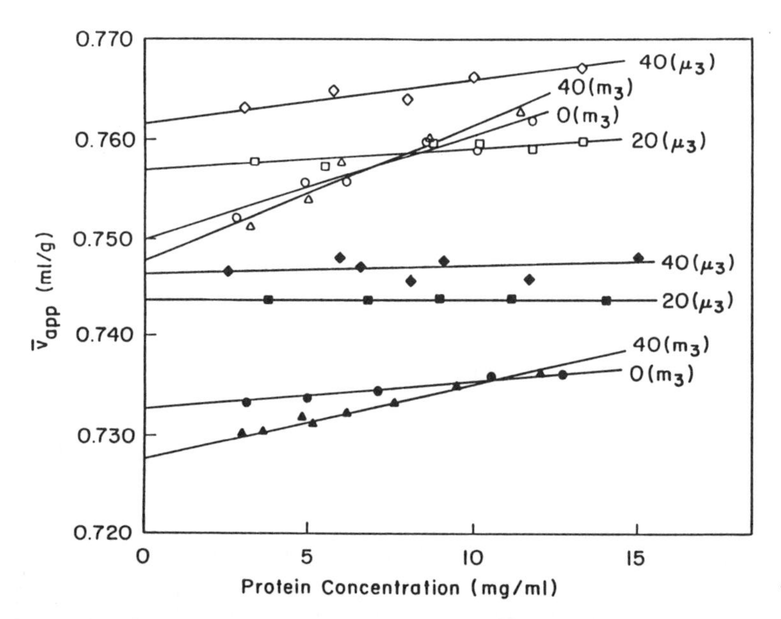

Fig. 5. Protein concentration dependence of the apparent partial specific volume of chymotrypsinogen A (◆, ■, ●, ▲) and β-lactoglobulin (◇, □, ○, △). The numbers in the figure show the glycerol concentration in vol%; μ_3 and m_3 in parentheses indicate the constant chemical potential and constant molality experiments, respectively. (Gekko and Timasheff [1981] *Biochemistry* **20,** 4667–4676; reprinted with permission from the American Chemical Society.)

References

1. Timasheff, S. N. (1993) *Annu. Rev. Biophys. Biomol. Struct.* **22,** 67–97.
2. Arakawa, T. and Timasheff, S. N. (1985) *Biophys. J.* **47,** 411–414.
3. Yancey, P. H., Clark, M. E., Hand, S. C., Bowlus, R. D., and Somero, G. N. (1982) *Science* **217,** 1214–1222.
4. Santoro, M. M., Liu, Y., Khan, S. M. A., Hou, L.-X., and Bolen, D. W. (1992) *Biochemistry* **31,** 5278–5283.
5. von Hippel, P. H. and Wong, K.-Y. (1965) *J. Biol. Chem.* **240,** 3909–3923.
6. Arakawa, T., Bhat, R., and Timasheff, S. N. (1990) *Biochemistry* **29,** 1914–1923.
7. Timasheff, S. N. (1994) in *Protein-Solvent Interactions* (Gregory, R., ed.) Marcel Dekker, New York, pp. 445–482.
8. Pittz, E. P. and Bablouzian, B. (1973) *Anal. Biochem.* **55,** 399–405.
9. Scatchard, G. (1946) *J. Am. Chem. Soc.* **68,** 2315–2319.
10. Stockmayer, W. H. (1950) *J. Chem. Phys.* **18,** 58–61.
11. Wyman, J., Jr. (1964) *Adv. Protein Chem.* **19,** 223–286.
12. Wyman, J. and Gill, S. J. (1990) *Binding and Linkage,* University Science Books, Mill Valley, CA.

13. Na, G. C. and Timasheff, S. N. (1981) *J. Mol. Biol.* **151,** 165–178.
14. Timasheff, S. N. and Kronman, M. J. (1959) *Arch. Biochem. Biophys.* **83,** 60–75.
15. Schellman, J. A. (1987) *Ann. Rev. Biophys. Biophys. Chem.* **16,** 115–137.
16. Schellman, J. A. (1990) *Biophys. Chem.* **37,** 121–140.
17. Timasheff, S. N. (1992) *Biochemistry* **31,** 9857–9864.
18. Timasheff, S. N. (1963) in *Electromagnetic Scattering* (Kerker, M., ed.), Pergamon, New York, pp. 337–355.
19. Inoue, H. and Timasheff, S. N. (1968) *J. Am. Chem. Soc.* **90,** 1890–1897.
20. Lee, J. C. and Lee, L. L. Y. (1987) *Biochemistry* **26,** 7813–7819.
21. Xie, G.-F. and Timasheff, S. N. (1994), in preparation.
22. Reisler, E., Haik, Y., and Eisenberg, H. (1977) *Biochemistry* **16,** 197–203.
23. Pace, C. N., Laurents, D. V., and Thomson, J. A. (1990) *Biochemistry* **29,** 2564–2572.
24. Greene, R. F., Jr. and Pace, C. N. (1974) *J. Biol. Chem.* **249,** 5388–5393.
25. Lee, J. C. and Timasheff, S. N. (1974) *Biochemistry* **13,** 257–265.
26. Lee, J. C., Gekko, K., and Timasheff, S. N. (1979) *Methods Enzymol.* **61,** 26–49.
27. Kupke, D. W. and Dorrier, T. E. (1978) *Methods Enzymol.* **48,** 155–162.
28. Casassa, E. F. and Eisenberg, H. (1964) *Adv. Protein Chem.* **19,** 287–395.
29. Cohen, G. and Eisenberg, H. (1968) *Biopolymers* **6,** 1077–1100.
30. Kielley, W. W. and Harrington, W. F. (1960) *Biochim. Biophys. Acta* **41,** 401–421.
31. Gekko, K. and Timasheff, S. N. (1981) *Biochemistry* **20,** 4667–4676.
32. Scatchard, G., Hamer, W. J., and Wood, S. E. (1938) *J. Am. Chem. Soc.* **60,** 3061–3070.
33. Bower, V. E. and Robinson, R. A. (1963) *J. Phys. Chem.* **67,** 1524–1527.
34. Stokes, R. H. (1967) *Aust. J. Chem.* **20,** 2087–2100.
35. Ellerton, H. D. and Dunlop, P. J. (1966) *J. Phys. Chem.* **70,** 1831–1837.
36. Barone, G., Rizzo, E., and Volpe, V. (1976) *J. Chem. Eng. Data* **21,** 59–61.
37. Bonner, O. D. (1976) *J. Chem. Thermodynamics* **8,** 1167–1172.
38. Schrier, M. Y. and Schrier, E. E. (1977) *J. Chem. Eng. Data* **22,** 73–74.
39. Scatchard, G. (1921) *J. Am. Chem. Soc.* **43,** 2406–2418.
40. Smith, E. R. B. and Smith, P. K. (1937) *J. Biol. Chem.* **117,** 209–216.
41. Robinson, R. A. and Stokes, R. H. (1955) *Electrolyte Solutions*, Academic, New York, p. 463.
42. Robinson, R. A. and Stokes, R. H. (1961) *J. Phys. Chem.* **65,** 1954–1958.
43. Robinson, R. A. and Stokes, R. H. (1955) *Electrolyte Solutions*, Academic, New York, pp. 461–489.
44. Smith, E. R. B. and Smith, P. K. (1940) *J. Biol. Chem.* **132**, 47–56.
45. Smith, P. K. and Smith, E. R. B. (1940) *J. Biol. Chem.* **132,** 57–64.
46. Cohn, E. J. and Edsall, J. T. (1943) *Proteins, Amino Acids and Peptides*, Reinhold, New York, pp. 218–228.
47. Ellerton, H. D., Reinfelds, G., Mulcahy, D. E., and Dunlop, P. J. (1964) *J. Phys. Chem.* **68,** 398–402; 403–408.
48. Bonner, O. D. (1982) *J. Chem. Eng. Data* **27,** 422–423.
49. Elworthy, P. H. and Florence, A. T. (1966) *Kolloid-Z. Z. Polym.* **208,** 157–162.
50. Rogers, J. A. and Tam. T. (1977) *Can. J. Pharm. Sci.* **12,** 65–70.

CHAPTER 12

Site-Directed Mutagenesis to Study Protein Folding and Stability

Philip N. Bryan

1. Introduction

In the early 1970s, recombinant DNA technology made in vitro manipulation of genetic information possible. In the early 1980s, advances in solid phase synthetic chemistry made the production of oligonucleotides of defined sequence rapid and economical. Since then, methods of site-directed mutagenesis (SDM) built on these advances have been refined to the point that unlimited variations of a gene sequence can be created given sufficient time. Furthermore, commercially available kits have opened the realm of site-directed mutagenesis to laboratories that are not specialists in DNA molecular biology. The ability to introduce specific mutations into a gene and then express and study the altered protein has provided a powerful experimental tool for studying the relationship between amino acid sequence and protein structure and stability.

The purpose of this chapter is to describe how site-directed mutagenesis can be useful to study protein stability and to discuss some principles for designing and analyzing site-directed mutants. The first issue to address is how site-direction helps in understanding protein stability. It is not necessarily obvious that creating new mutant versions of the multitude of naturally occurring proteins will simplify the protein folding problem. The principles that underpin the site-directed mutagenesis approach were largely discovered by biophysical chemists in the 1950s and early 1960s. The protein-folding reaction, which involves an enor-

From: *Methods in Molecular Biology, Vol. 40: Protein Stability and Folding: Theory and Practice*
Edited by: B. A. Shirley

mously complex set of noncovalent chemical interactions, was simplified into a two-state model. According to this model, the energetics of protein folding could be analyzed thermodynamically by considering the unfolding reaction to be a transition between a single native state *(N)* and many unfolded, but thermodynamically similar, forms *(U)*. The forces that contribute to the stability of proteins were described in terms of the free energy difference between the native and unfolded states *(1–3)*. The individual contributions to the total free energy change for the reaction, $N \Leftrightarrow U$, were described *(4)* by the relation:

$$\Delta G = \Delta G_{conf} + \Sigma_i^{\Delta g_{i,int}} + \Sigma_i^{\Delta g_{i,s}} + \Delta W_{el} \quad (1)$$

where ΔG_{conf} = configurational free energy (order/disorder term); $\Delta g_{i,int}$ = short range interactions (H-bonds, van der Waals interactions, salt bridges, cofactor binding, and so on); $\Delta g_{i,s}$ = short range interactions with solvent (hydrophobic effect, hydration of ions, and so on); and ΔW_{el} = long range electrostatic interactions (α-helix dipoles, and so on).

In this model for protein folding the symbol Δg_i represents contributions to the corresponding ΔG that are assigned to a localized portion of a protein molecule *(4)*. That is, the total free energy of all short range interactions within the ordered regions of a protein molecule is divided among those separate portions. Thus, the total ΔG between the states is expressed as the sum of contributions from all the physical and chemical factors that affect them *(5)*.

Considering protein stability as the sum of individual energetic contributions is a tremendous simplifying principle because all the complex interactions that result in protein folding need not be considered simultaneously. Site-directed mutagenesis has provided the means to alter a localized part of a protein. The two-state behavior of many proteins allows the energetic consequences of the localized alteration to be measured by its effect of the overall free energy of the unfolding reaction. Since the overall free energy of unfolding for proteins is small—typically 5–20 kcal/mol at 25°C, a change in $\Delta G_{\text{unfolding}}$ of 1 kcal/mol owing to a mutation is significant and easily measured.

The most informative research on the energetics of protein folding has concentrated on relatively small, monomeric, globular proteins. Because such proteins readily unfold and refold, they generally adhere well to the two-state approximation that underpins the site-directed mutagenesis approach. The primary forces driving protein folding as well as the basic features of folding pathways are being actively explored in these systems. This area has been extremely active and has been reviewed extensively *(6–9)*.

In spite of recent progress, understanding of the relationship between structure and stability is fragmentary. It is therefore impossible to write a protocol for the design of informative site-directed mutants. This chapter will instead illustrate some guidelines the author has followed in the design of a useful mutant of subtilisin BPN', the 275 amino acid serine protease from *Bacillus amyloliquefaciens*.

This example will illustrate five important elements of using SDM to understand protein stability.

1. Understand properties of the native protein.
2. Determine a specific question that SDM can help address.
3. Design a mutation.
4. Understand properties of the mutant protein.
5. Repeat steps 2–4.

2. Materials

2.1. Site-Directed Mutagenesis

The subtilisin gene from *Bacillus amyloliquefaciens* (subtilisin BPN') has been cloned, sequenced, and expressed at high levels from its natural promoter sequences in *Bacillus subtilis (10,11)*. Mutagenesis was performed according to the oligonucleotide-directed in vitro mutagenesis system, version 2 (Amersham International, Arlington Heights, IL). All restriction enzymes and T4 DNA ligase were purchased from New England Biolabs, Inc. (Beverly, MA), and alkaline phosphatase from Boehringer Mannheim Biochemicals (Mannheim, Germany). Single strand plasmid DNA was sequenced according to Sequenase™ (United States Biochemical, Cleveland, OH).

2.2. Expression

Mutant genes were cloned into a pUB110-based expression plasmid and used to transform *B. subtilis*. The *B. subtilis* strain used as the host contains a chromosomal deletion of its subtilisin gene and therefore produces no background wild-type activity *(12)*. L broth (100 mL) (10 g tryptone, 5 g yeast extract, 10 g NaCl/L) supplemented with 10 μg/mL kanamycin was inoculated with several kanamycin-resistant colonies in a 250-mL baffled flask. The culture was grown at 37°C 300 rpm until mid-log phase. This culture was used to inoculate 1.5 L of L broth buffered with 2.3 g KH_2PO_4, 12.5 g K_2HPO_4/L supplemented with 10 μg/mL kanamycin. The culture was grown at 34°C in a BioFlo Model C30 fermenter (New Brunswick Scientific, Hatfield, UK) for 18 h. The pH was maintained at 7.5 throughout the fermentation. S221C subtilisin was expressed at a level of ~150 mg of the correctly processed mature form per liter.

2.3. Protein Purification and Characterization

C221 subtilisins were purified as described *(13)*. In some cases the C221 mutant subtilisins were repurified on a sulfhydryl specific mercury affinity column (Affi-gel 501 BioRad, Richmond, CA). Protein purity was checked by discontinuous sodium dodecyl sulfate (SDS) polyacrylamide gel electrophoresis (PAGE) on 12% running gels (NOVEX, San Diego, CA). Assays of peptidase activity were performed by monitoring the hydrolysis of sAAPFna as described *(14)*. Protein concentration [P] was determined using $[P]^{0.1\%} = 1.17$ at 280 nm. For variants that contain the Y217K change, the $[P]^{0.1\%}$ at 280 nm was calculated to be 1.15 (or 0.96 × wt), based on the loss of one Tyr residue *(5)*.

2.4. Titration Calorimetry

The calorimetric titrations were performed with a Microcal Omega titration calorimeter as described in detail *(15)*. The titration calorimeter consisted of a matched reference cell containing the buffer and a solution cell (1.374 mL) containing the protein solution. Microliter aliquots of the ligand solution were added to the solution cell through a rotating stirrer syringe operated with a plunger driven by a stepping motor. After a stable baseline was achieved at a given temperature, the automated injections were initiated and the accompanying heat change per injection was determined by a thermocouple sensor between the cells. During each injection, a sharp exothermic peak appeared that returned to the baseline prior to the next injection occurring 4 min later. The area of each peak represented the amount of heat accompanying binding of the added ligand to the protein. The total heat, Q, was then fit by a nonlinear least squares minimization method *(15)* to the total ligand concentration $[Ca]_{total}$ according to the equation:

$$dQ/d[Ca]_{total} = \Delta H\{1/2 + [1 - (1 + r)/2 - Xr/2]/[Xr - 2Xr(1 - r) + 1 + r^2]^{1/2}\} \quad (2)$$

where $1/r = [P]_{total} \times K_a$ and $X_r = [Ca]_{total}/[P]_{total}$.

The protein concentrations ranged from 30–100 μ*M* whereas the concentration of the calcium solutions were about 20X the protein concentrations. Each binding constant and enthalpy were based on several titration runs at different concentrations. Titration runs were performed until the titration peaks were close to the baseline.

2.5. X-Ray Crystallography

Subtilisin crystals were obtained by dissolving 10 mg of lyophilized protein into 1 mL 50 m*M* glycine–NaOH, pH 9.0. Acetone was introduced by vapor diffusion into hanging drops *(16)*. Diffractive crystals of C221 subtilisin were obtained in several days. Several weeks were required to obtain suitable C221, Δ75–83 crystals.

Data were collected using X-rays generated by a Rigaku rotating anode and collected using a Siemens electronic area detector. The detector is mounted on a Supper oscillation camera controlled by a PCS microcomputer. Determination of crystal orientation and integration of reflective intensities were performed with XENGEN.

2.6. Spectroscopy

Absorption spectra were obtained using a Shimadzu UV-265 UV-visible recording spectrophotometer. Fluorescence spectra were obtained using a SPEX FluoroMax spectrofluorimeter. CD measurements were performed with a Jasco 720 spectropolarimeter using a water-jacketed cell. Temperature control was provided by a Neslab (Newington, NH) RTE-110 circulating water bath interfaced with a MTP-6 temperature programmer.

Unfolding/refolding reactions were followed by monitoring change in tryptophan fluorescence (excitation λ = 295, emission λ = 345) using a KinTek Stopped-Flow Model SF2001 interfaced with a Swan 486 PC. Temperature was controlled by a circulating water bath with the mixing syringes completely immersed in the thermostatted water. All solutions used in stop flow experiments were filtered by sterile 0.2 μm Acrodisc syringe filters (Gelman Sciences, Ann Arbor, MI) to remove any particulates. The filters were thoroughly washed with 50 mL of the appropriate buffer to rinse away any organic contaminants present on the membranes.

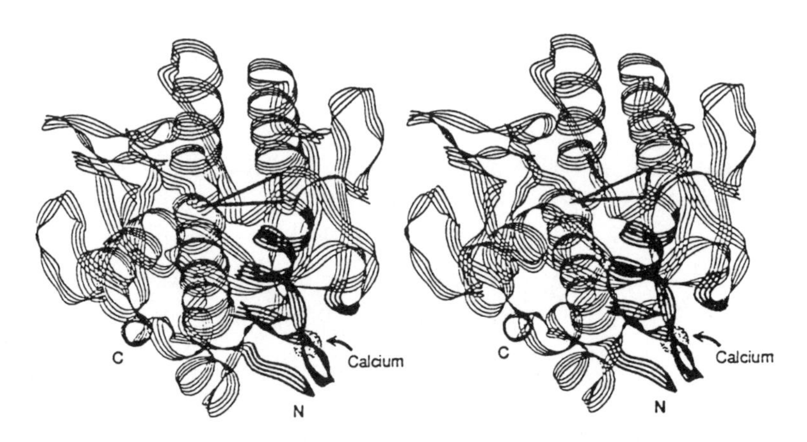

Fig. 1. Ribbon diagram depicting the structure of subtilisin BPN'. Solid sphere shows the position of the high affinity calcium-binding site A.

3. Methods

3.1. Understanding Properties of the Native Protein

Subtilisin is a very stable protein on the upper limit in size for a cooperative folding unit. Extracellular microbial proteases are typically robust since they must be capable of surviving an extracellular environment, which by virtue of their own presence is protease filled. Calcium binding sites are common features of extracellular proteases that make large contributions to their high stability. For example, thermolysin, from the thermophilic organism *Bacillus thermoproteolyticus*, contains four Ca^{2+} binding sites *(17)*. The thermophilic fungal proteases, proteinase K and thermitase, have two and three calcium binding sites, respectively *(18,19)*. One of the most intriguing features of these proteases is that their folding does not seem to occur spontaneously in vitro. In vivo biosynthesis is dependent on propeptide, which is eventually cleaved from the mature form of the enzyme *(20)*.

The subtilisin BPN' structure has been highly refined ($R = 0.14$) to 1.3 Å resolution and has revealed structural details for two metal ion binding sites *(21)*. One of these (site A) binds Ca^{2+} with high affinity, and is located near the *N*-terminus, whereas the other (site B) binds calcium and other cations much more weakly and is located about 32 Å away (Fig. 1).

Because of the dominant role of calcium binding in the stability of subtilisin, it seemed important to directly measure its contributions to the stability of the folded protein. Metal ions play an important role in stabilizing some proteins by binding at specific sites in their tertiary structure so that the free-energy of binding contributes to the stability of the native state. According to the principles outlined in Section 1., the unfolding reaction of subtilisin can be divided conceptually into two parts as follows:

$$N(\mathrm{Ca}) \overset{\Delta g_1}{\Leftrightarrow} N + \mathrm{Ca} \overset{\Delta g_2}{\Leftrightarrow} U \quad (3)$$

where N(Ca) is the native form of subtilisin with calcium bound to a high affinity calcium-binding site, N is the folded apoprotein, and U is the unfolded protein. The total free energy of unfolding is therefore equal to $\Delta g_1 + \Delta g_2$.

The energetics of calcium binding (Δg_1) were directly measured once the apoenzyme *(N)* was produced. Subtilisin after purification contains an equal molar amount of calcium tightly bound to the A site *(22)*. Complete removal of calcium from subtilisin is very slow, requiring 24 h of dialysis against EDTA at 25°C to remove all calcium from the protein and then an additional 48 h of dialysis in high salt *(13)* at 4°C to remove all EDTA from the protein. The apoenzyme was found to be unstable and is difficult to maintain in native form. The instability of the apoenzyme was addressed by eliminating the wild-type proteolytic activity of subtilisin to prevent autodegradation from occurring following removal of calcium from the enzyme. This was accomplished by converting the active-site serine 221 to cysteine. This mutant has become a vehicle through which the folding questions can be addressed without complications of proteolysis (*see* Note 1).

Using the inactive S221C subtilisin, both the association constant and kinetics of binding could be determined. The binding parameters obtained by titration calorimetry were $\Delta H = \sim\!-11$ kcal/mol; $K_a = 7 \times 10^6/M$ and a stoichiometry of binding of 1 calcium site/mol (Fig. 2). The standard free energy of binding at 25°C is 9.3 kcal/mol. The binding of calcium is therefore primarily enthalpically driven with only a small net loss in entropy ($\Delta S_{\mathrm{binding}} = 6.7$ cal/°mol). The kinetic barrier to calcium removal

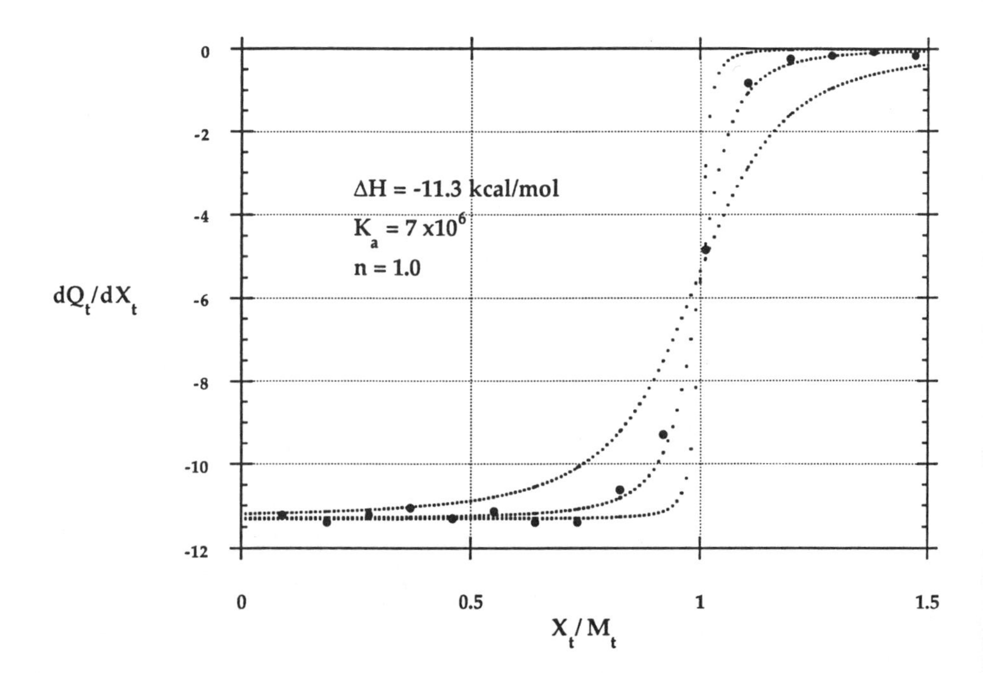

Fig. 2. Titration calorimetry of S221C subtilisin. The heat of calcium binding for successive additions of calcium are plotted vs the ratio of [Ca]/[P]. The data are best fit by a calculated binding curve using Eq. (1) in the text, assuming a binding constant of 7×10^6 and ΔH equal to –11.3 kcal/mol using Eq. (1) from the text. For comparison, calculated curves assuming $K_a = 1 \times 10^6$ and 1×10^8 are also shown. In this titration [P] = 100 µ*M*. Temperature was 25°C.

from the A site was found to be substantially larger than the standard free energy of binding. Thus the kinetics of calcium dissociation from subtilisin (e.g., in excess EDTA) are very slow ($t_{1/2}$= 1.3 h at 25°C).

3.2. Determining Questions Addressed by SDM

The in vitro folding of subtilisin is extremely slow, presumably because of a high kinetic barrier between the folded and unfolded states of subtilisin. The high activation barrier for calcium binding and dissociation suggested that formation of the calcium A-site during folding could be a difficult step in folding the mature form of the enzyme. Site-directed mutagenesis affords an opportunity to simplify the subtilisin folding reaction and test this hypothesis. It was hypothesized that a well designed calcium-free mutant subtilisin might fold much more readily in

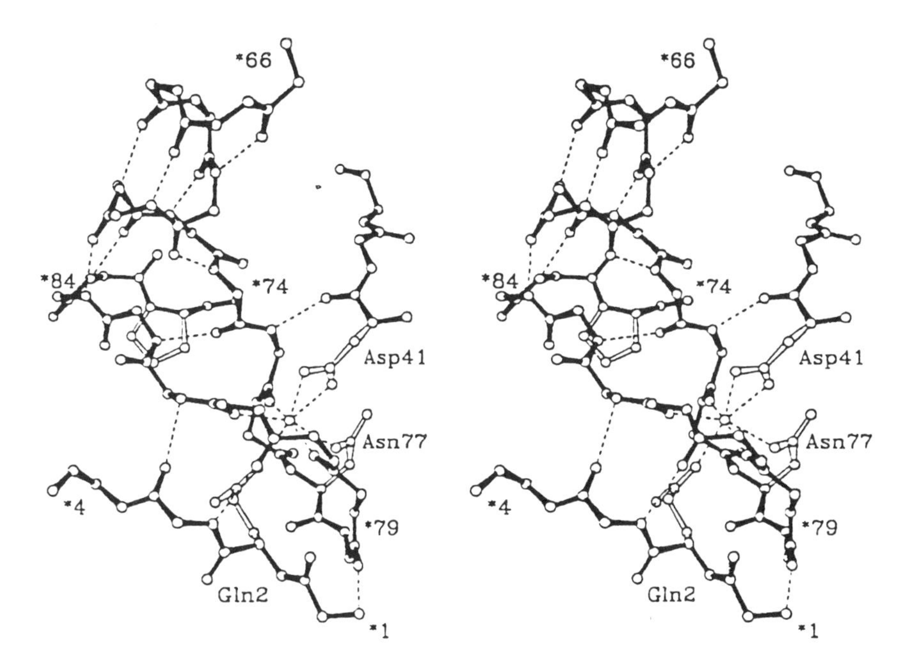

Fig. 3. Stereo closeup view of calcium A-site region. Dotted lines represent coordination bonds to calcium and hydrogen bonds under 3.2 Å. For clarity only side chains involved in calcium coordination and P86 are shown.

vitro than the wild-type protein. This calcium-free mutant should be unable to bind calcium but otherwise should be similar in structure to native subtilisin, allowing the $N \Leftrightarrow U$ reaction to be analyzed in detail.

3.3. Design of a Mutation to Eliminate Calcium Binding

The instability of the apoenzyme suggested that the initial prototypes of an engineered calcium-free subtilisin also would likely be unstable. The results also indicated, however, that if S221C subtilisin were used as a starting point, an engineered calcium-free subtilisin could be produced and maintained in native form.

A thorough inspection of the X-ray structure of native subtilisin was the next step in designing an informative mutation. Calcium at site A is coordinated by five carbonyl oxygen ligands and one aspartic acid. Four of the carbonyl oxygen ligands to the calcium are provided by a loop composed of amino acids 75–83 (Fig. 3). The geometry of the ligands is that of a pentagonal bipyramid whose axis runs through the carbonyl of 75 and 79. On one side of the loop is the bidentate carboxylate (D41), and on the other side is the N-terminus of the protein and the side chain of Q2. The seven coordination distances range from 2.3–2.6 Å, the shortest being to the aspartyl carboxylate. Three hydrogen bonds link the N-terminal segment to loop residues 78–82 in parallel-beta arrangement.

The most obvious strategy for eliminating calcium binding is to alter one or more of the calcium ligands. Since only the three side chains Q2, D41, and N77 are direct ligands to the calcium, the possibilities for single amino acid replacements is limited (*see* Note 2). Ultimately, a successful design was based around an unusual structural feature of subtilisin. Four of the ligands to the calcium are from a nine amino acid loop that creates a discontinuity in the last turn of a 14-residue α-helix involving amino acids 63–85. Inspection of the X-ray crystal structure suggested that the loop amino acids 75–83 could be removed to generate a continuous α-helix. Such a large alteration in structure would certainly eliminate calcium binding at the A-site. The obvious problems with this design are the lost and uncompensated interactions of Q2 and D41 with calcium, and amino acids 1–4 with the former loop residues. Since it seemed impossible to know in advance what compensating mutations to make, the initial goal was to produce a molecule stable enough to purify and further evaluate. Based on the characterization of aposubtilisin, however, it seemed reasonable that the deletion mutant would be stable enough to fold and purify.

3.4. Analyzing Mutant

3.4.1. Initial Characterization

Following purification of a mutant protein, any gross effects of a mutation on an enzyme's structure and function can be discovered in several ways that are both sensitive and rapid.

3.4.1.1. Determination of Enzymatic Activity

The Δ75–83 deletion in subtilisin was not expected to have any direct effects on catalytic function. Measurement of specific activity was therefore an easy way to determine whether the deletion had cataclysmic effects on native structure.

The specific activity of S221C subtilisins is low (~30,000× less than wild-type subtilisin) but can easily be measured with a sensitive assay.

Hydrolysis of the synthetic peptide substrate, succinyl-(L)-Ala-(L)-Ala-(L)-Pro-(L)-Phe-*p*-nitroanilide, can be measured spectrophotometrically. The release of the *p*-nitroanalide-leaving group is followed by the color change at 410 nm. Thus, it could be determined that Δ75–83, C221 subtilisin had the same specific activity as C221 subtilisin 0.0025 U/mg.

3.4.1.2. Spectroscopic Analysis

It could also be easily determined that Δ75–83, C221 subtilisin had very similar spectral properties to C221 subtilisin. The extinction and fluorescence properties of tyrosine and tryptophan are sensitive to their local environments, therefore careful inspections of spectra is one indicator of proper folding. Anomalous behavior of a mutant protein, such as aggregation, is evidenced by increased light scattering and is also easily detectable spectroscopically. Circular dichroism in the UV region is informative about changes in content of secondary structure. Since Δ75–83, S221C subtilisin has UV, fluorescence, and CD spectra indistinguishable from S221C subtilisin, one could be confident that the deletion had not dramatically altered the native folding topology.

3.4.1.3. Denaturation Behavior

Cooperative unfolding is a hallmark of a stable globular protein. Thus a mutant with a unique native conformation can be distinguished from a mutant with a disorganized or molten globule-like structure. The unfolding transition can be studied either with a chemical denaturant, such as urea, or by thermal denaturation. In either case, both the midpoint of denaturation and the breadth of the transition should be noted. A CD melting curve of Δ75–83, S221C subtilisin is shown in Fig. 4. A decrease in melting temperature as well as some loss of cooperative folding was observed relative to S221C subtilisin. These observations were consistent with expectations based on knowledge of the stability of S221C aposubtilisin.

After completing preliminary characterization, more detail and time-consuming structural characterization was undertaken.

3.4.2. Structural Analysis by X-Ray Crystallography

Large single crystal growth and X-ray diffraction data collection were performed essentially as previously reported *(16,23)*. The starting model

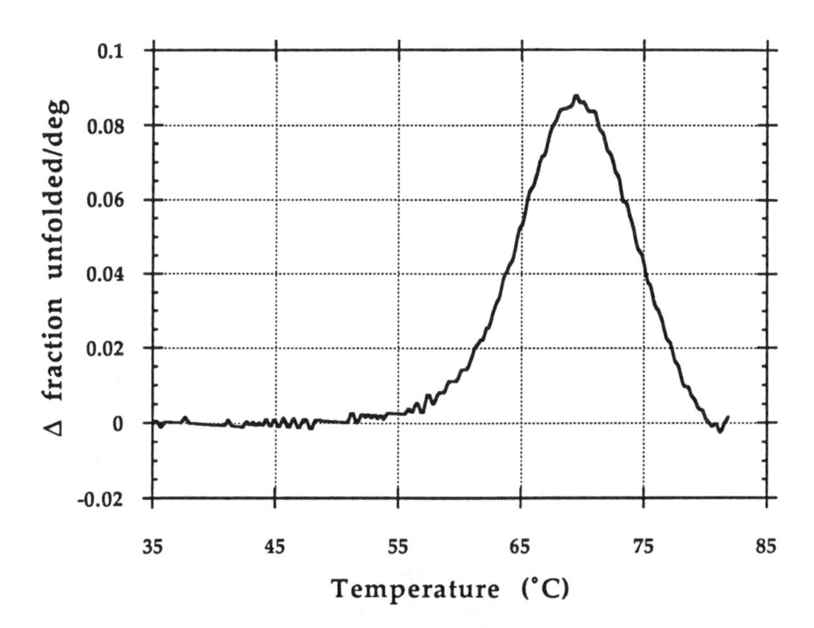

Fig. 4. CD melting of S221C, Δ75–83 subtilisin. Ellipticity at 222 nm was measured as a function of temperature. The derivative curve (Δ fraction unfolded/°C vs temperature) is shown. Subtilisin concentration was 1 μ*M*. Cell pathlength was 1 cm. The scan rate was 1°C/min.

for S221C, Δ75–83 subtilisin was made from S221C subtilisin (Protein Data Bank entry 1Sud.pdb).

Data sets with about 20,000 reflections between 8.0 and 1.8 Å resolution were used to refine the model using restrained least-squares techniques. Initial difference maps for S221C, Δ75–83 subtilisin, phased by a version of S221C subtilisin with the entire site A region omitted, clearly showed continuous density representing the uninterrupted helix, permitting an initial Δ75–83, S221C subtilisin model to be constructed and refinement begun. Each mutant was refined from R ~ 0.30 to R ~ 0.18 in about 80 cycles, interspersed with calculations of electron density maps and manual adjustments using the graphics modeling program FRODO *(24)*.

Except for the region of the deleted calcium-binding loop, the structures of S221C subtilisin and Δ75–83, S221C subtilisin are very similar, with a rms deviation of 0.18 Å between 262 α-carbons. The N-terminus of S221C subtilisin (as in the wild-type) lies beside the site A loop, furnishing one calcium coordination ligand, the side chain oxygen of Q2. In

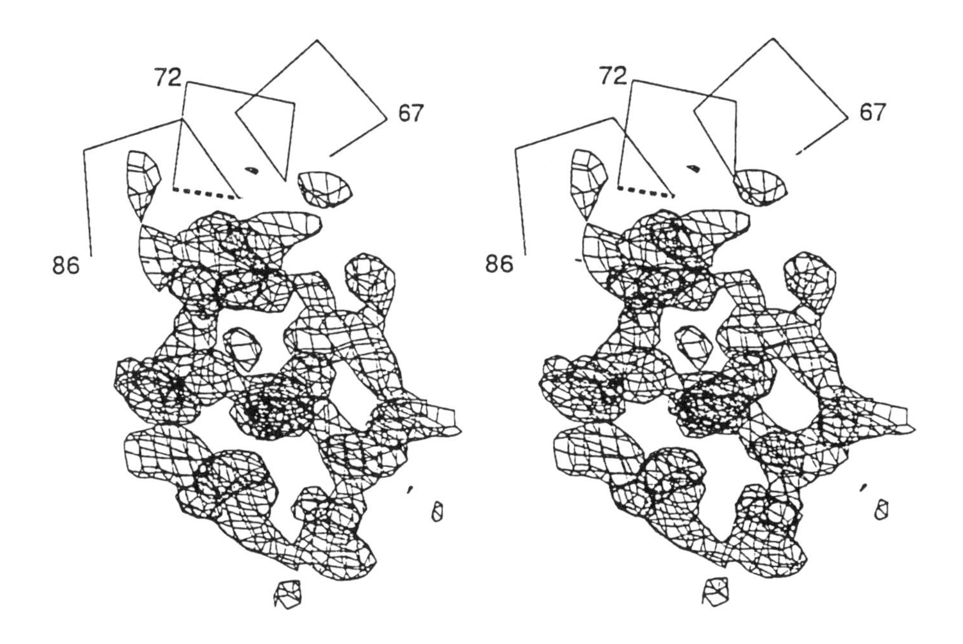

Fig. 5. Closeup view of the A-site deletion. The loop from S221C subtilisin is shown as a dotted line with the continuous helix of Δ75–83 subtilisin. Superimposed is the 3 sigma difference electron density (F_o – F_0[Δ75–83], phases from Δ75–83 subtilisin) showing the deleted A-site loop.

Δ75–83, S221C subtilisin the loop is gone, leaving residues 1–4 disordered. In S221C subtilisin (as in wild-type) the site A loop occurs as an interruption in the last turn of a 14-residue alpha helix; in Δ75–83, S221C subtilisin this helix is uninterrupted and shows normal helical geometry over its entire length. Diffuse difference density and higher temperature factors indicate some disorder in the newly exposed residues adjacent to the deletion (Fig. 5).

3.4.3. Energetic Analysis of the Folding Reaction

Having gained detailed structural information of Δ75–83 subtilisin through X-ray crystallographic analysis, the energetic consequences of the deletion could be rationally evaluated. The original hypothesis was that deleting the high affinity calcium binding site would eliminate a high kinetic barrier to the folding and unfolding reactions. A detailed examination of the energetic consequences of a mutation on the folding reaction should involve whatever tools are available (*see* Note 3). Five of the most useful analytical tools to study protein folding are reviewed in this volume: spectroscopic techniques (Chapters 3–6); urea and guanidine hydrochloride denaturation (Chapter 8); differential scanning calorimetry (Chapter 9); hydrogen/deuterium exchange techniques (Chapter 13); and kinetic analysis (Chapter 14). The results for some experiments measuring the kinetics of unfolding and refolding are presented below.

The kinetics of unfolding Δ75–83 subtilisin were measured at pH 2.15. CD spectra of Δ75–83 subtilisins taken after 1 min in 0.1*M* HPO_4^{2-}, pH 2.15 indicated that the protein was unfolded to predominantly random structure. Because the denaturation reaction for Δ75–83 subtilisins is rapid at this pH, kinetics of unfolding were measured by stopped-flow mixing methods. Unfolding of subtilisin was monitored by the 15% decrease in fluorescence of its three partially buried tryptophans. Subtilisin in 0.1*M* HPO_4^{2-}, pH 7.0 was mixed with 2 vol of 0.1*M* H_3PO_4. The resulting solution was 0.1*M* HPO_4^{2-}, pH 2.15. The denaturation reaction was fit to a single exponential equation. The residuals after subtracting the data from the calculated first order curve are plotted on a five-times expanded scale (Fig. 6). At 22°C, k = 14.3/s. The unfolding of S221C subtilisin under the same conditions occurs at a rate of ~0.001/s.

The folding rate of Δ75–83 subtilisin was measured at 0.1*M* HPO_4^{2-}. Refolding was initiated after a double pH jump. 3 μ*M* S221C, Δ75–83 subtilisin was dissolved in 0.05*M* HPO_4^{2-}, pH 7.0. Unfolding was induced by the addition of an equal volume of 0.1*M* H_3PO_4. The resulting solution has a pH of 2.15. The protein is allowed to denature for 1 s, which was shown above to be sufficient for complete denaturation. After the 1-s delay the protein is renatured by the addition of 1/2 vol of 0.15*M* HPO_4^{2-}, pH 12.0. The resulting solution is 0.1*M* HPO_4^{2-}, pH 7.2.

The refolding curve for C221, Δ75–83 subtilisin at 25°C is shown in Fig. 7A. The folding reaction was followed by the 20% increase in tryptophan fluorescence that occurs after refolding. Three refolding curves were averaged and fit to a single exponential equation. The calculated first order rate constant (k_f) equals 0.0027/s. The residuals after subtracting the data from the calculated first order curve are plotted on a five-times expanded scale (Fig. 7A). The folding reaction was also monitored by collecting a far UV CD spectrum (250–200 nm) at 5-min intervals. By 30 min after the second pH jump, the CD spectrum of the renatured S29 was identical to a spectrum of native S29 subtilisin. The decrease in

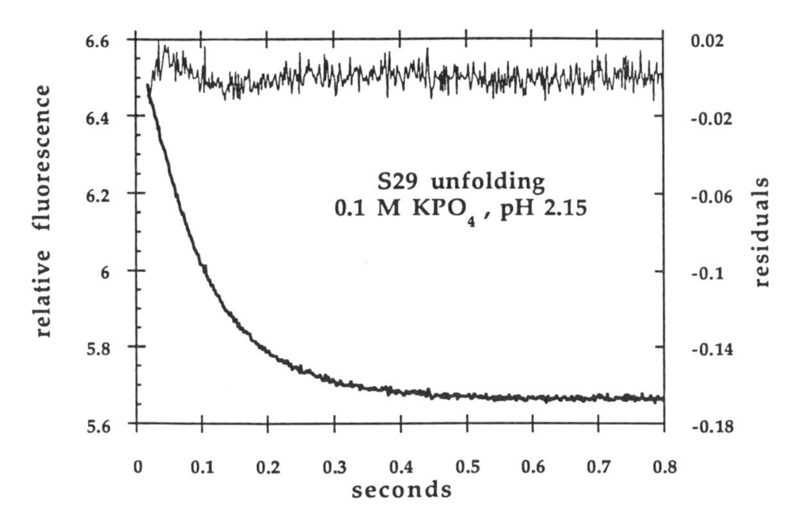

Fig. 6. Kinetics of unfolding S221C, Δ75–83 subtilisin. Unfolding in 0.1*M* HPO_4, pH 2.15 and 22°C is followed by the decrease in tryptophan fluorescence. The curve plotted is the average of 12 experiments. The residuals after subtracting the data from a single exponential fit (k = 10.4/s) are plotted on a five-times expanded scale.

ellipticity at 222 nm was plotted vs time and fit to a first order rate equation with $k_f = 0.0027/s$ (Fig. 7B). By comparison, unfolded C221 subtilisin, when returned to native conditions, collapses into an intermediate state that is not significantly converted to the native state after more than a week at 25°C.

The results of the kinetics for unfolding/refolding show that deletion of calcium binding loop does indeed lower the kinetic barrier between the folded and unfolded states of subtilisin.

3.5. Repetition of Steps 2–4

Having gained some additional understanding of the structure and stability folding of Δ75–83 subtilisin, further mutations can be contemplated. For example, one can explore ways to recoup the stability lost from calcium binding.

In our original design the deletion of the calcium binding loop was made without consideration of the fate of the former ligands to the calcium or to lost protein–protein interactions, because we thought it would be impossible to know in advance what compensating mutations to make.

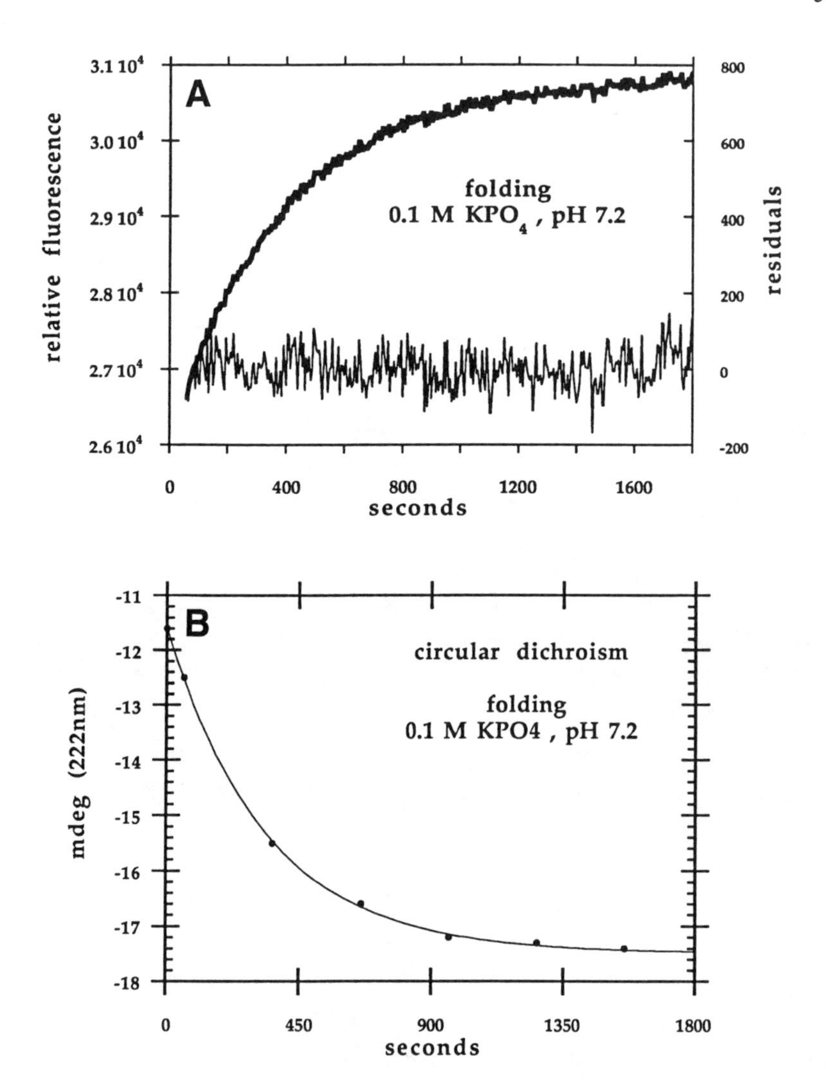

Fig. 7. Kinetics of refolding S221C, Δ75–83 subtilisin. (**A**) Refolding in 0.1*M* KPO_4, pH 7.2 and 25°C is followed by the increase in tryptophan fluorescence. The curve plotted is the average of three experiments. The residuals after subtracting the data from a single exponential fit ($k = 0.0027/s$) are plotted on a five-times expanded scale. (**B**) Refolding of S221C, Δ75–83 subtilisin in 0.1*M* KPO_4, pH 7.2 and 25°C is followed by CD. Spectra (250–200 nm) were collected at 5-min intervals. The decrease in ellipticity at 222 nm was plotted vs time and fit to a first order rate equation with $k = 0.0027/s$.

Our goal was to produce a molecule stable enough to purify and further evaluate. Although Δ75–83 subtilisin is much less stable than the wild-type, the mutant was sufficiently stable to allow its characterization and structure determination.

A major effect of the loss of the calcium A-site in Δ75–83 subtilisin is a partial loss of cooperative folding. This was observed in denaturation experiments using differential scanning calorimetry. In the wild-type protein, melting occurs cooperatively because the stability of the native state greatly exceeds that of any partially folded states. The breakdown of cooperativity in the Δ75–83 mutant results from the decrease in the stability of the native state relative to intermediate states. The key to engineering a stable subtilisin without the A-site is to restore cooperativity to the unfolding reaction.

An inspection of the X-ray structure of Δ75–83 subtilisin suggests reasons for the lack of cooperativity in unfolding. The regions of the structure most affected by the deletion are the N-terminal amino acids 1–4, the 202–219 β-ribbon, and the omega loop containing D41. The first four residues in Δ75–83 subtilisin are disordered in the X-ray structure since all its interactions were with the calcium loop. Other than the N-terminus, there are three other residues whose side chain conformations are distinctly different from wild-type. Y6 swings out of a surface niche into a more solvent-exposed position, as an indirect effect of the destabilization of the N-terminus. D41, a former calcium ligand, and Y214 undergo a coordinated rearrangement, forming a new hydrogen bond.

Based on knowledge of the structure and stability of the deletion mutant, compensating for the loss of the calcium A-loop by modifying the N-terminal part of the protein should be possible.

4. Notes

1. Eliminating the problem of autohydrolysis: The problem of autohydrolysis in folding studies is unique to proteases and is trivial to eliminate by site-directed mutagenesis of an active site amino acid. The active site nucleophile of subtilisin is serine 221. Substituting this amino acid with any other will largely eliminate proteolytic activity. The main consideration in choosing a substitute is to avoid altering the conformational stability. A secondary consideration is to retain sufficient enzymatic activity to serve as a convenient measure of native conformation. We chose to mutate serine 221 to cysteine. This substitution reduces proteolytic activity by a factor of

~3×10^4, a level that is measurable but no longer problematic for folding studies. This mutation also has little effect of the thermal denaturation temperature of subtilisin.

2. Q2R mutation: We constructed a Q2R mutant in the hope that the positively charged quanidinium group would displace calcium and possibly form a salt bridge with D41. It was thought that if the protein were stable enough to fold, that the mutant protein would be unable to bind calcium at the A-site. The mutant protein was in fact able to fold, but contrary to expectations, the mutant protein still bound calcium with a dissociation constant in the μ*M* range. Apparently, the Q2R mutation causes the N-terminal residues to swing away from the calcium site while calcium remains coordinated to the remaining six ligands.
3. Mutant design: Although it is tempting to pontificate about the mechanics of mutant design, my experience in designing and analyzing many uninformative mutants prevents me from doing so. My experience has been that the most important and time consuming step in the design process is step one—understanding properties of the native protein.

Acknowledgments

I wish to thank Travis Gallagher for preparing the computer graphic figures of subtilisin.

References

1. Brandts, J. F. (1964) *J. Am. Chem. Soc.* **86,** 4302–4314.
2. Kauzmann, W. (1959) *Adv. Protein Chem.* **14,** 1–63.
3. Tanford, C. (1962) *J. Am. Chem. Soc.* **84,** 4240–4247.
4. Tanford, C. (1970) *Adv. Protein Chem.* **24,** 1–95.
5. Pantoliano, M. W., Whitlow, M., Wood, J. F., Dodd, S. W., Hardman, K. D., Rollence, M. L., and Bryan, P. N. (1989) *Biochemistry* **28,** 7205–7213.
6. Leatherbarrow, R. J. and Fersht, A. R. (1986) *Protein Eng.* **1,** 7–16.
7. Matthews, B. W. (1987) *Biochemistry* **26,** 6885–6888.
8. Goldenberg, D. P. (1988) *Ann. Rev. Biophys. Biophys. Chem.* **17,** 481–507.
9. Alber, T. (1989) *Ann. Rev. Biochem.* **58,** 765–798.
10. Wells, J. A., Ferrari, E., Henner, D. J., Estell, D. A., and Chen, E. Y. (1983) *Nucleic Acids Res.* **11,** 7911–7925.
11. Vasantha, N., Thompson, L. D., Rhodes, C., Banner, C., Nagle, J., and Filpula, D. (1984) *J. Bacteriol.* **159,** 811–819.
12. Fahnestock, S. R. and Fisher, K. E. (1987) *Appl. Environ. Microbiol.* **53,** 379–384.
13. Bryan, P., Alexander, P., Strausberg, S., Schwarz, F., Wang, L., Gilliland, G., and Gallagher, D. T. (1992) *Biochemistry* **31,** 4937–4945.
14. DelMar, E., Largman, C., Brodrick, J., and Geokas, M. (1979) *Anal. Biochem.* **99,** 316–320.

15. Wiseman, T., Williston, S., Brandts, J. F., and Lin, L.-N. (1989) *Anal. Biochem.* **179,** 131–137.
16. Gallagher, T. D., Bryan, P., and Gilliland, G. (1993) *Proteins: Str. Funct. Gen.* **16,** 205–213.
17. Matthews, B. W., Weaver, L. H., and Kester, W. R. (1974) *J. Biol. Chem.* **249,** 8030–8044.
18. Betzel, C., Teplyakov, A. V., Harutyunyan, E. H., Sänger, W., and Wilson, K. S. (1990) *Protein Eng.* **3,** 161–172.
19. Gros, P., Kalk, K. H., and Hol, W. G. J. (1991) *J. Biol. Chem.* **266,** 2953–2961.
20. Ikemura, H., Takagi, H., and Inouye, M. (1987) *J. Biol. Chem.* **262,** 7859–7864.
21. Finzel, B. C., Howard, A. J., and Pantoliano, M. W. (1986) *J. Cell. Biochem. Suppl.* **10A,** 272.
22. Pantoliano, M. W., Whitlow, M., Wood, J. P., Rollence, M. L., Finzel, B. C., Gilliland, G., Poulos, T. L., and Bryan, P. N. (1988) *Biochemistry* **27,** 8311–8317.
23. Bryan, P. N., Rollence, M. L., Pantoliano, M. W., Wood, J., Finzel, B. C., Gilliland, G. L., Howard, A. J., and Poulos, T. L. (1986) *Prot. Str. Funct. Gen.* **1,** 326–334.
24. Jones, T. A. (1978) *J. Appl. Crystallogr.* **11,** 268–272.

Chapter 13

Hydrogen Exchange Techniques

J. Martin Scholtz and Andrew D. Robertson

1. Introduction

Hydrogen exchange, in conjunction with NMR spectroscopy, is the only method with which one can, in principle, monitor protein folding and stability at nearly every individual amino acid residue in a protein. The unique power of hydrogen exchange is attributable to the ubiquity and the chemistry of the peptide NH group in proteins: Peptide NHs are labile to exchange with hydrogens of the aqueous solvent, and exchange often is slowed by many orders of magnitude in native proteins. If the hydrogen isotope in water is deuterium, then the exchange event in a vast excess of D_2O can be summarized in the following scheme:

$$-\underset{\displaystyle H}{N}-\overset{\displaystyle O}{C}- + D_2O \rightarrow -\underset{\displaystyle D}{N}-\overset{\displaystyle O}{C}- + HOD$$

Deuterons are silent in a proton NMR experiment, so the resonances corresponding to the peptide NH will diminish in intensity as exchange proceeds. If NH resonance assignments have been made for the peptide or protein of interest, then exchange can be monitored at known sites throughout the sequence and structure.

For small peptides, the three major determinants of the observed exchange rate are pH, temperature, and peptide sequence. The contributions of all three factors are fairly well understood, to the extent that one can predict the exchange behavior in a peptide with reasonable accuracy *(1–3)*. In native proteins, however, NH exchange may be slowed many

From: *Methods in Molecular Biology, Vol. 40: Protein Stability and Folding: Theory and Practice*
Edited by: B. A. Shirley

orders of magnitude; this is a consequence of participation in hydrogen-bonded structure and the resultant exclusion of solvent *(4,5)*. The extent to which individual sites show diminished exchange varies over a remarkable range within a given protein, suggesting that hydrogen bonding and solvent exclusion, as identified from protein crystal structures, are not the only factors responsible for the observed exchange. The heterogeneity in the kinetics of hydrogen exchange is most likely a consequence of the unequal contribution of different regions within a protein to its overall stability.

Even though the precise details concerning the mechanism of hydrogen exchange from proteins are not known, the practical consequence of slowed exchange is a large number of probes distributed throughout the three-dimensional structure of a protein that are sensitive to the stability of the protein. These probes permit one to identify hydrogen bonds and assess their stability, and to monitor the microscopic consequences of events such as ligand binding, protein–protein interactions, or changes in solvent conditions. All this information may also be obtained for equilibrium folding intermediates and, in slightly more complicated experiments, kinetic folding intermediates as well. It should be emphasized at the outset that, in most hydrogen exchange experiments, only the subset of peptide NH showing slow exchange in the native protein is observed. This subset typically represents anywhere from 15–50% of all peptide NH in proteins studied to date.

This chapter is directed toward those who will employ proton NMR spectroscopy to follow the exchange of protons and deuterons at individual peptide NH in proteins for which proton NMR assignments are available. Spectral assignment methods and strategies are covered in a number of excellent reviews and monographs *(6–9)*. We will focus on studies of protein folding and the stability of monomeric globular proteins. We refer those interested in studies of ligand binding and protein–protein interactions to a number of recent publications *(10,11)*. Proton–tritium exchange and related low- to intermediate-resolution methods for following NH exchange have been discussed elsewhere *(5,12,13)*. In addition, some of the material covered here has been treated in varying detail in a number of recent reviews *(14,15)*.

2. Methods

2.1. Acquisition and Analysis of NMR Data

2.1.1. Acquisition of NMR Data

All of the experiments described in this chapter use proton NMR spectroscopy to measure proton occupancy at individual NHs. We assume that resonance assignments have been made for at least some of the NHs, although such assignments are not essential for obtaining valuable information. We also assume that the reader has, or is in the process of acquiring, some basic knowledge of NMR spectroscopy.

The choice of 1D vs 2D NMR depends on a number of factors, including the size of the peptide or protein, the number of NHs to be observed, chemical shift dispersion in the NH region of the spectrum, protein availability and solubility, acquisition-time requirements (e.g., limited instrument time or relatively fast NH exchange), and the characteristics of the NMR spectrometer. At present, it is not uncommon to have access to a 500 MHz spectrometer, and our discussion will apply to such instruments. For those with access to spectrometers operating at frequencies other than 500 MHz, all considerations concerning sensitivity and resolution must be scaled accordingly.

In almost every instance, 1D NMR is a good choice for pilot studies and establishing experimental conditions, although newer instruments make 2D data acquisition and processing nearly as easy as 1D methods. Alternatively, if access to a spectrometer is limited or if the protein is particularly valuable, tritium-exchange experiments may be appropriate for the initial experiments *(12,16)*. Most spectrometers observe proton frequencies on samples contained in 5 mm NMR tubes and a typical sample volume is about 0.6 mL. Both larger and smaller sizes are available, but are not nearly as common. The performance characteristics of the spectrometer, the available instrument time, and the rate of NH exchange dictate the desired protein concentration; a 1 m*M* solution of a 10 kDa protein (0.6 µmol or 6 mg in 0.6 mL) will permit acquisition of a good quality 1D spectrum in <1 min on a modern 500 MHz instrument. As sensitivity scales with protein concentration and the square root of the total acquisition time, a reduction of the protein concentration to 0.5 m*M* will require four times the acquisition time to match the sensitivity obtained with the 1 m*M* sample.

Two-dimensional proton NMR experiments typically require at least 100 times the total acquisition time of the 1D experiments. A variety of 2D proton experiments can be used to measure NH occupancies, the most common of which is the absolute value (or magnitude) mode COSY *(17)*. For studies of NH exchange in ribonuclease A, 2 m*M* samples of protein were used for COSY experiments taking approx 3 h and typical acquisition parameters were 256 data blocks (or t_1 increments) each of which consisted of 32 summed scans, 2048 data points, a 1 s recycle delay, and spectral widths of 6000 Hz in both dimensions. If the protons of interest have spin-lattice relaxation times (T_1 values) that are appreciably longer than 2 s, longer recycle delays may be necessary. In general, the optimal acquisition conditions are determined empirically. The following is a reasonable scheme for identifying conditions that minimize the total acquisition time for a COSY experiment.

1. Start with a relatively long acquisition time to ensure good sensitivity and resolution. For a sample containing 1 m*M* protein, good acquisition parameters are 512 blocks containing 32 scans each, 2 K data points per block, a 1.5 s recycle delay, and a spectral width just broad enough (typically 5000–6000 Hz on a 500 MHz spectrometer) to encompass the entire proton spectrum. Under these conditions, it will take about 8.5–10 h to obtain the data.
2. Working with this data set, one can then test the effect of reducing the acquisition time by simple manipulations of the data using the processing software. For example, a series of 2D Fourier transforms can be done using an ever smaller subset of the total number of blocks (i.e., ever smaller values of t_1); this permits identification of the minimum number of blocks needed to achieve the desired resolution in the COSY spectrum. Similarly, one can also reduce the acquisition time, t_2, by simply performing the Fourier transform after artificially reducing this number with a weighting function. Once these t_1 and t_2 have been minimized, one can assess the effects on sensitivity of reducing the recycle delay and the number of scans. The aim here is to achieve a level of sensitivity that will just permit the weakest crosspeak to be accurately measured when proton occupancy is 0.25–0.5 full occupancy. This approach can be used with any 2D NMR experiment.

Phase-sensitive COSY and NOESY experiments may also be used to acquire exchange data *(18,19)*; each experiment has its strengths and weaknesses. The magnitude-mode COSY is the simplest to implement and process, but resolution suffers relative to the phase-sensitive experiments. The NOESY experiment is appealing because of the good resolution and the ease with which peak identities can be confirmed, but changes in peak intensities during exchange experiments may result from both a change in proton occupancy as well as perturbations of relaxation pathways as nearby NH undergo exchange.

An increasingly popular 2D experiment that is readily implemented on many modern spectrometers is the proton detected or inverse-detected heteronuclear NMR experiment (for example, *see* ref. *20*). For hydrogen exchange studies, ^{15}N is the heteronucleus of choice and data are acquired using the Heteronuclear Multiple Quantum Coherence (HMQC) or Heteronuclear Single Quantum Coherence (HSQC) pulse sequence; the latter usually provides better resolution *(21)*. The impetus for these experiments is excellent sensitivity and resolution relative to the 2D proton experiments. The strong one-bond coupling (ca. 100 Hz) between the nitrogen and proton of each peptide NH results in intense crosspeaks after a short acquisition time, comparable to that needed for the 2D proton experiments. High resolution is achieved because of the added chemical shift dispersion from the ^{15}N nucleus. The result is that one is able to examine proteins that exceed 20 kDa in size; proton NMR experiments on proteins larger than 10–15 kDa are difficult because of the small coupling constants involved (2–12 Hz). The limitations to the heteronuclear approach are:

1. The need for ^{15}N-labeled protein, which is usually obtained through overproduction in bacteria;
2. The cost of ^{15}N labeled precursors; and
3. Access to necessary hardware, software, and technical expertise.

These experiments are, however, rapidly becoming routine.

Hydrogen exchange may, in some cases, be followed using ^{13}C NMR *(22,23)*. The chemical shift of the peptide carbonyl carbon is sensitive to the hydrogen isotope attached to the neighboring nitrogen and, depending on the rate of exchange and the relative concentrations of H_2O and D_2O, one may observe two distinct carbonyl resonances whose intensities reflect the relative populations of NH and ND. Although accurate rate determinations can only be measured over a narrow range of exchange rates, these rates are more rapid than those accessible with the other methods described in this chapter and, thus, this method may complement the standard hydrogen exchange methods.

2.1.2. Analysis of NMR Data

Resonance intensities in 1D NMR spectra may be quantified by:

1. Measuring peak height;
2. Integration of peak areas;
3. Multiplication of peak height by linewidth at half-height; and
4. Line-shape analysis *(24)*.

For hydrogen exchange studies, the first method generally works as well or better than any of the other methods *(25)*. All of the methods are dependent on good baselines and phasing; the integration method is especially sensitive, so a judicious choice of phasing parameters and baseline correction is necessary. The weighting or apodization function by which the FID is multiplied prior to Fourier transformation should be strong enough to minimize any noise in the spectrum, since spurious noise can result in scatter in the intensity measurements. Intensity measurements for partially overlapping peaks may require line–shape analysis. Alternatively, intensities may be measured on the sides of peaks most distant from the overlap. Most NMR data processing software packages include routines for performing line–shape analysis; these permit the user to model in a number of resonances with varying chemical shift, intensity, and linewidths. If real-time measurements of exchange are being made with a single sample in the spectrometer, the peak intensities may not need to be normalized to a nonexchanging peak. If, however, a number of different samples are being used to obtain exchange rates, then a normalization is essential. Aromatic peaks and upfield-shifted methyl resonances are good controls for intensity.

In 2D NMR spectra, one may measure peak heights *(26)* or peak volumes *(27)*; routines for these measurements are usually available with the processing software. With volume measurements, it is important to integrate "blank" regions in the 2D spectrum, which provide a measure of the significance of peak intensities to determine baseline values; peak intensities approaching values for blanks should be viewed with caution. Flat baseplanes, which are most easily obtained with the magnitude-mode COSY experiment, greatly facilitate these measurements.

Many NMR software packages include routines for the determination of first-order rate constants. These may be used for fitting exchange data from a series of spectra. Alternatively, nonlinear least squares fitting of peptide NH intensities vs time may be done with any of the large number of commercially available software packages with curve-fitting capabilities. With any of the fitting procedures, it is best to let the program fit the value for intensity at infinite time, because this may not in fact be exactly zero. With partially or completely overlapping peaks, one may be able to fit the exchange data to a sum of exponentials, especially if rates for the two peptide NH differ by more than about threefold.

2.1.3. Hydrogen Exchange in Model Compounds

NH exchange is catalyzed by specific acid and base, with possibly a very small contribution from water-catalyzed exchange:

$$k_{ex} = k_H[H_3O^+] + k_{OH}[OH^-] + k_w$$

Under most conditions, NH exchange exhibits pseudo-first order kinetics, which permits facile measurement of exchange rates. In addition to pH, the rate of NH exchange from small peptides depends on temperature, the nature of neighboring side chains, and the electrostatic environment *(1,14,28)*. Most of these effects are well characterized in small peptides and one can generally predict exchange rates in peptides and denatured proteins with reasonable accuracy *(2,3,20,29)*. Knowledge of these rates is often very useful in studies of exchange from native proteins, because they serve as the basis for assessing quantitatively the extent to which exchange is perturbed in the protein. Methods for predicting rates as a function of sequence, temperature, and pH have been published *(1,16,30)*. We have incorporated these methods into simple BASIC and FORTRAN programs, which are available from the authors. Some simple approximations that are useful in evaluating potential experimental conditions are: Above pH 3 or so, intrinsic rates of exchange increase tenfold with each unit increase in pH, and the rates of exchange increase ≈ threefold with every 10°C increase in temperature.

2.2. Exchange from Native Proteins and Equilibrium Folding Intermediates

2.2.1. Native Proteins

Almost all studies of NH exchange from proteins involve the subset of NHs exhibiting very slow exchange, typically showing rates <100× slower than that for small peptides under the same solution conditions. This set of NH may comprise over 50% of all of the backbone NH in a protein *(20)*. The study of NH exchange from native proteins can be used to identify hydrogen-bonded secondary structure *(6)*, to monitor the consequences of ligand binding on highly localized features of protein structure *(4)*, and to determine binding constants *(31)*. Moreover, studies of exchange from native proteins are essential for any type of exchange study, since all NMR measurements are made with native protein and knowledge of native exchange behavior is used to evaluate the efficacy of any quench used in studies of dilute protein solutions or very rapid exchange *(see below)*.

The rate of NH exchange-out will depend on the "intrinsic" factors, pH, temperature, and sequence, as well as hydrogen bonding and solvent exclusion. These contributions from protein structure will, in turn, depend on local features of structure as well as the overall stability of the protein. One may estimate the slowest possible rate of exchange if the free energy of unfolding, ΔG_u, is known. In the simplest model for exchange, the "protection factor" is the reciprocal of an equilibrium constant for exposure of an NH to solvent. It then follows that the slowest possible rate of exchange will occur from sites that are only exposed to solvent after complete unfolding of the protein. In this case, the protection factor will be $\exp(\Delta G_u/RT)$, and the slowest possible rate of NH exchange is $k_{ex} \cdot [\exp(-\Delta G_u/RT)]$. This estimate can be extremely helpful in designing HX experiments.

The simplest exchange experiments involve dissolving or diluting protein into D_2O and then following the progress of exchange with proton NMR spectroscopy. A typical experiment is performed as follows:

1. One µmol of protein is dissolved in 0.6 mL of H_2O containing 0.1*M* KCl. The solution may be buffered with a nonprotonated buffer such as phosphate, with a low concentration ($\leq$ 5 m*M*) of protonated buffer, or with deuterated buffer.
2. The sample pH is adjusted to the desired value and the sample is lyophilized.
3. The dried sample is redissolved in D_2O in which the pH* (the pH measured with a standard electrode without correction for the deuterium isotope effect) has been adjusted to the desired value. Alternatively, buffer and salt can be omitted from the first step and added with the D_2O.
4. The sample is rapidly cleared of particulates, if necessary, by filtration using a low dead-volume syringe or by centrifugation.
5. The sample is transferred to the NMR tube, which is placed in the spectrometer magnet.

6. The spectrometer can be prepared for the sample to minimize shimming and tuning by using a mock sample with similar characteristics. This is a useful step if relatively fast exchange events ($\tau_{ex} \leq 5$ min) are anticipated.
7. Since exchange from individual NHs exhibits pseudo-first order kinetics, the choice of a "zero" time is somewhat arbitrary; it is usually easiest to use the start of NMR data acquisition as zero time. For very slow exchange (days), one can keep track of the elapsed time using watches and clocks. Between measurements, the sample can be removed from the magnet and stored in a water bath. If the exchange is to be measure over the course of minutes or hours, then one may be able to collect data using the "kinetics" routines that are standard on many spectrometers. In this instance, one must take into account the finite data acquisition time in order to calculate accurately the delays between time points.

Steps 2 and 3, in which exchange is initiated, may be replaced by gel filtration on a small desalting column equilibrated in D_2O. A dilution of approx threefold is anticipated during chromatography so, in favorable cases, one may load the concentrated protein in a small volume (≤0.2 mL) onto the column (1 × 4 cm) and perform the NMR analysis directly on the eluted protein. If, after chromatography, the protein is too dilute for the NMR experiment, then a concentration step *(see below)* will be necessary.

Measurement of resonance intensities can be made in "real-time" as described above, if the exchange rates of interest are slow enough (k_{ex} ≤0.3/min for 1D NMR, and $k_{ex} \leq 0.03$/min for 2D NMR). In favorable cases, the rates measurable with 2D methods may approach those of the 1D methods. With concentrated solutions of protein, for example, the lower limit for the total acquisition time may be determined not by the sensitivity but rather by phase-cycling and resolution requirements. In some cases, 2D NMR experiments may be performed without phase-cycling, in which case NMR data may be collected in as little as 5 min *(32)*.

NH exchange experiments may also be performed under conditions where exchange is very fast ($k_{ex} > 0.3$/min), the protein very dilute (< 0.1 m*M*), or both. Such experiments entail the use of a quench that, after a known time of exchange, arrests or minimizes further exchange during subsequent manipulation of the sample and acquisition of NMR data. Exchange may be quenched by reducing the pH and temperature to decrease the intrinsic rate of exchange and by increasing protein stability through changes in pH, temperature, and ionic strength. The influence of pH and temperature on protein stability can be investigated by 1D NMR methods, or with more sensitive spectroscopic methods such as UV-absorbance (Chapter 4), spectrofluorimetry (Chapter 3), and circular dichroism spectropolarimetry (Chapter 5). Quenching may also be achieved by adding a ligand. For a protein with a single binding site for ligand *X*, the free energy of unfolding will be increased by a factor of $RT^*\ln(1 + K_a[X])$, where K_a is the bimolecular association constant for *X*. Regardless of the quench conditions, it should be emphasized that no quench is complete; some exchange will occur during sample preparation and acquisition of NMR data. As a consequence, all samples from a given experiment must be handled in an identical fashion so that the extent of "post-quench" exchange is the same for all samples; the apparent first-order rate constants determined in these experiments will thus be unaffected by any residual exchange.

The following is a sample protocol for measuring exchange with dilute protein solutions; the precise conditions for exchange are likely to differ from protein to protein:

1. A 50-mL plastic conical tube, containing 4 mL of 100 μ*M* protein in H_2O and the desired buffer, is placed in a water bath set to the experimental temperature. The pH may be set to the value required for the exchange experiment, but this is not necessary since the pH can be adjusted in the following dilution step.
2. NH exchange is initiated by the addition of 36 mL of buffered D_2O equilibrated at the experimental temperature. The pH* of the buffered D_2O solution is adjusted to give the desired pH* for the exchange experiment.
3. Further exchange is quenched using the procedure previously established by the investigator. For example, the quench may be achieved through rapid adjustment of pH and temperature or addition of a stabilizing ligand or cofactor.
4. The sample volume is reduced to 0.5 mL by ultrafiltration.
5. The sample is then transferred to a 5 mm NMR tube for spectroscopic measurements or stored frozen for analysis at a later time.

If the protein sample needs to be concentrated for NMR spectroscopy, then the simplest method is lyophilization followed by dissolution of the sample in D_2O that has been adjusted to the desired pH*. Some artifactual exchange has been observed with lyophilized protein *(33)*, but this is likely to be of little consequence in most studies. Ultrafiltration may also be used to concentrate the sample and change the buffer composition.

Filtration usually will require at least 2 h to complete, so care should be taken to keep the samples as cool as possible in order to minimize further exchange; filtration may, for example, be performed in a cold room or with the filtration unit immersed in a bucket filled with ice water. The sample volume dictates the type of filtration apparatus to be used: Samples over 15 mL will require stirred cells whereas smaller samples may be concentrated with filtration units designed for use in high-speed centrifuges.

2.2.2. Interpretation of Equilibrium HX Studies

Hydrogen exchange from native proteins and equilibrium unfolding intermediates may be interpreted both qualitatively and quantitatively. One important question to consider early on is what constitutes slow exchange. Given the probable uncertainties in predicted rates of exchange, derived from data for small peptides, as well as possible perturbations of the pH dependence of chemical exchange in proteins *(3)*, it is probably safest to conclude that NH exchange in proteins is slowed significantly only if the rate is decreased greater than five- to tenfold. Lower levels of protection may be measured with confidence in well-behaved systems, such as small peptides, where one can control for electrostatic effects *(3,28)*.

In the absence of sequence-specific proton NMR assignments, slowed exchange is good evidence for hydrogen-bonded structure. This information may be of value to those investigating proteins of unknown conformation, where slowed exchange can provide evidence of a folded structure. Slowed exchange at known NH resonances can aid in the identification of hydrogen-bonded secondary structure *(6)*, and, in some instances, has provided conformational constraints suitable for inclusion in high-resolution structure determinations *(34)*. Slowed exchange from equilibrium intermediates in unfolding has provided valuable information about residual native-like hydrogen bonded structure in these intermediates *(35–37)*.

Quantitative evaluation of NH exchange data has been the subject of intense discussion for a number of years *(4,5)*. Much of the controversy centers on the nature of the fluctuations in protein structure that permit exchange with solvent, but most investigators agree that under conditions where native protein is marginally stable, exchange occurs by way of unfolding *(38)*. A simple two-state model for exchange at individual NHs in proteins is:

$$\mathrm{C(H)} \underset{k_{-1}}{\overset{k_1}{\rightleftharpoons}} \mathrm{O(H)} \underset{D_2O}{\overset{k_{ex}}{\longrightarrow}} \mathrm{O(D)} \underset{k_1}{\overset{k_{-1}}{\rightleftharpoons}} \mathrm{C(D)} \quad (1)$$

where *C* and *O* refer to the closed and open sites, respectively *(39)*. Exchange of a proton (H) with a solvent deuteron (D) takes place only from the *O* state and occurs at the rate k_{ex}, characteristic of model compounds. The rate constants k_1 and k_{-1} refer to the opening and closing reactions, respectively. In the presence of a vast excess of D_2O, the reaction need not be considered beyond the chemical exchange step because all *O*(D) is converted to *C*(D) and there is no detectable back reaction. Assuming a rapid equilibrium between *C* and *O*, the overall rate of exchange, k_{obs}, is described by the following equation:

$$k_{obs} = (k_1(k_{ex})/(k_{-1} + k_{ex}) \quad (2)$$

Under the solution conditions employed in most studies of exchange from native proteins, $k_{ex} \ll k_{-1}$, and Eq. (2) reduces to:

$$k_{obs} = k_{ex}(k_1/k_{-1}) = k_{ex}K_o \quad (3)$$

where K_o refers to the equilibrium constant for the opening reaction. Measurement of k_{obs} and knowledge of k_{ex} from model compound data permits determination of K_o. From this one can describe the free energy of the opening reaction ΔG_o, which is simply $-RT\ln K_o$. In using this appealing approach, one must recognize that it involves a number of assumptions, some of which can be verified by experiment *(38)*. The model, nevertheless, has proved to be a fruitful framework for interpreting NH exchange from proteins.

Two mechanisms have been proposed to explain NH exchange under more strongly native conditions: local unfolding and solvent penetration. Recent studies of NH exchange in RNase A suggest that examination of exchange as a function of low concentration of chemical denaturants may provide a measure of the contribution of the two mechanisms to exchange *(40)*. The results suggest that both mechanisms are operative in native proteins.

2.3. Hydrogen Exchange in Kinetic Studies of Protein Folding

2.3.1. Experimental Design

The general experiment uses hydrogen exchange as a method of monitoring the protein folding process by looking at the ability of structure acquisition to retard exchange with bulk solvent. The basic premise of the experiment is that a fully unfolded protein will exhibit hydrogen exchange with the chemical exchange rate, k_{ex}, observed for unstructured peptides. As the protein refolds, hydrogen-bonded secondary structure develops and regions of the protein become less exposed to solvent. This results in a decrease in the observed hydrogen exchange rates for the amide protons that are buried and/or involved in stable structure. The degree to which exchange is decreased is related to the stability of the structure present during the exchange process.

The two critical variables to consider in the design of an experiment using hydrogen exchange to follow the kinetics of protein refolding are pH and temperature. The chemistry of hydrogen exchange is such that the effects of pH and temperature are known for the chemical exchange (k_{ex}) of unstructured peptides. Refolding is generally initiated by a dilution of a chemical denaturant; fortunately, the effects of low concentrations of denaturants on the chemical exchange rate are small and have been measured for some model compounds *(41)*. Temperature changes are generally not used to initiate refolding or the exchange process, but are quite useful in quenching hydrogen exchange. For the HX step, high pH is used, whereas the quench is generally performed at low pH. The solubility, stability, and the kinetics of refolding for the protein under all of these conditions must be known in order to design a successful experiment and to interpret the resulting data.

Since protein refolding generally occurs on the millisecond time scale (*see* Chapter 14), specialized equipment is required to afford the necessary mixing and dilutions for a kinetic analysis of HX. There are several sources of rapid-mixing devices, including BioLogic, Update Instruments, and Hi-Tech Scientific. In addition to these commercial sources, several rapid-mixing devices have been constructed by individual investigators *(42,43)*. The general features of these devices are all similar in that they allow for several mixing events separated by various time delays. For many of the kinetic refolding experiments, it is desirable to have three independent mixing events separated by variable delays. The mixing events need to accommodate both large (~10-fold) and small dilutions, whereas the delay times range from very short (ms) to rather long times (min). Regardless of the type of instrument employed, it is critical to calibrate the instrument in order to determine dead times, mixing times, dead volumes, and dilution volumes. A general calibration procedure has been described *(42)*.

For the techniques described below, we will assume that NMR is to be used to quantify the proton occupancy of individual amide protons. Therefore, the same considerations about sample amount, volume, and solution conditions for the NMR experiment that were discussed above apply here as well. In addition, concentration of the sample is a necessary component of kinetic experiments, since one usually has a large volume of dilute protein. The sample concentration methods discussed above are also used in the kinetic experiment.

2.3.2. Competition Method

In this experiment, HX and refolding are allowed to occur simultaneously. The basic experiment is to take a solution of unfolded, fully deuterated protein *(see* Section 2.3.5.*)* in concentrated chemical denaturant and initiate refolding and exchange by a dilution into H_2O. Usually a 1:10 dilution is employed so as to increase the H_2O:D_2O ratio to ~10:1. This dilution is also used to decrease the concentration of chemical denaturant so folding will commence. A change in pH along with a dilution of the denaturant can also be used to initiate refolding. The key is to jump through the unfolding transition and arrive at refolding conditions that are very native-like so as to increase the probability of detecting folding intermediates (*see* Chapter 14). After a variable refolding time, a quench is applied to minimize further exchange. The quench is identical to that described above in terms of temperature and pH; the only other consideration is the necessity for a rapid quench afforded by the rapid-mixing device. The sample is then concentrated and prepared for NMR analysis in a fashion similar to that described above.

Since refolding and HX occur simultaneously, a complete understanding of the data requires a knowledge of both of these processes at the selected pH and temperature:

$$\begin{array}{ccc} U^* & \xrightarrow{k_f} & N^* \\ k_{ex}\downarrow & & \\ U & \xrightarrow{k_f} & N \end{array} \qquad \frac{N^*}{N} = \frac{k_f}{k_{ex}}$$

N and *U* are the native and unfolded forms of the protein, respectively, and the asterisk represents deuterated protein. Both the exchange rate constant, k_{ex}, and the folding rate constant, k_f, depend on the solution conditions, but usually not to the same extent. For a given temperature, we know how k_{ex} varies with pH, so pH is used to balance k_f/k_{ex} to arrive at a suitable experimental time range. Generally, rather long times are used for the refolding and HX steps to assure that hydrogen exchange has gone to completion. The basic method, along with several applications, has been reviewed *(13,14,38)* and the experiment has been applied to the study of a variety of proteins *(44,45)*.

The simplest application of this method is to use a single refolding/HX pH and vary the refolding/HX time in order to look at the time course for exchange of individual amides. This method has been described in detail by Roder *(38)*. Noncoincidence of the exchange data for individual amide protons indicates the presence of kinetic intermediates on the folding pathway. An inspection of the structural disposition of amides with similar kinetics of exchange may indicate which elements of secondary structure are forming first in the folding process.

The stability of any kinetic intermediates that are observed can be probed by changing the pH of the refolding/HX step or by varying the amount of chemical denaturant present during the refolding step. Again, a successful interpretation of the data requires accurate values for k_{ex} and k_f under the refolding conditions since pH and the presence of chemical denaturants will affect both kinetic processes. A potential problem in this second approach is that the mechanism or pathway for folding may change depending on the solution conditions employed. Therefore, it is desirable to look at the time course for exchange of all of the individual amides under all refolding conditions in order to determine if the pathway for folding is changing with solution conditions.

Since refolding and HX occur simultaneously in the competition method, this method is especially good for detecting early refolding intermediates. The principal advantages of the method over the Pulsed HX methods *(see* Section 2.3.3.*)* are:

1. Fewer mixing events are used;
2. One can look at early events in refolding; and
3. It is possible to determine if the pathway for folding changes with solution conditions.

The chief disadvantages of the method are:

1. One is generally limited to studying refolding at lower pH where k_{ex} is small;
2. It is difficult to study late intermediates in refolding;
3. Changing the pH affects both k_{ex} and k_f;
4. The refolding pathway may change with pH; and
5. The stability of refolding intermediates may change with pH.

2.3.3. Pulsed HX Methods

In contrast to the competition method, the pulsed HX methods can take "snapshots" of the protein at any time during folding. The basic experiment has been reviewed in a number of excellent articles *(14,38)* and the technique has been applied to a variety of different proteins *(20,37,46–52)*. Refolding is allowed to occur for a variable period of time, under any solution conditions, before the labeling pulse is applied. The labeling pulse is generally of short duration but at pH values that are high enough to label any hydrogens that are not involved in stable structure. The duration and pH of the labeling pulse can be adjusted to achieve the desired pulse intensity as defined by Englander and Mayne *(14)*. The labeling step is terminated by another mixing event that quenches exchange by lowering the pH and often the temperature, and folding is allowed to go to completion. There are two basic modifications of this technique; one that allows a description of the time course of the development of stable structure and one that looks at the stability of structure that is observed in kinetic refolding intermediates.

In the first modification, the kinetics of the acquisition of protection are determined by applying the labeling pulse at variable times after refolding has been initiated. The intensity of the HX pulse is held constant at a value that assures labeling of all free hydrogens without affecting the exchange of amide protons involved in stable structure. By varying the delay before the application of the labeling pulse, one can determine the time course for the development of protection. Different amide protons might show different kinetics, suggesting a pathway for folding (*see* Section 2.3.4.)

In order to determine the stability of kinetic intermediates that form at a given time after refolding has been initiated, the intensity of the HX pulse can be varied. In this method, the HX pulse is applied at a constant time after refolding has been initiated, but the pH or the duration of the HX pulse is varied. The stability of protected hydrogens can be determined from the variation in exchange rate with the intensity of the labeling pulse. In combination with the previous experiment, one can look at the time course for acquisition of protection as well as the stability of the intermediate that is giving rise to the observed protection.

The general procedure is similar in many ways to that outlined for the competition method, however, an additional mixing step is needed and, therefore, accurate calibration of the instrument is critical. The protein is unfolded in denaturant and D_2O, thus deuterating all the peptide amides. In the first mixing step, a dilution of the chemical denaturant is performed to initiate folding. This dilution can be with H_2O, provided the pH and temperature are low enough to avoid exchange. If higher refolding pH values are necessary, or if you want to look at the HX behavior after longer refolding times, the refolding dilution must be with D_2O. After a variable refolding time, an exchange pulse is applied in the form of a sharp increase in pH in H_2O. The short labeling pulse is terminated by quenching to low pH and temperature. Refolding is allowed to go to completion and the sample is prepared for NMR analysis.

The principal advantage of the pulsed HX method is that refolding can be performed under solution conditions that are more likely to stabilize folding intermediates (low temperatures, moderate pH, and so on). The HX pulse (at higher pH) is short enough (10–50 ms) that it is not likely to affect k_f significantly. Also, one can look at refolding in the presence of additional cofactors, ligands, chaperones, and so on under conditions where they are more likely to have an effect on protein folding. This method is also well-suited to the study of intermediates that form late in the folding process, in contrast to the competition method described in Section 2.3.2.

The chief disadvantage of the pulsed HX method is that it is difficult to look at very short refolding times since two mixing events must occur prior to the labeling step. The earliest times are dictated by the dead time of the instrument and the efficiency of the mixing events, but are typically in the 10–20 ms range. This is becoming a problem for many proteins that appear to achieve a significant level of protection in the dead time of this experiment. Since more mixing and dilution steps are required, one often ends up with a large volume of protein solution that must be concentrated prior to NMR analysis. Care must be taken to minimize further exchange during the preparation of the sample for NMR analysis.

2.3.4. Data Analysis and Interpretation

After quantification of the proton occupancy of each of the time points of interest for all of the amide protons, one determines the rate constants and amplitudes for each observed kinetic phase. The observed rate constants are then normalized using the chemical exchange rate (k_{ex}) for the particular residue in question under the solutions conditions used in the HX step. If all the protons show a single kinetic phase with a relative amplitude of unity, there are no kinetic intermediates populated and the refolding reaction is a simple two-state reaction.

If, however, one observes different protection curves for separate amide protons, a kinetic intermediate must be involved in the reaction. This is analogous to the ratio test for equilibrium folding intermediates where different probes of structure reveal different unfolding transitions. By investigating the structural disposition of amide protons that show similar protection curves, one can determine if specific elements of secondary structure, or subdomains of the protein, are behaving as structure units. The stability of each of these structural units can be determined by comparing the protection curves generated from experiments performed with different HX intensities *(14)*.

A more complicated picture emerges when one observes multiple kinetic phases for an individual amide proton. This multiphasic behavior usually results from heterogeneity in the unfolded state with independent parallel folding pathways. A comparison of the relative amplitudes and protection profiles for different amide protons will determine if kinetic intermediates are observed. If the unfolded population is heterogeneous, and each of the unfolded conformations refold in a concerted fashion, then all of the amide protons will exhibit identical protection curves in terms of both the amplitudes and time constants for all of the observed phases. If, on the other hand, there is an intermediate on one or more of the parallel folding pathways, the protection curves for separate amide protons will not be coincident. The stability of these kinetic intermediates can be determined by comparing the protection profiles generated from experiments performed with different HX intensities.

2.3.5. Example of the Pulsed HX Method

As an example of the pulsed HX method, it is instructive to look at the approach used by Udgaonkar and Baldwin *(50,52)* for ribonuclease A (RNase A):

1. Fully deuterated RNase A was prepared by dissolving the protein in D_2O at pH* 3 and heating the solution to 60°C for 20 min. The deuteration was repeated after lyophilization to achieve complete exchange.
2. The deuterated RNase A was unfolded in 2.65*M* GuHCl, 40 m*M* glycine in D_2O at pH* 2.0 at a protein concentration of 60–70 mg/mL.
3. Refolding was initiated by a 10.5-fold dilution of 0.1 mL of the unfolded RNase A solution into a refolding buffer containing 0.442*M* sodium sulfate, 0.055*M* sodium formate, in H_2O at pH 4.25.
4. After various refolding times, an exchange pulse is applied in the form of a 1.5-fold dilution into the HX buffer containing 0.4*M* sodium sulfate, 0.25*M* GuHCl, 0.1*M* glycine in H_2O so the final pH was 9 or 10.
5. The 37 ms exchange pulse was terminated by diluting 1.33-fold into a quench buffer containing 0.4*M* sodium sulfate, 0.25*M* GuHCl, and 0.1*M* sodium formate so the final pH was 2.9.
6. Refolding was allowed to proceed to completion (≈10 min) at this pH. For all three mixing steps, the deadtime was approx 5 ms and the entire experiment was performed at 10°C.
7. For the zero time point, the HX pulse was applied directly to the unfolded protein solution. The samples from 5 or 6 different experimental trials were combined and the protein solution was concentrated and prepared for NMR analysis using ultrafiltration methods.

This protocol represents a general approach for the pulsed HX method used to determine the time course of acquisition of protection. The protein is initially deuterated and unfolded with GuHCl at low pH. Folding is initiated by a large dilution into H_2O at pH 4.0 where the chemical exchange rate is slow compared with refolding. The sodium sulfate is used to provide conditions that favor the formation of kinetic intermediates during this refolding step. Since refolding is performed in H_2O, the HX pulse only needs to increase the pH to 9 or 10 with a small volume of concentrated glycine buffer. Exchange is quenched by the addition of a concentrated sodium formate solution at low pH, such that the resulting pH is 2.9, the minimum for chemical exchange of free peptide amide protons. Refolding is allowed to proceed to completion at this pH. The HX and quench buffers also contain the same concentrations of GuHCl and sodium sulfate present in the refolding step.

Whereas the exact conditions required for all of the steps of a refolding HX experiment will depend on the particular protein involved, the basic ideas are all similar. For some proteins, it may not be possible to use the extremes of pH outlined above and alternate methods for the quench and HX steps will be required. In spite of the apparent complexities in the kinetic refolding experiments, these are extremely powerful techniques; techniques that can provide information on the protein folding pathway at the level of individual residues. As further advances in NMR spectroscopy and rapid-mixing technology become available, these hydrogen exchange experiments will prove valuable in shedding more light on the protein folding problem.

References

1. Molday, R. S., Englander, S. W., and Kallen, R. G. (1972) *Biochemistry* **11**, 150–158.
2. Roder, H., Wagner, G., and Wüthrich, K. (1985) *Biochemistry* **24**, 7407–7411.
3. Robertson, A. D. and Baldwin, R. L. (1991) *Biochemistry* **30**, 9907–9914.
4. Woodward, C. K. and Hilton, B. D. (1979) *Ann. Rev. Biophys. Bioeng.* **8**, 99–127.
5. Englander, S. W. and Kallenbach, N. R. (1984) *Q. Rev. Biophys.* **16**, 521–655.
6. Wüthrich, K. (1986) *NMR of Proteins and Nucleic Acids*. Wiley, New York.
7. Robertson, A. D. and Markley, J. L. (1990) *Biological Magnetic Resonance* **9**, 155–176.
8. McIntosh, L. P. and Dahlquist, F. W. (1990) *Q. Rev. Biophys.* **23**, 1–38.
9. Clore, G. M. and Gronenborn, A. M. (1991) *Annu. Rev. Biophys. Chem.* **20**, 29–63.
10. Paterson, Y., Englander, S. W., and Roder, H. (1990) *Science* **249**, 755–759.
11. Englander, S. W., Englander, J. J., McKinnie, R. E., Ackers, G. K., Turner, G. J., Westrick, J. A., and Gill, S. J. (1992) *Science* **256**, 1684–1687.
12. Rosa, J. J. and Richards, F. M. (1981) *J. Mol. Biol.* **145**, 835–851.
13. Kim, P. S. (1986) *Methods Enzymol.* **131**, 136–156.
14. Englander, S. W. and Mayne, L. (1992) *Annu. Rev. Biophys. Biomol. Struct.* **21**, 243–265.
15. Tao, F., Fuchs, J., and Woodward, C. (1993) in *Techniques in Protein Chemistry* (Angeletti, R., ed.), Academic, New York, pp. 513–521.
16. Englander, S. W. and Poulsen, A. (1969) *Biopolymers* **7**, 379–393.
17. Aue, W. P., Bartholdi, E., and Ernst, R. R. (1976) *J. Chem. Phys.* **64**, 2229–2246.
18. Pedersen, T. G., Thomsen, N. K., Andersen, K. V., Madsen, J. C., and Poulsen, F. M. (1993) *J. Mol. Biol.* **230**, 651–660.

19. Chyan, C.-L., Wormald, C., Dobson, C. M., Evans, P. A., and Baum, J. (1993) *Biochemistry* **32**, 5681–5691.
20. Lu, J. and Dahlquist, F. W. (1992) *Biochemistry* **31**, 4749–4756.
21. Bax, A., Ikura, M., Kay, L. E., Torchia, D. A., and Tschudin, R. (1990) *J. Magn. Reson.* **86**, 304–318.
22. Kainosho, M. and Tsuji, T. (1982) *Biochemistry* **21**, 6273–6279.
23. Henry, G. D., Weiner, J. H., and Sykes, B. D. (1987) *Biochemistry* **26**, 3626–3634.
24. Ferretti, J. A. and Weiss, G. H. (1989) *Methods Enzymol.* **176**, 3–11.
25. Dempsey, C. E. (1986) *Biochemistry* **25**, 3904–3911.
26. Wagner, G. and Wüthrich, K. (1982) *J. Mol. Biol.* **160**, 343–361.
27. Wand, A. J. and Englander, S. W. (1985) *Biochemistry* **24**, 5290–5294.
28. Kim, P. S. and Baldwin, R. L. (1982) *Biochemistry* **21**, 1–5.
29. Radford, S. E., Buck, M., Toping, K. D., Dobson, C. M., and Evans, P. A. (1992) *Proteins: Struct. Funct. Genet.* **14**, 237–248.
30. Englander, J. J., Calhoun, D. B., and Englander, S. W. (1979) *Anal. Biochem.* **92**, 517–524.
31. Schreier, A. A. and Baldwin, R. L. (1976) *J. Mol. Biol.* **105**, 409–426.
32. Marion, D., Ikura, M., Tschudin, R., and Bax, A. (1989) *J. Magn. Reson.* **85**, 393–399.
33. Robertson, A. D., Purisima, E. O., Eastman, M. A., and Scheraga, H. A. (1989) *Biochemistry* **28**, 5930–5938.
34. Meadows, R. P., Nettesheim, D. G., Xu, R. X., Olejniczak, E. T., Petros, A. M., Holzman, T. F., Severin, J., Gubbins, E., Smith, H., and Fesik, S. W. (1993) *Biochemistry* **32**, 754–765.
35. Baum, J., Dobson, C. M., Evans, P. A., and Hanley, C. (1989) *Biochemistry* **28**, 7–13.
36. Hughson, F. M., Wright, P. E., and Baldwin, R. L. (1990) *Science* **249**, 1544–1548.
37. Jeng, M. F., Englander, S. W., Elöve, G. A., Wand, A. J., and Roder, H. (1990) *Biochemistry* **29**, 10,433–10,437.
38. Roder, H. (1989) *Methods Enzymol.* **176**, 446–473.
39. Hvidt, A. and Nielsen, S. O. (1966) *Adv. Prot. Chem.* **21**, 287–386.
40. Mayo, S. L. and Baldwin, R. L. (1993) *Science* **262,** 873–876.
41. Loftus, D., Gbenle, G. O., Kim, P. S., and Baldwin, R. L. (1986) *Biochemistry* **25**, 1428–1436.
42. Cash, D. J. and Hess, G. P. (1981) *Anal. Biochem.* **112**, 39–51.
43. Fersht, A. R. and Jakes, R. (1975) *Biochemistry* **14**, 3350–3356.
44. Bycroft, M., Matouschek, A., Kellis, J. J., Serrano, L., and Fersht, A. R. (1990) *Nature* **346**, 488–490.
45. Dobson, C. M. (1991) *Ciba Found. Symp.* **161**, 167–181.
46. Briggs, M. S. and Roder, H. (1992) *Proc. Natl. Acad. Sci. USA* **89**, 2017–2021.
47. Mullins, L. S., Pace, C. N., and Raushel, F. M. (1993) *Biochemistry* **32**, 6152–6156.
48. Matouschek, A., Serrano, L., Meiering, E. M., Bycroft, M., and Fersht, A. R. (1992) *J. Mol. Biol.* **224**, 837–845.
49. Radford, S. E., Dobson, C. M., and Evans, P. A. (1992) *Nature* **358**, 302–307.
50. Udgaonkar, J. B. and Baldwin, R. L. (1988) *Nature* **335**, 694–699.
51. Roder, H., Elöve, G. A., and Englander, S. W. (1988) *Nature* **335**, 700–704.
52. Udgaonkar, J. B. and Baldwin, R. L. (1990) *Proc. Natl. Acad. Sci. USA* **87**, 8197–8201.

CHAPTER 14

Protein Folding Kinetics

Thomas Kiefhaber

1. Introduction

The aim of protein folding studies is to understand the relationship between the information encoded in the linear amino acid sequence of a polypeptide chain and its three-dimensional structure. The elucidation of the kinetic folding mechanism of a protein is the first step on the way to characterize its complete folding pathway. Subsequent steps comprise the characterization of transiently formed intermediates and of the transition states between the various states of the protein. A folding pathway is understood when all transient intermediates and the transition states between them are characterized. This chapter deals with several aspects of the process of determining folding pathways. The elucidation of kinetic folding mechanisms under various solvent conditions will be described and methods used in the characterization of the transition states for individual folding reactions will be discussed. The detailed characterization of folding intermediates is described in Chapters 13 and 15.

In recent years the investigation of folding reactions has mainly focused on small, single domain proteins as simple model systems. In most cases folding and unfolding of these proteins is reversible and devoid of side reactions such as aggregation processes that compete with proper folding. This reversibility is an obligatory prerequisite for the mechanistic interpretation of folding kinetics. Furthermore, the small size of the investigated proteins allows the molecular interpretation of observed folding steps. This has been strongly facilitated in recent years by the use of site-directed mutagenesis to determine the influence of specific parts of a polypeptide chain on individual folding steps. The folding

From: *Methods in Molecular Biology, Vol. 40: Protein Stability and Folding: Theory and Practice*
Edited by: B. A. Shirley

of large, multi-domain proteins has recently been addressed in reviews by Jaenicke *(1)* and Garel *(2)*.

Folding transitions of small proteins are usually highly cooperative and partially folded intermediates are not detected under equilibrium conditions (*3*; exceptions are discussed in Chapter 15). Therefore, the investigation of partially folded protein structures has mainly focused on kinetic studies of refolding and unfolding reactions with the aim of detecting transient intermediates and of characterizing the reactions that lead to their formation and their subsequent conversion to native protein.

Protein folding kinetics are often complex for two major reasons: The unfolded states of proteins can be heterogeneous, which leads to different forms of unfolded proteins that refold on parallel pathways, and folding intermediates can be formed under certain solvent conditions and are subsequently converted to native protein in a sequential process.

A heterogeneity in unfolded proteins was first observed by Garel and Baldwin *(4)*, who discovered that fast and slow refolding molecules exist in unfolded ribonuclease A (RNase A). As a molecular interpretation the *cis-trans* isomerization of Xaa–Pro peptide bonds was proposed in 1975 *(5)*. Unlike other peptide bonds, where the trans isomer is strongly favored over the *cis* conformation, in Xaa–Pro peptide bonds the amount of *cis* peptide conformers in random polypeptides is about 10–40% under equilibrium conditions *(6)*. The activation energy for *cis-trans* isomerization is large (~20 kcal/mol), giving rise to low rates of interconversion of the two isomers. This has major implications for the protein folding process. In native states of proteins Xaa–Pro peptide bonds are fixed in one of the two isomers (although exceptions to this rule have been reported). After the protein unfolds, these bonds become free to isomerize and will establish an equilibrium between *cis* and *trans* isomers. Refolding of the polypeptide chains will thus start from a heterogeneous mixture of unfolded molecules, each of the species having a different set of Xaa–Pro isomers. The number of different unfolded species consequently rises with the number of Xaa–Pro peptide bonds in a protein. In refolding the different unfolded species can reach the native state on parallel pathways. Unfolded molecules with the same set of prolyl isomers as the native protein can usually refold rapidly (fast folding molecules; U_F), whereas unfolded molecules with one or more incorrect prolyl isomers have to undergo slow *cis-trans* isomerization reactions before they can reach the native conformation (slow folding molecules; U_S).

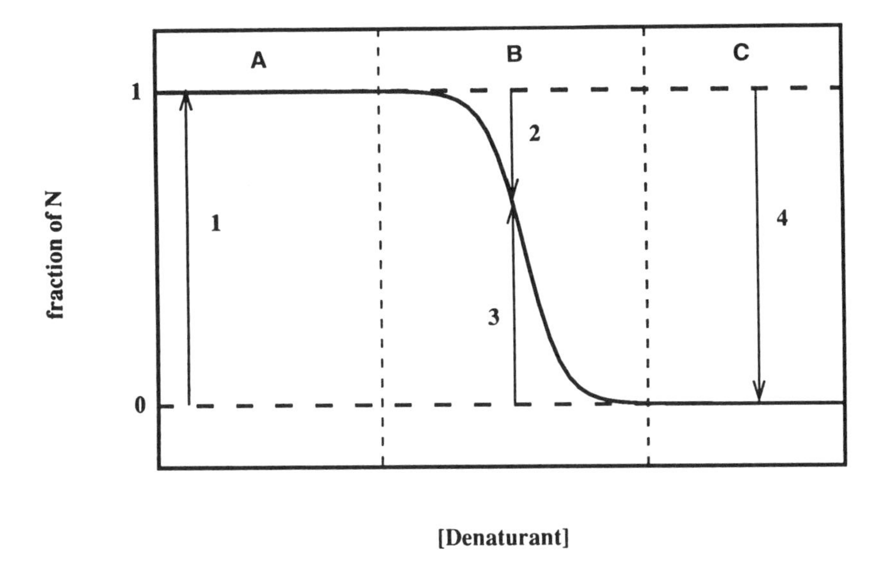

Fig. 1. Schematic denaturant-induced unfolding transition of a small globular protein that follows two-state behavior. The solid line represents the denaturant dependence of the fraction of native protein; dashed lines are the extrapolated baselines for the native and the unfolded state. The regions of the transition representing the native baseline (**A**), the transition region (**B**), and the unfolded baseline (**C**) are indicated. The arrows represent the different kinds of folding experiments discussed in the text. (1) represents refolding kinetics under strongly native conditions, (2) and (3) represent unfolding and refolding kinetics under identical conditions in the transition region, respectively, and (4) are unfolding kinetics under unfolding conditions.

These isomerization reactions can represent late or early steps in the refolding process, depending on the refolding conditions. The role of prolyl isomerization reactions in the refolding of several proteins has meanwhile been demonstrated *(7–10)*. Methods of detecting the influence of prolyl isomerization reactions on the refolding kinetics under various solvent conditions will be discussed in detail in Section 2.

The folding pathway of a protein is strongly dependent on the experimental conditions. A denaturant-induced unfolding transition of a protein can be divided into three different regions of stability (Fig. 1). The different stability regions determine the properties of the observed protein folding reactions and each of them allows the characterization of different aspects of the folding mechanism.

1. In the region of the native baseline the native folded structure of a protein is favored and consequently also partially folded structures are stabilized. Therefore, refolding kinetics under these conditions are usually complex reactions involving transient folding intermediates. A common picture for folding of small proteins under native solvent conditions has emerged in the last few years. In a very rapid step intermediates can be formed that already contain part of the native secondary structure but are not stabilized by specific tertiary side chain interactions. The formation of the specific tertiary contacts occurs in subsequent slower folding steps.
2. In the region of the unfolding transition the stability of the native state is only marginal and thus partially folded intermediates, which are by definition less stable than the native conformation, are usually not observed. Under these conditions the native and the unfolded state of a protein are both present in significant amounts. A special feature of this region is that both unfolding and refolding kinetics can be monitored under identical final conditions, depending on whether folding is started from native or from unfolded molecules.
3. The unfolded form of a protein is the most stable state in the region of the unfolded baseline. Here, unfolding can be monitored after transferring native protein into solutions of high concentrations of denaturants where the denatured state is thermodynamically stable.

It has been a point of discussion in recent years whether protein folding is under kinetic or under thermodynamic control, i.e., whether the final state of a folding reaction is the thermodynamically most stable state or whether some other folding pathway might reach a state of lower free energy. Evidence from folding experiments clearly indicate that the final state of refolding experiments is in most cases independent of the refolding conditions and thus independent of the folding pathway. The same native conformation is reached under strongly native conditions where transient intermediates are formed and under marginally native conditions where no partially folded structures are detected. Additionally, there is an obvious need for a protein to be able to reach the native state very rapidly in order to prevent competing side reactions such as aggregation or degradation by proteases in the cell. It is therefore plausible to assume that most protein pathways have evolved that lead to the thermodynamically lowest state of energy and also allow folding in a very short time *(11)*. Interesting exceptions are secreted proteases that require a pro sequence (that is cleaved off when the mature enzyme is formed) to refold *(12,13)* and protease inhibitors of the serpin family that refold to a metastable, active form that is slowly converted to a more stable, but inactive state *(14,15)*. During in vivo folding additional proteins (molecular chaperones) are present to protect refolding proteins from competing side reactions (*see* Chapter 16). The fact that the native states of in vivo and in vitro folding are identical shows, however, that these helper proteins do not direct the folding pathway and folding still leads to the thermodynamically most stable conformation.

Various kinetic and thermodynamic treatments of protein folding reactions have been discussed in classical reviews by Tanford *(16)* and have been followed up in reviews by Baldwin *(17)*, Kim and Baldwin *(18,19)*, Jaenicke *(1)*, and Schmid *(11)*. Several practical aspects of protein folding kinetics have been described in vol. 131 of *Methods in Enzymology (20)* and by Rudolph and Jaenicke *(21)*.

2. Methods

It should be kept in mind that kinetic methods used to characterize chemical reactions can never prove a mechanism but rather can exclude mechanisms that are not consistent with the observed data. Kinetic folding studies will thus help to find the simplest folding mechanism that is in agreement with the experimental data.

2.1. Analysis of Kinetic Data

2.1.1. General Treatment of Kinetic Data

A major advantage in the analysis of folding reactions of small, single domain proteins is that all reactions are first order, which simplifies the data analysis. From the analysis of kinetic experiments the number *(n)* of observable kinetic phases, their relaxation times ($\tau_i = 1/\lambda_i$; where λ_i is the apparent rate constant) and their respective amplitudes (A_i) can be determined. In the case of unimolecular reactions the observed change in signal with time can be represented as a sum of *n* exponentials:

$$B_t - B_\infty = \sum_{i=1}^{n} A_i e^{-\lambda_i t} \tag{1}$$

where B_t is the observed signal at the time t_i and B_∞ is the signal after the reaction is complete. For the comparison of kinetic data the amplitudes are often normalized and given as the fraction (α_i) of total observed signal change between the native and the unfolded state of a protein, which can be measured in equilibrium unfolding transitions:

$$\alpha_i = A_i / [B(N) - B(U)] \quad (2)$$

where B(N) is the signal of the native state and B(U) the signal of the unfolded state under the conditions where the folding experiments have been performed. This procedure allows the comparison of kinetic data monitored with different probes.

Generally, folding pathways can be represented as consecutive, reversible, first order reactions with x different intermediate states:

$$U \underset{k_{21}}{\overset{k_{12}}{\rightleftharpoons}} I_1 \underset{k_{32}}{\overset{k_{23}}{\rightleftharpoons}} \cdots \rightleftharpoons I_x \underset{k_{z(z-1)}}{\overset{k_{(z-1)z}}{\rightleftharpoons}} N \quad (3)$$

A pathway with z different species is characterized by $2(z - 1)$ microscopic rate constants (k_{ij}) and $z - 1$ nonzero apparent rate constants (λ_i). The apparent rate constants are determined by the final conditions of the system but are independent of the initial conditions. They are complex functions of the microscopic rate constants, which can be determined only in special cases. Their respective amplitudes, however, are functions of both the initial and the final conditions. The rigorous solution of a system of reversible first order reactions is described elsewhere *(22,23)*.

The kinetic treatment of kinetic schemes with off-pathway intermediates of the scheme:

$$\begin{array}{ccc} U & \underset{k_{21}}{\overset{k_{12}}{\rightleftharpoons}} & N \\ k_{13} \upharpoonleft\downharpoonright k_{31} & & \\ I & & \end{array} \quad (4)$$

is similar to the kinetic mechanisms described above and has been discussed in detail by Ikai and Tanford *(24)* and by Utijama and Baldwin *(25)*.

2.1.2. Determination of Time Constants and Amplitudes of Folding Reactions

Experimental kinetic curves can be analyzed by linearization of the data. A plot of ln ($B_t - B_\infty$) as a function of the refolding time yields a straight line for monophasic unimolecular reactions *(26)*. An easier and more precise way to analyze kinetic data is, however, the use of commercially available nonlinear least squares fitting programs. These programs allow fitting the data to different mechanisms and the quality of a fit can be judged by looking at the residuals, i.e., the deviation of the experimental data from the fitted values. This procedure is definitely preferable if more than one folding phase is detected.

2.2. Folding Reactions in the Native Baseline Region

The native baseline region represents solvent conditions that favor folded structures. As a consequence, partially folded transient intermediates can often form rapidly under these conditions. The presence of transient intermediates leads to complex refolding kinetics, where typically several fast and slow reactions can be distinguished. Another origin for the complexity of refolding reactions in the native baseline region can arise from a heterogeneous unfolded state caused by *cis-trans* equilibria of Xaa–Pro peptide bonds. These different unfolded species can refold on parallel pathways. Under strongly native conditions, unfolded molecules with nonnative prolyl isomers can often form native-like intermediates that can be partially active and differ from the native state mainly in the isomerization state of a Xaa–Pro peptide bond *(23,27)*. Their rate of conversion to native protein is very slow, since it is limited by a prolyl isomerization reaction. Thus, prolyl residues can act as effective traps to slow down late steps in folding and thus facilitate the characterization of transient folding intermediates. The formation of structure prior to proline isomerization can have opposite effects on the isomerization rate. In the case of RNase A the *trans* to *cis* isomerization of Pro 93 is accelerated in a native-like folding intermediate *(28)*, whereas in RNase T1 the rate of *trans* to *cis* isomerization of Pro 39 is decelerated in a partially folded form *(29)*.

A general kinetic scheme for protein folding reactions under native solvent conditions, considering several parallel pathways with x unfolded species, is given by Eq. (5):

$$
\begin{array}{c}
U_{\mathrm{F}} \rightleftharpoons I^{1}{}_{\mathrm{F}} \rightleftharpoons \ldots \rightleftharpoons I^{n}{}_{\mathrm{F}} \rightleftharpoons N \\
U_{\mathrm{S(I)}} \rightleftharpoons I^{1}{}_{\mathrm{S(I)}} \rightleftharpoons \ldots \rightleftharpoons I^{n}{}_{\mathrm{S(I)}} \rightleftharpoons N \\
\cdot \\
\cdot \\
\cdot \\
U_{\mathrm{S(x)}} \rightleftharpoons I^{1}{}_{\mathrm{S(x)}} \rightleftharpoons \ldots \rightleftharpoons I^{n}{}_{\mathrm{S(x)}} \rightleftharpoons N
\end{array}
\qquad (5)
$$

where U_F are fast-folding molecules and U_S are slow-folding molecules that differ in the isomerization state of prolyl residues, and I^1 to I^n are partially folded intermediates.

A special feature of folding reactions in the native baseline region of a folding transition is that the folding rates are much higher than the unfolding rates so that the reverse reactions can usually be neglected for late steps on the folding pathway. Rapidly formed intermediates are, however, often only marginally stable and in rapid equilibrium with the unfolded state.

Usually, not all unfolded species contribute to the observed folding reactions to the same extent. Prolyl peptide bonds that are *cis* in the native state isomerize to a large extent to their nonnative *trans* form in the unfolded state and thus generally have a much larger amplitude in refolding reactions than the isomerization of trans prolyl peptide bonds. Thus, if the native state of a protein contains *cis* prolyl peptide bonds, the contribution of *trans* proline residues to the refolding kinetics under native conditions can often be neglected.

A general procedure to determine the kinetic mechanism of folding of a protein under native conditions should comprise the following steps:

1. Choose the experimental conditions and denaturant.
2. Perform refolding studies and monitor the resulting kinetics with various spectroscopic and functional probes and determine the apparent rate constants and their amplitudes.
3. Measure equilibrium unfolding transition curves under the same conditions and with the same probes and compare the maximum difference between *N* and *U* to the amplitude of the kinetic experiments in order to detect fast reactions that occur within the dead time of mixing.
4. Perform experiments to distinguish between parallel and sequential pathways and to detect transient folding intermediates.
5. Check for the presence of multiple unfolded states caused by proline isomerization reactions.
6. Determine possible folding mechanisms using the kinetic data and design further experiments to exclude alternative mechanisms.

2.2.1. The Choice of Experimental Conditions

Usually, high concentrations of denaturants, such as GuCl or urea, are used to produce unfolded protein. In most cases 6*M* GuCl or 8*M* urea is sufficient for complete unfolding. However, some proteins are extremely resistant to denaturants and cannot be unfolded even at high concentrations of GuCl or urea at room temperature. The concentration of denaturant needed for complete unfolding should be determined in equilibrium unfolding transitions (*see* Chapter 8). GuCl is a stronger denaturant than urea but is a salt. Thus, if a protein contains binding sites for salt, GuCl can have unwanted side effects. Also, aggregation of proteins during refolding can be salt dependent (in either way), so that in preliminary experiments the best suited denaturant should be evaluated.

Many proteins unfold at low pH values. In many cases, however, residual structure has been detected in acid-unfolded proteins, so that chaotropic reagents should preferably be used for complete unfolding of proteins. A further advantage of denaturant-induced unfolding over pH-induced unfolding is the ability of denaturants to solubilize unfolded or partly folded polypeptide chains and thus prevent aggregation. Often, a combination of high denaturant concentrations and low pH is used to efficiently unfold proteins.

Since the goal of performing folding experiments in the native baseline region is usually to detect partially folded intermediates, the chosen refolding conditions should strongly favor folded conformations. It is therefore important to keep the residual concentrations of denaturant as low as possible in refolding kinetics in order to allow transient intermediates to form. The influence of the residual denaturant concentration on the observed folding reactions should be tested by performing folding kinetics in the presence of various residual denaturant concentrations.

2.2.2. Refolding Experiments Monitored with Different Spectroscopic and Functional Probes

In standard experiments refolding is initiated by diluting completely unfolded protein into native conditions in an optical cell and monitoring resulting refolding reactions directly in the respective spectrometer. It is crucial to control the temperature of the refolding solution during the whole experiment since folding rates are usually very temperature sensitive. To establish a homogenous temperature in the optical cell in long term experiments, it is useful to stir the solution permanently. This additionally reduces photodestruction of the samples, which is particularly important if tryptophan fluorescence emission is monitored.

The concentration dependence of the observed refolding kinetics should be measured to test the influence of the protein concentration on the refolding kinetics. The observed refolding reactions have to be concentration independent if only first-order reactions are involved.

The mixing step is very important in performing folding experiments and the applied mixing technique should depend on the folding rates. The rates of folding reactions in proteins can vary from the time range of a few milliseconds to several hours. For fast folding reactions that are complete within a few seconds or less, it is necessary to use stopped-flow mixing. This method has been reviewed in detail by Utijama and Baldwin *(25)*. Slower folding reactions should be monitored in standard optical instruments, since here the signal is much more stable over extended periods than in specially designed stopped-flow machines and usually less concentrated protein solutions are required.

Following the refolding process with different spectroscopic probes allows one to monitor different regions of a protein. Far and near-UV CD, UV-absorbance, and fluorescence emission spectroscopy are particularly well-suited spectral probes. Far UV-CD monitors ordered structure in the polypeptide backbone, whereas the other methods detect changes in the environment of aromatic amino acid residues and are therefore more sensitive to tertiary contacts in a protein *(30; see also* Chapter 3–5). During refolding of many proteins under native conditions intermediates with high contents of secondary structure are formed very early, resulting in a rapid and large change in the far-UV CD signal of the refolding protein with only little or no effects on near UV-CD, fluorescence emission, or UV-absorbance.

Complementary to spectroscopic methods, it is advisable to probe the regain of functional properties during refolding. For this purpose the binding of inhibitors (or activators) and the recovery of the enzymatic activity are commonly used probes. Both these methods are sensitive to the spatial reconstitution of specific regions of the protein. Folding studies in the presence of inhibitors should be monitored with probes that do not detect any signal change for refolding of the protein itself. A commonly applied method is the use of inhibitors that show spectral changes after binding to the protein.

Measuring reactivation during refolding requires that the folding reactions of interest are rather slow, since activity assays usually involve a transfer of sample from the refolding solution to the conditions of the activity assay and, further, the activity assays themselves are usually slow. Typically, unfolded protein is refolded under native conditions and, after various times of refolding (t_i), aliquots of the refolding solutions are transferred into a standard activity assay. To prevent unfolded or partially folded molecules from further refolding, it is advisable to add protease to the activity assays. It has to be tested, however, that the protease used does not affect the activity of the native protein under the conditions of the activity assays.

2.2.3. Detection of Reactions that Occur in the Dead-Time of Mixing

Since early steps in protein folding can be extremely fast, it should always be tested whether folding reactions occur in the dead-time of the applied mixing technique. An easy way to do this is to measure an equilibrium unfolding transition of the protein (*see* Chapter 8) and extrapolate the unfolded baseline of the transition into the conditions where the refolding experiments were performed. This allows the determination of the maximum signal difference between the native and the unfolded states under the conditions of the refolding experiments (*see* Fig. 1). The comparison of this maximum amplitude to the sum of the amplitudes of the observed refolding reactions allows detection of missing amplitudes.

2.2.4. Distinction Between Parallel and Sequential Reactions and Detection of Transient Folding Intermediates

If complex refolding kinetics are observed during refolding, it is important to determine whether the different reactions are part of a sequential pathway (Eq. [6a]) with transient intermediates or caused by parallel channels that lead to native protein (Eq. [6b]).

$$U \xrightarrow{k_1} \mathrm{I} \xrightarrow{k_2} N \tag{6a}$$

$$\begin{aligned} U^{\mathrm{I}} &\xrightarrow{k_1} N \\ U^{\mathrm{II}} &\xrightarrow{k_2} N \end{aligned} \tag{6b}$$

2.2.4.1. Direct Optical Measurements to Detect Transient Folding Intermediates

Comparing the results of refolding kinetics monitored with different probes and under different final conditions is one way to determine the presence of partially folded transient intermediates *(16)*. If a pathway from the unfolded to the native state does not contain any intermediate structures, the amplitude of this reaction is only determined by the number of molecules that use this pathway. Therefore, in this case the amplitudes of the observed refolding reaction should be independent of the probes used to monitor folding. If, on the other hand, the amplitude of a certain reaction depends on the probe used to monitor folding, this reaction is probably part of a sequential pathway. To perform this kind of comparison, probes should be chosen that monitor different regions or properties of the protein *(see* Section 2.2.2.*)*.

If native protein is formed on a sequential pathway and the rate constants of the involved reactions are of the same order of magnitude, the time course of the measured signal should be characterized by an initial lag phase. Since spectral signals usually also change after the formation of partially folded intermediates, this lag in the appearance of native protein is commonly overlaid by the signal change that is accompanied with the formation of intermediates.

2.2.4.2. Forward and Backward Jump Experiments to Monitor the Formation of Native Molecules or Transient Intermediates

The method of choice for detecting the formation of native protein during folding is unfolding assays. These assays make use of the fact that native protein is usually separated from any partially folded conformations by a high energy barrier and therefore unfolds much slower than any intermediates (Fig. 2). Consequently, if a mixture of native molecules and folding intermediates is unfolded at high denaturant concentrations, only the unfolding kinetics of the native molecules can be detected since the unfolding of the intermediates occurs too fast. In unfolding assays protein is refolded under native conditions. After various times (t_i), aliquots of the refolding solution are transferred to conditions where complete unfolding of native protein occurs (e.g., at high concentrations of denaturant). The amplitude of the detected unfolding reaction is proportional to the amount of native protein that was present in the refolding solution after the time t_i. This amplitude thus increases with the same kinetics as the native state is formed (*see* Fig. 3) and thus any lag phases in the formation of native molecules can be detected.

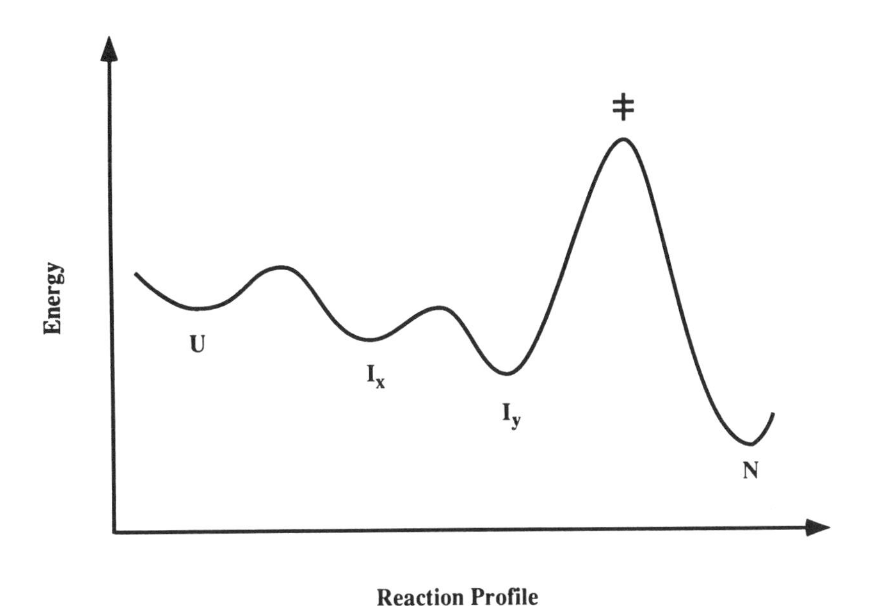

Fig. 2. Schematic energy diagram for a hypothetical folding reaction. *U* represents the unfolded state, *N* is the native state and I_x and I_y are intermediate states on the folding pathway. The transition state for the overall folding reactions is indicated by ‡. It is probably structurally close to the native state.

A modified version of this assay can be used to monitor the formation and decay of folding intermediates. In these experiments refolding and the drawing of the aliquots is essentially carried out as in the assays described above but the unfolding step is performed under conditions where the native protein is still stable and only partially folded intermediates unfold. Here the amplitude of the observed unfolding reaction is thus a measure for the amount of the particular folding intermediate that was present at t_i.

Figure 3 shows the time course of formation of native molecules and a folding intermediate during refolding of ribonuclease T1 S54G,P55N measured with the methods described above. Here, no lag phase in the appearance of native protein is observed, since the formation of the intermediate is too fast (in the ms time region) compared to the slow interconversion of the intermediate to the native state.

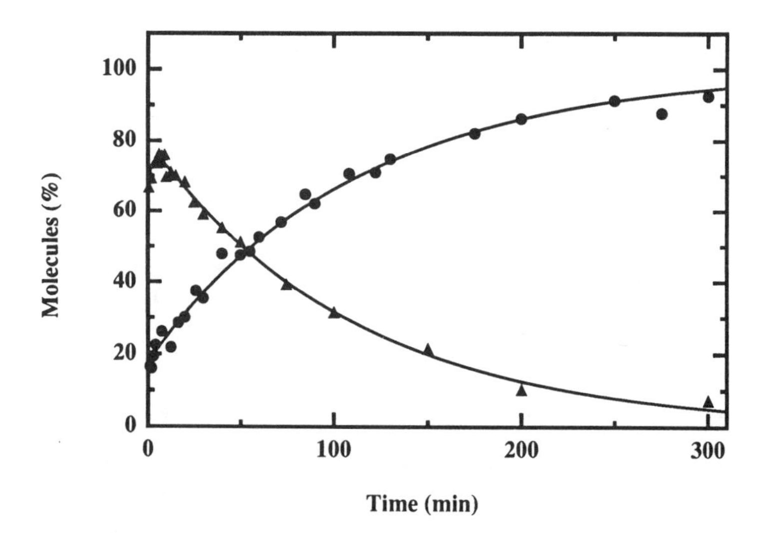

Fig. 3. Time course of formation and decay of a folding intermediate (▲) that is formed during refolding of RNase T1 S54G,P55N under strongly native conditions and time course of formation of native molecules (●) under the same conditions. Completely unfolded protein (in 6.0*M* GuCl, 0.1*M* glycine, pH 2.0) was refolded in 0.15*M* GuCl, 0.1*M* NaAc, pH 5.0 at 10°C. After various times of refolding (ti), aliquots were withdrawn and transferred into 2.8*M* GuCl, 0.1*M* NaAc, pH 5.0 (to monitor unfolding of the intermediate) or into 5.2*M* GuCl, 0.1*M* glycine, pH 2.0 (to monitor unfolding of the native molecules). The amplitudes of the resulting unfolding reactions are plotted against the refolding time. The solid lines represent fitted curves *(44)*.

2.2.4.3. Additional Methods of Detecting Folding Intermediates

An excellent method of detecting and characterizing folding intermediates at a molecular level is pulsed hydrogen exchange experiments. The use of this method has led to a variety of information on rapidly formed transient folding intermediates in recent years. The method is discussed in detail in Chapter 13.

Indirect information on the presence of folding intermediates on a folding pathway can come from the determination of the activation energy for a folding step. In the case of RNase A a rapidly formed folding intermediate on the pathway of the U_F molecules was postulated since a negative activation energy was observed for this reaction *(31)*. This effect is caused by a rapid equilibrium between a folding intermediate and unfolded molecules:

$$U \underset{k_2}{\overset{k_1}{\rightleftharpoons}} I \xrightarrow{k_3} N \quad K_I = k_1/k_2 \tag{7}$$

The apparent rate constant for this mechanism is given by

$$k_{obs} = K_I * k_f \tag{8}$$

Thus, a negative activation energy can be observed if the equilibrium between U and I has a negative enthalpy, i.e., if the intermediate becomes more stable with decreasing temperature.

2.2.5. *Detection of Proline Isomerization Reactions*

If slow reactions are observed during refolding, it must be examined whether these reactions are caused by *cis-trans* isomerization of Xaa–Pro peptide bonds. A direct way to determine whether the rate of a folding step is limited by prolyl isomerization is to find out whether it is accelerated in the presence of the enzyme peptidyl–prolyl *cis-trans* isomerase (PPIase). This enzyme is known to catalyze prolyl isomerization reactions both in model peptides *(32)* and in proteins *(33)*. Catalysis of a slow step in protein folding by PPIase is strong evidence for the involvement of prolyl residues in this step. It was shown, however, that the efficiency of catalysis by PPIase is strongly dependent on the accessibility of the involved proline residue *(34)*. Thus, if a slow step is not catalyzed by PPIase, this could either mean that prolines are not involved in this reaction or that the responsible proline residue is not accessible to the enzyme because of rapidly formed intermediates.

Another way to detect proline isomerization reactions is double jump assays. They were first suggested by Brandts and coworkers *(5)* and have since then been used to identify refolding reactions limited by proline isomerization in many proteins. These assays make use of the fact that, after unfolding of a native protein, at first only molecules are formed that have all prolyl peptide bonds in the same conformation as the native protein (U_F-molecules). The following establishment of the *cis-trans* equilibrium at each Xaa–Pro peptide bond is a very slow process at low temperature and yields the equilibrium distribution of U_F and U_S molecules. This method has been reviewed in detail by Schmid *(35)*.

In double jump experiments, native protein is unfolded at high concentrations of denaturant, where the $N \longrightarrow U_F$ reaction is very fast (it should typically be complete within a few seconds) compared to proline isomerization. A good separation of the rates of the unfolding reaction and of the proline isomerization step can be achieved in two different ways. An increase in denaturant concentration during unfolding increases the rate of unfolding, whereas the rate of proline isomerization is not influenced. A decrease in the temperature at which unfolding is carried out slows down proline isomerization more than the unfolding reaction. This is caused by the high activation energy of the proline isomerization reaction. A combination of both possibilities should be used to get a good separation of the rates of the two reactions.

After various times of unfolding (t_i), aliquots are drawn from the unfolding solution and transferred into native conditions to initiate refolding. After short times of unfolding, most of the molecules will still be in the U_F state and thus the slow refolding reactions will have little amplitude. This amplitude will increase with increasing time of incubation (t_i) in the unfolding solution until the equilibrium distribution of U_F and U_S is established. The plot of the amplitudes of the slow folding reactions versus the unfolding time gives the kinetics of formation of the slow-folding species in the unfolded chain (Fig. 4). The observed relaxation times for proline isomerization reactions in unfolded proteins are usually about 10–100 s at 25°C and have an activation energy of about 20 ± 2 kcal/mol. So, if a slow equilibration reaction in the unfolded chain is observed that shows the above characteristics, it is likely that the respective slow refolding reactions are limited by prolyl isomerization.

2.2.6. Determination of Possible Folding Mechanisms

In many cases the folding mechanisms of proteins are not purely sequential or parallel processes but comprise both features. Scheme 1 shows the refolding mechanism of ribonuclease T1 (RNase T1) as an example. RNase T1 contains 2 *trans* and 2 *cis* prolyl residues in the native state. After unfolding, each of these prolyl peptide bonds adopts a *cis-trans* equilibrium. In refolding experiments 3.5% of the unfolded molecules refold fast (U_F-molecules), whereas the remaining unfolded

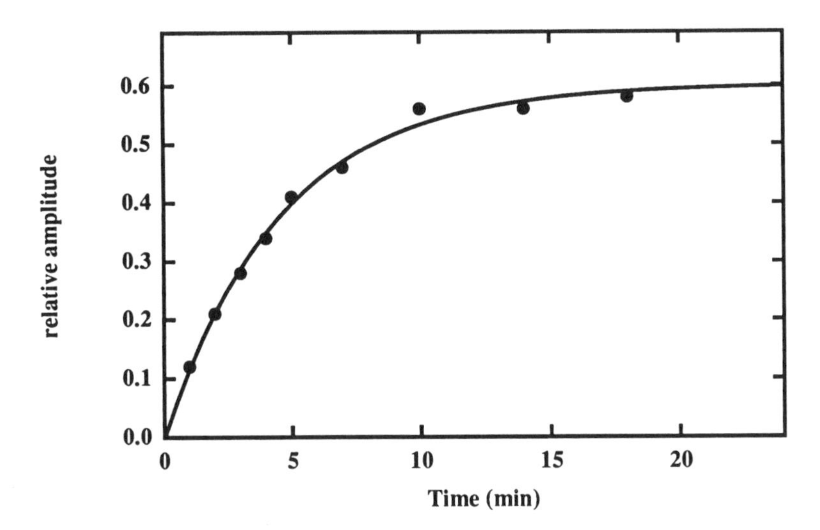

Fig. 4. Double-jump experiments to monitor proline isomerization in the unfolded state of RNase T1 S54G,P55N. The protein was unfolded rapidly and completely in 6.0*M* GuCl, 0.1*M* glycine, pH 1.6 at 10°C. After various times of incubation (ti), aliquots were withdrawn from the unfolding solution and refolded in 0.15*M* GuCl, 0.1*M* NaAc, pH 5.0 at 25°C. The amplitude of the observed slow refolding reaction is plotted against the incubation time t_i. The solid line reprsents the curve fitted to a single exponential.

molecules refold in the time region of minutes to hours. The two prolyl residues that are *cis* in the native state were shown to be the major source of the slow folding reaction. The mechanism indicates that all slow folding species can form intermediates very rapidly. These intermediates are slowly converted to native molecules in reactions that are limited by the *trans* to *cis* isomerization reaction of at least one Xaa–Pro peptide bond.

The refolding mechanism shown in Scheme 1 was developed on the basis of a combination of several spectroscopically detected refolding kinetics and kinetic jump experiments as described above. The use of replacements of individual amino acid residues with the help of site-directed mutagenesis further allowed the assignment of individual proline residues to specific folding reactions *(10,29,34,36)*. The molecular interpretation of the data was greatly facilitated by the availability of the three-dimensional structure of the molecule *(37)*.

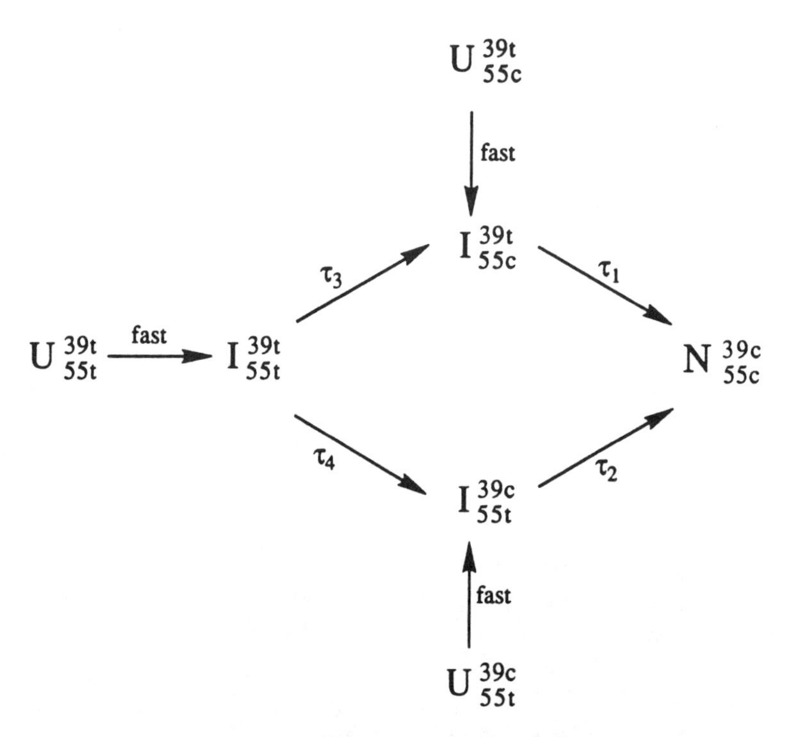

Scheme 1. Kinetic mechanism for the slow refolding reactions of wild-type ribonuclease T1. Four major unfolded species exist that differ in the isomerization states of the two *cis*-proline residues. Unfolded molecules with all prolyl residues in the same configuration as the native state can refold rapidly (U_F molecules) whereas the three unfolded species with at least one nonnative prolyl–peptide bond can form partially folded intermediates rapidly under strongly native conditions. The rate of interconversion of the intermediates to native protein is limited by proline isomerization reactions. The relaxation times for the various reactions are $\tau_1 = 3{,}000$ s, $\tau_2 = 500$ s, $\tau_3 = 100$ s, and $\tau_4 = 190$ s at pH 8 and 10°C; and $\tau_1 = 7{,}000$ s, $\tau_2 = 250$ s, $\tau_3 = 170$ s, and $\tau_4 = 400$ at pH 5.0 and 10°C. Additionally, unfolded molecules with all prolyl peptide bonds in their native conformation (U_F-molecules) refold in the ms time region (reprinted in part with permission from ref. *10* copyright 1990 American Chemical Society).

2.3. Folding and Unfolding Reactions in the Transition Region

Folding in the transition region of small monomeric proteins is usually simplified by the absence of detectable intermediates. This is caused by the fact that partially folded structures are usually much less stable than the native state and are thus not formed under conditions where native molecules are at the edge of their stability. Therefore only native and completely unfolded molecules have to be considered in folding reactions in the transition region:

$$U \underset{k_u}{\overset{k_f}{\rightleftharpoons}} N \tag{9}$$

where k_f and k_u are the rate constants for the folding and unfolding reactions, respectively.

The observed macroscopic rate constant (λ) in this case is given by the sum of the forward and reverse rate constants:

$$\lambda = k_f + k_u \tag{10}$$

If, additionally, the equilibrium constant (K_{eq}) of the system at given conditions is known:

$$K_{eq} = k_f/k_u \tag{11}$$

k_u and k_f can be determined using Eqs. (10) and (11).

2.3.1. Transition State Analysis

2.3.1.1. Transition State Theory

If both k_f and k_u can be determined for a system and the principle of microscopic reversibility is fulfilled, the activated state of the folding process can be located energetically. The complete energy profile of the folding/unfolding reaction (*see* Fig. 5) can be obtained by using the Eyring formalism:

$$\ln k = (\ln (\kappa\, k_b T))/h - \Delta G^{\ddagger}/RT \tag{12}$$

where κ represents a transmission factor and $\Delta G^{\ddagger}$ is the free energy of activation. The value of κ has an upper limit of 1 *(26)*. Recent theoretical studies suggest that the preexponential factor may not be $k_b T/h$ in the rate equation, so that an analysis of experimental data according to Eq. (12) may not yield the actual free energy of the transition state *(38)*. The data obtained by using the Eyring formalism can, however, still be useful in the comparison of mutant forms of the same protein if it is warranted that both κ and the preexponential factor do not change (for a review *see* ref. *39*). The absolute values for the energy of the transition state, however, will strongly depend on the value of κ and the correct preexponential factor.

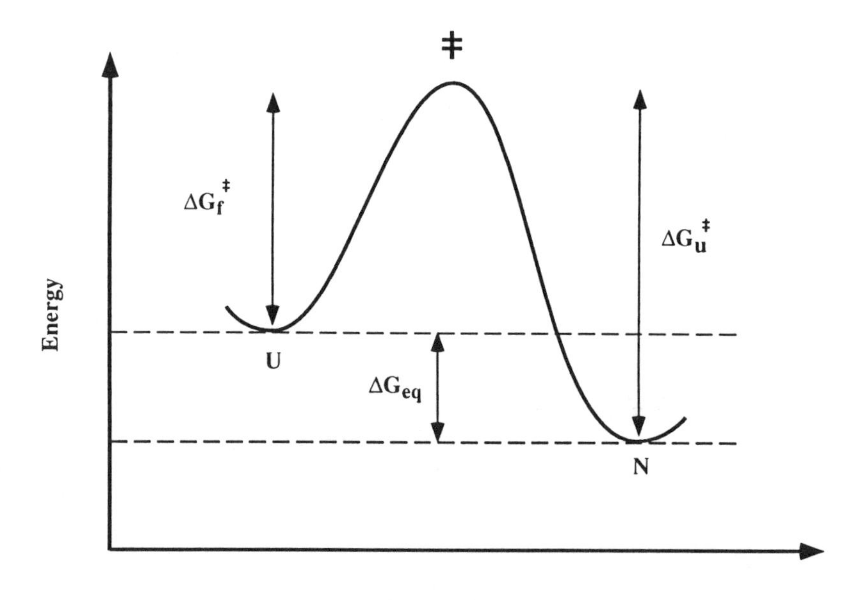

Fig. 5. Schematic energy diagram for a hypothetical folding reaction under conditions where no transient intermediates are populated. The diagram shows the relationship between the free energies of activation for the refolding and unfolding reaction ($\Delta G_f^\ddagger$ and $\Delta G_u^\ddagger$, respectively) and the difference in free energy between the native and unfolded state (ΔG_{eq}).

The activated state can be further characterized by measuring the influence of the denaturant concentration on the rate constants of the unfolding and refolding reactions. This analysis is based on the experimental observation that log k_u and log k_f are often linear functions of the denaturant concentration in the transition region and under unfolding conditions *(16)*.

$$\log k_u = \log k_u\,(H_2O) + m_u\,[\text{denaturant}]$$
$$\log k_f = \log k_f\,(H_2O) + m_f\,[\text{denaturant}] \qquad (13)$$

Thus, according to Eq. (13), the plot of log λ vs denaturant concentration yields a V-shaped curve, that is determined by k_f at low denaturant concentrations and by k_u at high denaturant concentrations. Near the center of the V the apparent rate constant is determined both by k_f and k_u.

From the obtained values for k_f and k_u, $\Delta G^\ddagger_f$ and $\Delta G^\ddagger_u$ can be calculated using Eq. (12). These values can be compared to the free energy for unfolding of the protein as determined by equilibrium transition curves (*see* Chapter 8):

$$\Delta G_{eq} = \Delta G_{eq}(H_2O) + m_{eq}\,[\text{denaturant}] \quad (14)$$

From

$$\Delta G_{eq} = \Delta G_f^{\ddagger} - \Delta G_u^{\ddagger} \quad (15)$$

in combination with Eqs. (13) and (14) follows that

$$m_{eq} = m_f - m_u \quad (16)$$

A schematic diagram of the relationship between ΔG_{eq}, $\Delta G_f^{\ddagger}$, and $\Delta G_u^{\ddagger}$ is shown in Fig. 5.

Equation (16) can be used as a test for the validity of the two-state assumption. If both refolding and unfolding are two-state processes, the difference between m_f and m_u has to equal m_{eq}. Deviations hereof indicate conditions where the two-state assumption is not valid *(16,40)*.

Comparison of the *m*-values of the folding and unfolding reactions with the *m*-value for the equilibrium unfolding has been further used to draw conclusions about the degree of organization of the transition state of folding *(16)*. The *m*-value of a particular reaction is believed to be proportional to the change in accessible surface area in this step. Thus, comparison of denaturant dependence of the refolding and unfolding rate constants with the denaturant dependence of K_{eq} is a measure of the amount of surface area that becomes accessible in the activated state. Kuwajima and coworkers have performed this kind of analysis to characterize the transition state of folding of α-lactalbumin and apo-α-lactalbumin and concluded that the structural organization of the activated state in this protein is about 35% compared to the native state in respect to its exposure to GuCl. Additionally, the activated state was shown to be able to bind Ca^{2+} ions *(41)*.

2.3.1.2. Limitations of Transition-State Theory

There are various factors that limit the use of transition state theory for protein folding. As mentioned above, the values obtained for the activation free energy are strongly dependent on the value of the transmission factor κ and of the preexponential term. Neither value is accessible for protein folding reactions. Thus, the values obtained for $\Delta G^{\ddagger}$ should be denoted as apparent activation free energy. Further, the data analysis discussed here is only valid under conditions where no transient intermediates are formed. Thus the experiments are restricted to a very narrow concentration range. Extrapolation to low denaturant concentrations is usually not warranted, since the mechanism of folding changes as the stability of partially folded intermediates increases. This behavior is represented by a nonlinearity of the denaturant dependence of the refolding rate constant at low concentrations of denaturant.

Also, it is not accepted by everyone that transition state theory is applicable to protein folding. Especially physicists have pointed out that the analysis of phase transitions in polymers may have to be analyzed in a different way *(42)*.

2.3.1.3. Influence of Coupled Prolyl Equilibria on the Observed Folding Rates

An important factor that influences the denaturant dependence of the apparent rate constants of the folding and unfolding reaction under equilibrium conditions is the heterogeneity of the unfolded state, e.g., caused by linked *cis-trans* equilibria at prolyl peptide bonds. It has been shown that both *cis* and *trans* prolyl residues can influence the observed folding rates and their denaturant dependence in the transitions region in the absence of transient intermediates. The magnitude of the effect depends strongly on the relative rate constants of the actual folding step of the protein ($U_F \rightleftharpoons N$) and of the proline isomerization reaction as well as on the number of coupled prolyl equilibria and their equilibrium distribution. In general the effects of prolines are more pronounced if the actual folding reaction is slow and if the prolines are *cis* in the native state *(23,43)*. The coupling of prolyl residues gives rise to additional kinetic phases (cf., Eq. [1]). For example, the coupling of one prolyl equilibrium to a folding reaction:

$$U_S \underset{k_-}{\overset{k_+}{\rightleftharpoons}} U_F \underset{k_u}{\overset{k_f}{\rightleftharpoons}} N \tag{17}$$

gives rise to one additional kinetic phase. The apparent rate constants for this mechanism are given by:

$$\lambda_{1/2} = 1/2\ \{-A_2 \pm (A_2^2 - 4A_1)^{1/2}\} \tag{18}$$

with

$$A_2 = -(k_+ + k_- + k_f + k_u) \tag{19}$$

and

$$A_1 = k_+(k_f + k_u) + k_-k_u \tag{20}$$

In contrast to the rate constants for the folding and unfolding reactions, the rate of prolyl isomerization is independent of the denaturant concentration. This leads to characteristic profiles for the denaturant dependences of the apparent rate constants. Figure 6A,B shows the influence of one coupled *cis* prolyl residue when protein folding is faster or slower than the actual folding step, respectively. In the case that folding is faster than prolyl isomerization, the coupled proline equilibrium merely leads to a second well separated kinetic phase. The rate of this additional reaction is equal to:

$$\lambda = k_+ \tag{21}$$

under refolding conditions and to

$$\lambda = k_+ + k_- \tag{22}$$

under unfolding conditions.

The amplitudes of the fast and slow observed reactions in refolding correspond to the equilibrium distribution of U_F and U_S molecules in the unfolded protein, respectively. This kind of refolding behavior was observed in refolding of the C_L domain of immunoglobulins *(7)*.

The effects of coupled proline equilibria on the observed folding rates are more complex if the folding reaction and proline isomerization have similar rate constants or when folding is slower than proline isomerization. Figure 6B shows that, under these conditions, one coupled prolyl equilibrium decreases the observed folding rate in the transition region and that the denaturant dependence of the refolding rate apparently gets weaker. Additionally, a second kinetic phase is present. The amplitude of this phase is, however, strongly dependent on the initial conditions. If folding is started from the equilibrium mixture of U_F and U_S molecules, often only the slowest phase is detected in refolding in the transition region. The additional kinetic phase can be detected, however, if refolding is started after short times of unfolding when most of the molecules are still in the U_F conformation. In these experiments both kinetic phases have measurable amplitudes *(23)*.

As a consequence of the influence of prolyl equilibria on the observed folding rates, the apparent folding rate constants (λ_i) often can not be

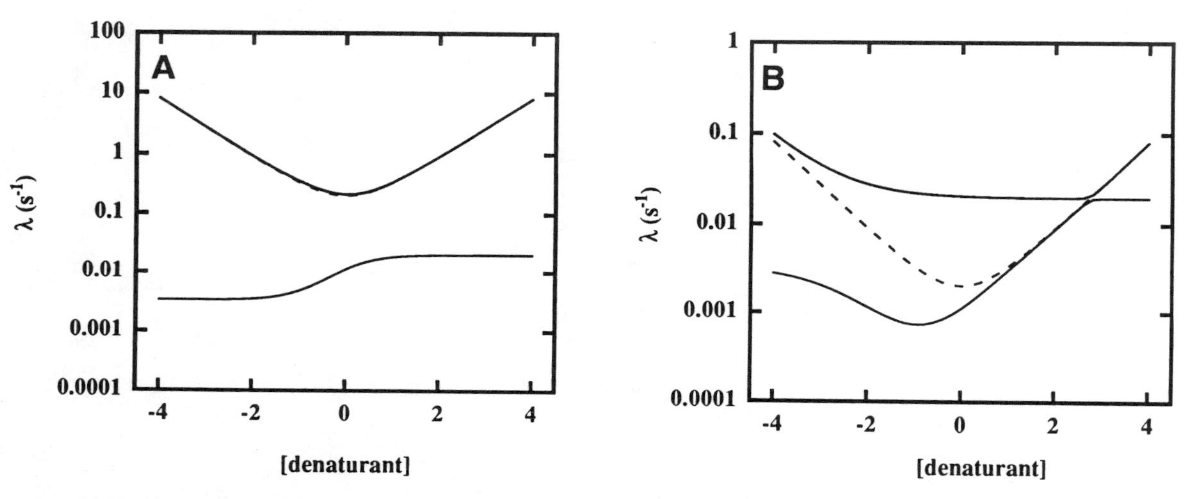

Fig. 6. Influence of kinetic coupling between proline isomerization and protein folding for the case that the folding reaction ($U_F \rightleftharpoons N$) is ten times faster (**A**) and ten times slower (**B**) than the proline isomerization reaction at the midpoint of the $U_F \rightleftharpoons N$ transition. The data shown resemble the case of the coupling of one proline residue that is cis in the native state and has an equilibrium constant ($K_{eq} = f_{(cis)}/f_{(trans)}$) of 0.20. The solid lines represent the values of the resulting apparent rate constants (λ_i) and the dashed lines represent the rate constant of the actual folding reaction ($k_f + k_u$) that would be observed in the absence of coupled proline equilibria. The denaturant concentrations are given in arbitrary units relative to the midpoint of the $U_F \rightleftharpoons N$ transition (adopted from ref. *23*). For an amplitude analysis of the cases presented and further simulations *see* ref. *23*.

used for transition state analysis since they are a complex function of all microscopic rate constants (k_{ij}) Rather, all microscopic rate constants have to be determined according to Eqs. (17) and (18) and the denaturant dependence of k_u and k_f can then be used for transition state analysis.

2.3.2. Experimental Procedure

1. Measure the denaturant dependence of refolding and unfolding kinetics under conditions where folding is completely reversible and no refolding intermediates are detected.
2. Compare the *m*-values of the denaturant dependence of the refolding and unfolding reaction to the *m*-value obtained in equilibrium unfolding studies to detect deviations from two-state behavior.
3. Check for additional kinetic phases caused by prolyl equilibria in the unfolded chain.
4. Analyze the denaturant dependence of the refolding and unfolding reactions according to transition state theory.

2.3.2.1. Map the Conditions Suitable for Performing the Reversible Folding Studies

Methods for determining conditions where no transient folding intermediates are observed are described in Section 2.2.4. Stock solutions with identical concentrations of native and completely unfolded protein should be prepared and folding should be initiated by diluting either native or completely unfolded protein into the final conditions. It is important to perform unfolding and refolding studies under identical final conditions in the transition region. If the system obeys microscopic reversibility, the observed apparent rate constants must be identical for refolding and unfolding under identical final conditions. The amplitudes of the observed kinetic phases depend, however, both on the initial and on the final conditions.

2.3.2.2. Validity of the Two-State Assumption

An easy way to detect deviations from the two-state assumption for folding under equilibrium conditions is the comparison of the denaturant dependence of the observed refolding and unfolding rate constants to the denaturant dependence of the equilibrium constant. According to Eq. (16) a comparison of the *m*-value for the equilibrium unfolding reaction with the *m*-values for the refolding and unfolding reactions allows the detection of deviations from the two-state mechanism. This method was used to demonstrate non-two state behavior in the folding of barnase *(40)*.

Methods for detecting partially folded intermediates have been discussed in Section 2.2.4. They are also applicable to detecting transient intermediates in the transition region. If the two-state assumption is valid, the amplitudes of all observed kinetic phases have to be independent of the probes used to measure refolding under identical initial and final conditions.

2.3.2.3. Detection of Additional "Silent" Kinetic Phases

Additional kinetic phases caused by kinetic coupling between prolyl isomerization and protein folding can have little or no amplitude if refolding is started from the equilibrium mixture of fast and slow folding molecules. These coupled equilibria can nevertheless strongly influence the observed folding rates in the transition region *(23)*. Refolding experiments should therefore be performed starting from predominantly fast-folding molecules. In these experiments the native protein is unfolded rapidly at a very high concentration of denaturant and at low temperature (cf., the double jump assays described in Section 2.2.5.). Because of the slow rate of proline isomerization nearly all molecules will still have all prolyl peptide bonds in their native conformations (U_F molecules) after short times of unfolding. Refolding of these short-time unfolded molecules often allows the detection of additional kinetic phases. Refolding experiments starting from U_F molecules should be performed under various denaturant concentrations throughout the transition region to obtain the denaturant dependence of the apparent rate constants of all detectable kinetic phases and their respective amplitudes. With the help of these experiments an additional fast reaction in the transition region of RNase A and RNase T1 was detected *(43)*.

In many cases the detection of additional kinetic phases and their amplitudes allows the determination of all microscopic rate constants. In general, for a system with n microscopic rate constants n independent experimental observables have to be determined for a complete characterization of the system. If, for example, in a system with two unfolded species that do not differ in their spectral properties the two macroscopic rate constants and their amplitudes can be measured, all four microscopic rate constants can be determined. If the two unfolded species differ in some spectral property, the additional determination of the equilibrium constant allows one to calculate the extent by which the unfolded species differ in this property *(24)*.

2.3.2.4. Data Analysis

Plotting the log of the apparent rate constant vs the denaturant concentration usually gives a characteristic V-shaped profile. Since the apparent rate constant in a two-state transition is the sum of the unfolding and the refolding rate constants, the resulting plot is dominated by the refolding rate at low denaturant concentrations and by the unfolding rate at high denaturant concentrations. Near the center of the V, both rate constants contribute to the apparent rate constant. The data can be analyzed according to Eq. (13): m_f and m_u and the folding and unfolding rate constants extrapolated to zero denaturant ($k_f[H_2O]$ and $k_u[H_2O]$) can be obtained by fitting the data to Eq. (23):

$$\log \lambda = \log(k_f\,[H_2O]\,\exp[m_f\{\text{denaturant}\}] + k_u[H_2O]\,\exp[m_u\{\text{denaturant}\}]) \tag{23}$$

It should be noted that $k_f\,(H_2O)$ and $k_u(H_2O)$ are hypothetical rate constants at zero denaturant. As mentioned above the mechanism of folding usually changes at low concentrations of denaturant which makes an extrapolation of the experimental data to native solvent conditions invalid.

2.4. Unfolding Kinetics Under Strongly Destabilizing Conditions

At high concentrations of denaturant, where the unfolded state is thermodynamically favored, usually single first-order unfolding reactions are observed. Since the transition state for unfolding is probably structurally very close to the native state (Fig. 2), no transient intermediates are populated in unfolding reactions.

There are two possible molecular origins if multiple unfolding reactions are observed under unfolding conditions. One is the presence of multiple states in the native protein and the other is kinetic coupling between the folding reaction and prolyl equilibria. The two possibilities can easily be distinguished by measuring the denaturant dependence of the unfolding phases. If the observed multiple unfolding reactions originate in a heterogeneity of the native state, their amplitudes should be independent of the final unfolding conditions. In this case the equilibrium distribution of the individual species in the native state determines the relative amplitudes. If, however, the multiple unfolding reactions arise from a kinetic coupling between prolyl isomerization and the actual folding reaction, the amplitudes should depend on the final conditions and at very high denaturant concentrations only one unfolding reaction should be visible *(23)*. In the case of very stable proteins these conditions might, however, not be experimentally accessible.

3. Conclusion

Several aspects are important in the determination of the folding mechanism of a protein. First of all the stability regions of a protein have to be determined and the conditions under which folding intermediates can be detected have to be characterized. Two major experimental tools have proven to be very useful in the characterization of folding pathways. One is the use of multiple probes to monitor folding, the other is the use of interrupted folding and unfolding experiments. Especially the second method gives valuable information on the origin of observed folding reactions and can be used to detect "silent" folding or isomerization steps. Proline isomerization equilibria influence protein folding kinetics under various different conditions and can have unexpected effects on the observed folding rates. In order to investigate the actual folding steps ($U_F \rightleftharpoons N$) it is therefore advisable to investigate proteins that contain only few or no proline residues. Especially the presence of *cis* prolyl residues in the native state leads to a large population of slow-folding molecules in the unfolded state; consequently they have more prominent effects on the observed kinetics in the transition region than trans proline residues.

References

1. Jaenicke, R. (1987) *Progr. Biophys. Mol. Biol.* **49**, 117–237.
2. Garel, J.- R. (1992) in *Protein Folding* (Creighton, T. E., ed.), Freeman, New York, pp. 405–454.
3. Privalov, P. L. and Khechinashvili, N. N. (1974) *J. Mol. Biol.* **86**, 665–684.
4. Garel, J.-R. and Baldwin, R. L. (1973) *Proc. Natl. Acad. Sci. USA* **70**, 3347–3351.
5. Brandts, J. F., Halvorson, H. R., and Brennan, M. (1975) *Biochemistry* **14**, 4953–4963.
6. Grathwohl, C. and Wüthrich, K. (1976) *Biopolymers* **15**, 2025–2041.
7. Goto, Y. and Hamaguchi, K. (1982) *J. Mol. Biol.* **156**, 91–910.
8. Kelley, R. F. and Richards, F. M. (1987) *Biochemistry* **26**, 6765–6774.
9. Wood, L. C., White, T. B., Ramdas, L., and Nall, B. T. (1988) *Biochemistry* **27**, 8562–8568.
10. Kiefhaber, T., Grunert, H. P., Hahn, U., and Schmid, F. X. (1990) *Biochemistry* **29**, 6475–6480.
11. Schmid, F. X. (1992) in *Protein Folding* (Creighton, T. E., ed.), Freeman, New York, pp. 197–239.
12. Zhu, X., Ohta, Y., Jordan, F., and Inouye, M. (1989) *Nature* **339**, 483–484.

13. Baker, D., Sohl, J. L., and Agard, D. A. (1992) *Nature* **356**, 263–265.
14. Franke, A. E., Danley, D. E., Kaczmarek, F. S., Hawrylik, S. J., Gerard, R. D., Lee, S. E., and Geoghegan, K. F. (1990) *Biochim. Biophys. Acta* **1037**, 16–23.
15. Carrell, R. W., Evans, D. L., and Stein, P. E. (1992) *Nature* **353**, 576–578.
16. Tanford, C. (1968) *Advan. Prot. Chem.* **23**, 121–282.
17. Baldwin, R. L. (1975) *Annu. Rev. Biochem.* **44**, 453–475.
18. Kim, P. S. and Baldwin, R. L. (1982) *Annu. Rev. Biochem.* **51**, 459–489.
19. Kim, P. S., and Baldwin, R. L. (1990) *Annu. Rev. Biochem.* **59**, 631–660.
20. Hirs, C. H. W. and Timasheff, S. N., eds. (1986) *Methods in Enzymology, vol. 131 Enzyme Structure,* Academic, Orlando, FL.
21. Jaenicke, R. and Rudolph, R. (1989) in *Protein Structure. A Practical Approach* (Creighton, T. E., ed.), IRL, Oxford, pp. 191–224.
22. Szabo, Z. G. (1969) in *Comprehensive Chemical Kinetics* (Bamford, C. H. and Tipper, C. F. H., eds.), Elsevier, Amsterdam, pp. 1–80.
23. Kiefhaber, T., Kohler, H. H., and Schmid, F. X. (1992) *J. Mol. Biol.* **224**, 217–229.
24. Ikai, A. and Tanford, C. (1973) *J. Mol. Biol.* **73**, 145–163.
25. Utijama, H. and Baldwin, R. L. (1986) *Methods Enzymol.* **131**, 51–70.
26. Moore, J. W. and Pearson, R. G. (1981) *Kinetics and Mechanism* (3rd ed.), Wiley, New York.
27. Cook, K. H., Schmid, F. X., and Baldwin, R. L. (1979) *Proc. Natl. Acad. Sci. USA* **76**, 6157–6161.
28. Schmid, F. X. (1986) *FEBS Lett.* **198**, 217–220.
29. Kiefhaber, T., Grunert, H. P., Hahn, U., and Schmid, F. X. (1992) *Prot. Struct. Funct. Genet.* **12**, 171–179.
30. Schmid, F. X. (1989) in *Protein Structure. A Practical Approach* (Creighton, T. E., ed.), IRL, Oxford, pp. 251–286.
31. Hagerman, P. J. and Baldwin, R. L. (1976) *Biochemistry* **15**, 1462–1473.
32. Fischer, G., Bang, H., and Mech, C. (1984) *Biomed. Biochim. Acta* **43**, 1101–1111.
33. Lang, K., Schmid, F. X., and Fischer, G. (1987) *Nature* **329**, 268–270.
34. Kiefhaber, T., Quaas, R., Hahn, U., and Schmid, F. X. (1990) *Biochemistry* **29**, 3061–3070.
35. Schmid, F. X. (1986) *Methods Enzymol.* **131**, 70–82.
36. Kiefhaber, T., Quaas, R., Hahn, U., and Schmid, F. X. (1990) *Biochemistry* **29**, 3053–3061.
37. Heinemann, U. and Saenger, W. (1982) *Nature* **299**, 27–31.
38. Chen, B.-L., Baase, W. A., Nicholson, H., and Schellman, J. A. (1992) *Biochemistry* **31**, 1464–1476.
39. Matthews, C. R. (1987) *Methods Enzymol.* **154**, 498–511.
40. Matouschek, A., Kellis, J. J., Serrano, L., Bycroft, M., and Fersht, A. R. (1990) *Nature* **346**, 440–445.
41. Kuwajima, K., Mitani, M., and Sugai, S. (1989) *J. Mol. Biol.* **206**, 547–561.
42. Kramer, H. A. (1940) *Physica* **4**, 284–304.
43. Kiefhaber, T. and Schmid, F. X. (1992) *J. Mol. Biol.* **224**, 231–240.
44. Kiefhaber, T., Schmid, F. X., Willaert, K., Engelborghs, Y., and Chaffotte, A. (1992) *Protein Sci.* **1**, 1162–1172.

Chapter 15

Molten Globules

Anthony L. Fink

1. Introduction

The term molten globule refers to a compact denatured state: Both experimental and theoretical descriptions have been proposed. Recent reviews include those of Ptitsyn *(1,2)*, Kuwajima *(3)*, Dill and Shortle *(4)*, Christensen and Pain *(5)*, and Baldwin *(6)*. There is considerable controversy regarding the details of exactly what constitutes a molten globule. This is occasioned in part by apparently contradictory experimental evidence for, on the one hand, an unstructured collapsed species with a hydrophobic core, and on the other a structured species with fixed specific secondary structure (summarized by Baldwin *[6]*). Baldwin proposed that the solution to this apparent paradox was the existence of two classes of molten globule, the "true" molten globule, corresponding to the structured form, and a "collapsed unfolded form" for the unstructured species *(6)*. As discussed below there is now clear experimental data to indicate that two classes of compact denatured states of proteins exist: compact intermediates, in the thermodynamic sense (i.e., a minimum in the free energy profile for the reaction), which may be considered to be molten globules, and compact forms of the unfolded state *(7)*. It is important to note that it is often experimentally difficult to distinguish between the two forms, especially by spectral methods. Theoretical models for the existence of two classes of denatured states have been presented by Dill and coworkers *(8)*, by Ptitsyn *(1,2)*, and by Finklestein and Shakhnovich *(9)*. Transient molten globule-like intermediates have also been observed during the refolding of unfolded proteins *(10,11)*.

From: *Methods in Molecular Biology, Vol. 40: Protein Stability and Folding: Theory and Practice*
Edited by: B. A. Shirley

For the purpose of this chapter the major focus will be on equilibrium molten globules, rather than "kinetic" ones.

That denatured proteins, especially when denatured by heat or pH, may have residual structure was demonstrated by Aune et al. *(12)*, who showed that the addition of urea or guanidine hydrochloride to some thermally denatured proteins led to a further cooperative transition. The existence of distinct denatured states was questioned by Privalov *(13)* who demonstrated, using calorimetry, that there was little or no difference in the heat capacity between the apparent intermediate and the "fully" unfolded state, indicating the full solvent exposure of the hydrophobic residues in the "intermediate." More recent investigations suggest small differences in the heat capacity and enthalpy of some molten globules and unfolded states. For example, Xie et al. *(14)* reported an intrinsic enthalpy difference between the molten globule and unfolded states of α-lactalbumin at neutral pH, however, this has recently been disputed *(15)*, and it is likely that there is no *significant* enthalpic difference between the two states in most cases. The most reasonable explanation is that in the molten globule the majority of the hydrophobic residues, buried in the native state, have become exposed to solvent.

Although the evidence has been in the literature for a long time, it is only recently that there has been general recognition of the fact that denatured proteins can exist in a wide range of structures in terms of compactness and residual secondary structure. The fact that some of these are almost as compact as the native state, and contain essentially as much secondary structure, raises the question of the distinction between intermediate states and unfolded states (both being denatured states in the sense of being nonnative conformations). Confusion exists, in part, because of semantic problems. Many terms have been used rather loosely, and also to mean different things by different authors. For example, "intermediate" has been used both in the sense of a thermodynamic intermediate state, and to indicate a structure or species with properties between two other conformations. The term "state" is often not used in the thermodynamic sense. In many cases in which compact denatured states have been reported, insufficient experimental evidence was available to determine whether the species was a true intermediate in the thermodynamic sense, or just a compact substate of the unfolded state.

The term molten globule was first used in the literature to describe the acid-denatured state of cytochrome c by Ohgushi and Wada *(16)*, which resembled a compact state with native-like secondary structure but fluctuating tertiary structure, which had recently been reported by Ptitsyn and coworkers for α-lactalbumin *(17)*. It has now been used to describe a number of proteins denatured under a variety of conditions in which residual structure was observed. In some cases the compactness of the species was measured, in others not. Thus the actual usage of the term has been rather general, at least by most authors, and has been, in effect, used to describe denatured states with significant residual structure, especially secondary structure. In cases where the size was measured, some of these were quite compact; others were considerably expanded relative to the native state. Some are undoubtedly true intermediate states, others are just compact forms (substates) of the unfolded state. The distinction between a true (thermodynamic) intermediate and a compact substate of the unfolded state can only be determined by experiments in which the existence of an energy barrier between the putative intermediate and the native and unfolded states is shown. Spectroscopic experiments that show residual structure, and spectral transitions that appear cooperative, in the absence of thermodynamic or kinetic data, cannot distinguish the two classes of denatured states.

Two things are now clear: Not only can the denatured state (in the sense of the nonnative state) exist as a compact conformation, but for a given protein it may exist in different compact conformations, depending on the external environment; and for both intermediates and unfolded states the degree of folding and compactness can vary continuously over a wide range depending on the external pressures for collapse (folding) or expansion (unfolding). A particularly good example of this is the heat shock protein or molecular chaperone DnaK. In the presence of low concentrations of GuHCl the native state is converted into a compact intermediate whose hydrodynamic radius increases monotonically with increasing denaturant; when first detected at low guanidine hydrochloride (0.25*M*) the Stokes radius (determined by SEC HPLC or dynamic light scattering) is 43 Å, when last observed at higher denaturant (1.4*M*) the R_s is 55 Å *(7)*. Similarly, the unfolded state, when first detected at relatively low denaturant concentration, has R_s = 53 Å, whereas at concentrations of GuHCl above 4*M* its R_s is 69 Å. The compact intermediate is sufficiently stable, and sufficiently well separated in terms of energy barriers from the native and unfolded states, that it is readily observed under equilibrium conditions by SEC HPLC. In the vicinity of 0.6–0.9*M*

GuHCl it is the only species present and thus could be characterized readily by dynamic light scattering and other techniques *(7)*.

2. Theoretical Models of the Molten Globule

Dill and coworkers have used statistical mechanics to model protein stability; by incorporating electrostatics to their model, they were able to accurately predict the conformational phase diagram for apomyoglobin, showing the boundaries between native, molten globule, and unfolded states as a function of pH and ionic strength *(18)*. Their model, which predicts protein stability as a function of temperature, pH, and salt concentration, taking into account the physical properties of the constituent amino acids, predicts two stable forms of denatured states, a compact form, the molten globule, and an expanded denatured state. Among other critical predictions of this theory is that there is a major loss of hydrophobic contacts in the molten globule relative to the native state; that the main electrostatic contribution to stability is entropic; and that more hydrophobic polypeptides will have more prominent compact intermediate states.

Finklestein and Shakhnovich *(9)*, using polymer solution theory, predict that two forms of the molten globule, in addition to the unfolded state, will exist, namely a "wet," more compact, and a "swollen," more expanded, form of the molten globule. By applying mean-field theory to a denatured protein, which they considered as a stochastic polymer, they determined a theoretical phase diagram for stable protein states; particular attention was paid to the density and solvent penetration of the different states. The key disruption in going from the native state to intermediate states was the breakdown of the tight packing in the core of the native state.

Ptitsyn's models assume a nonuniform expansion of the native state on denaturation, which allows retention of the "frame" of the native state in the molten globule *(2)*. The frame is defined to include the central parts of helices and sheets that carry nonpolar side chains contributing to the hydrophobic core of the native state. According to Ptitsyn's current model, water does not significantly penetrate this hydrophobic core in the molten globule, although the packing may become looser permitting greater mobility of these side chains. This model, in which the hydrophobic core is retained, whereas much of the rest of the molecule is relatively unfolded, is claimed to reconcile experimental data that, on the one hand, indicates some native-like properties, and on the other, reveals unfolded-like properties *(2)*.

It has been pointed out that whenever an unfolded protein is placed under native-like conditions it will rapidly collapse to a species with many of the properties associated with a molten globule *(4)*. There is some controversy regarding whether a hydrophobic collapse occurs before or after the initial formation of secondary structure during refolding, or as proposed by Dill *(19)* the formation of secondary structure is a consequence of the collapse. Strong correlation between the ellipticity at 222 nm and the volume of a protein in different conformational states *(7)* favors the latter, whereas the observation that apocytochrome c at neutral pH, low ionic strength, is relatively compact but has no ordered secondary structure as judged by CD *(20)*, favors the former model (this appears to be the only reported such case, and may be a unique situation).

In the majority of, if not all, cases of molten globules or compact denatured states the conformational transitions are manifested as either two- or three-state transitions, where the three states involved are: $N \Leftrightarrow$ molten globule $\Leftrightarrow U$. Depending on the specific experimental conditions, the observed transitions may be between N and molten globule or molten globule and U, or N and U. The particular transition is governed by the underlying conformational phase diagram *(21)*.

Under both equilibrium and transient refolding kinetics conditions, for a given protein and specific experimental conditions, it seems likely that there will be one major stable intermediate state, corresponding to the molten globule, i.e., on the potential energy surface for the reaction one intermediate energy well will predominate. This species will be an ensemble of substates, but for the given conditions, clustered around a rather narrow range of conformations. For different conditions the properties of this molten globule will be different. The more native-like it is (i.e., more compact, more secondary structure) the more narrow the range of substates.

3. Properties of the Molten Globule

The following properties are necessary to fully characterize a species as a molten globule:

1. A compactness much closer to that of the native state than the unfolded state, measured by techniques such as intrinsic viscosity, dynamic light scattering, size exclusion chromatography, or small-angle X-ray scattering;

2. A substantial amount of secondary structure as revealed by techniques such as CD or FTIR;
3. Little or no native-like tertiary structure as evidenced by the lack of a near-UV CD signal or a 1D NMR spectrum indicative of substantial side-chain equivalence; and
4. Significant exposed hydrophobic surface area compared to the native state leading to a propensity for aggregation and to the ability to bind hydrophobic dyes such as ANS.

In addition, based on the limited number of extant studies, it is anticipated that there will be a small, or even no, enthalpy and heat capacity difference between the molten globule and the fully unfolded state, i.e., the thermal unfolding of the molten globule is effectively noncooperative. In the absence of a complete characterization it is not possible to unambiguously identify a denatured state as a molten globule.

The most characterized molten globule is that of α-lactalbumin (summarized in refs. *2,3,22*), and its properties will be used to illustrate the typical experimental manifestations of a molten globule. α-Lactalbumin has been shown to adopt a molten globule conformation under a variety of conditions, including pH < 4, pH > 9.5, [GuHCl] = 2*M* (pH 7, 20°C), thermal unfolding (90°C), and removal of Ca^{2+} at pH 7. The concentration of Ca^{2+} is critical in determining the conformational state of α-lactalbumin: In the presence of Ca^{2+} the native state is stabilized relative to the molten globule and the latter is not observed *(23)*. The α-lactalbumin molten globule exhibits native-like secondary structure, as indicated by far-UV CD, IR, and ORD spectra, whereas the side chains appear in a solvent-exposed environment as determined by NMR, near-UV CD, and aromatic absorbance and fluorescence, although two of the four tryptophans are solvent-inaccessible based on solvent-perturbation experiments. The molten globule is only slightly less compact than the native state (approx 10–20% expansion as measured by small-angle X-ray scattering, size exclusion chromatography, and electrophoresis). The α-lactalbumin molten globule binds hydrophobic fluorescent dyes, such as ANS, and has a strong propensity to aggregate. No cooperative unfolding is observed for thermal denaturation of the A state (acidic pH) form of the α-lactalbumin molten globule, however, some cooperativity is seen with the guanidine-induced molten globule *(14)*, although the significance of this observation has been questioned *(15)*. From fluorescence polarization studies the mobility of the Trp side chain in the molten globule is similar to that of the native state. H/D exchange kinetics indicate that exchange is much faster in the molten globule compared to the native state. NMR analysis of the α-lactalbumin molten globule reveals that some regions of secondary structure (helices) are native-like, and that some tertiary interactions are present *(24)*. Low-angle X-ray scattering also revealed that the molten globule had a relatively dense core.

Dobson and coworkers have examined the structure of the α-lactalbumin molten globule using NMR techniques *(24)*. The NMR spectrum of the molten globule is clearly different from those of the native and unfolded states. Because of the widespread overlap and substantial linewidths of many of the resonances in the molten globule, techniques such as magnetization transfer (with the native state) and amide protection were necessary to obtain structural information. Similar NMR studies have now been reported for the molten globules of apomyoglobin *(25)*, cytochrome c *(26)*, barnase *(27)*, and ubiquitin *(28)*.

4. Experimental Observations of Molten Globules

Compact denatured states are most frequently observed under mildly denaturing conditions, such as low concentrations of denaturant, extremes of pH, high concentrations of certain salts, or high temperatures. The existence of such a species, for a given protein, and its degree of folding and compactness, are determined by a combination of factors: intrinsic ones reflecting the basic structure of the particular protein, i.e., its sequence, and external ones reflecting the environment, such as pH, solutes, and temperature. Other factors that can lead to formation of molten globules or similar species include truncation of the end of the polypeptide chain (e.g., fragments of staphylococcal nuclease in which several residues have been removed from the C-terminus *[29]* or removal of the N-terminal region as with tryptophan synthase *[30]*), reduction of disulfides *(31)*, and removal of metal ions *(23)*. Compilations of proteins in which molten globules have been observed have been made by Ptitsyn *(2)* and Christensen and Pain *(5)*.

Techniques that can be used to detect and characterize a molten globule will now be discussed. It is important to note that there are a *mini-*

mum of four criteria necessary to confirm that a denatured state is indeed a molten globule, namely:

1. Significant secondary structure;
2. Little or no native-like tertiary structure;
3. Compactness closer to the native state than the unfolded state; and
4. A significant energy barrier between the putative molten globule and the native and unfolded states (in order to distinguish the molten globule from a compact form of the unfolded state).

Secondary structure can be monitored by far-UV CD, FTIR, Raman, and NMR spectroscopy; tertiary structure can be measured by near-UV CD and NMR NOEs; side chain mobility can be determined by fluorescence polarization and esr of spin-labeled derivatives; size can be measured by size exclusion chromatography, intrinsic viscosity, dynamic light scattering, and small angle X-ray scattering; energy barriers can be ascertained by thermodynamic measurements, such as DSC or titration calorimetry, and kinetics experiments measuring the rates of interconversion between conformational states.

Frequently the binding of fluorescent hydrophobic dyes, such as ANS, can be used to suggest the presence of molten globules, owing to the preferential binding of the dye to the intermediate compared to native and unfolded states. This method of detection of molten globules has been used at both extremes of pH and in the presence of denaturants *(7,32–34)*, however, in the absence of the other criteria listed above it is insufficient to characterize a species as a molten globule. It is important to note that the presence of ANS tends to increase the propensity of molten globules and compact denatured states to aggregate, and that aggregation increases the ANS fluorescence emission.

5. Experimental Conditions to Generate Equilibrium Stable Molten Globules

Compact denatured states or molten globules have now been observed for a significant number of proteins, and it is assumed that most if not all proteins can form such species if the appropriate experimental conditions can be found; in other words, molten globules are general intermediates in protein folding, and are found both as stable equilibrium intermediate states, as well as transient kinetic intermediates during the refolding process *(11)*.

Molten globules have been most widely observed under acidic denaturation conditions. Many proteins, when brought to conditions of pH <3 and moderate ionic strength, will exhibit the properties of a compact denatured state. The compact species observed at low pH are usually known as A states, and have many molten globule-like properties. Fink and coworkers investigated the acid denaturation of several proteins and found that their behavior could be classified into three types *(35)*, although these are all manifestations of the same underlying conformational phase diagram *(21)*, reflecting different phase boundary positions. On HCl titration of a salt-free solution, some proteins, such as ubiquitin and T4 lysozyme, remain native-like to pH 1 or lower, others, such as α-lactalbumin, transform into the A state (the molten globule state) at pH 3–4, whereas a substantial number, such as cytochrome c and β-lactamase, unfold in the vicinity of pH 3–4, to the acid-unfolded state, and then, below pH 2, refold into the A state *(36)*. In the case of apomyoglobin the transition is more complicated, namely $N \rightarrow A \rightarrow U \rightarrow A$ reflecting the details of the conformational phase diagram *(21)*. Whether or not a molten globule can be observed at low pH is determined by the conformational phase boundaries. Conformational phase diagrams as a function of pH, ionic strength, temperature, and denaturant concentration have been constructed *(21,37)*.

The acid-unfolded state may be converted into the A (molten globule) state by the addition of anions *(38)*. Different anions vary in both their effectiveness (i.e., concentration required) as well as in the degree of folding they induce in the A state. For those proteins whose conformational phase diagrams permit observation of the A state at low pH and moderate ionic strength, the typical conditions necessary to form the molten globule are as follows: The pH should be adjusted to pH = 2, using HCl (this will form the acid-unfolded state if the ionic strength is low). Sodium or potassium salts of the desired anion should be present at their characteristic required concentrations (0.2–0.5*M* chloride; 10–50 m*M* perchlorate; 10–50 m*M* trifluoroacetate; 5–40 m*M* trichloroacetate; 2–10 m*M* sulfate; 50–100 μ*M* ferricyanide) *(38)* (higher concentrations may be used, but frequently lead to aggregation or precipitation). The manner in which the final conditions are generated is usually not important, however, depending on the specific protein, there are sometimes intermediate combinations of pH, ionic strength, and protein concentration that lead to aggregation or precipitation.

Some specific examples of the conditions necessary to form molten globules will now be presented. This list is by no means comprehensive, but does serve to illustrate the range of situations under which these species are found.

5.1. α-Lactalbumin

The molten globule of α-lactalbumin is formed in the absence of Ca^{2+} (i.e., from the apo-protein). Typical conditions for formation are: pH 7, 30°C, no calcium (10 m*M* EDTA); pH < 3, 0.05*M* KCl; pH 7, 2*M* GuHCl. Reduction of the disulfides in α-lactalbumin leads to an unfolded conformation, however, the addition of salts, especially at low temperatures, induces a transition to a molten globule-like conformation *(39)*.

5.2. Cytochrome c

Recently there have been several investigations of the molten globule conformation of cytochrome c *(26,40–42)*. The A state of cytochrome c is readily formed at acidic pH (pH < 3) in the presence of moderate ionic strength, as discussed above *(35,36,38)*. A similar molten globule can be made from apocytochrome c at low pH and moderate salt concentrations *(20)*.

5.3. DnaK

The heat shock protein DnaK, the hsp70 family member from *E. coli*, and other hsp70 proteins, form a molten globule at pH 7 on raising the temperature to 45°C *(43,44)*. In the case of the mammalian cytosolic hsp70s the molten globule forms large soluble aggregates; DnaK, on the other hand, remains monomeric in its molten globule state. The DnaK molten globule has also been observed under conditions of low denaturant (midpoint of the transition at 20°C is 0.45*M* GuHCl), as discussed above *(7)*. Nucleotides, such as ATP and ADP, which bind tightly to hsp70 stabilize the native state and shift the midpoint of the thermal transition to the molten globule to 60°C, and the denaturant-induced molten globule to 0.9*M* GuHCl.

5.4. β-Lactamases

The Class A β lactamases also form stable molten globules. The enzyme from *Staphylococcal aureus* forms two compact denatured states, state I, which is more compact, and state H, which is more expanded. Both are found in moderate concentrations of denaturant

(11,45–47). This enzyme shows similar acid pH behavior to the *Bacillus* enzymes, except the molten globule is more prone to aggregate. The enzyme from *Bacillus cereus* forms molten globules at both low and high pH at moderate ionic strength *(34)*. The molten globule has a strong propensity to aggregate, which can be minimized by low protein concentration and low temperature. The β-lactamase from *B. licheniformis* behaves similarly to the enzyme from *B. cereus*: e.g., the A state is stable at pH 2 in 8 m*M* trichloroacetate (20°C) at 15 μ*M* protein concentration (sufficient for esr investigations *[48]*). At higher concentrations of protein or salt aggregation occurs.

5.5. Staphylococcal Nuclease

Fragments of staphylococcal nuclease in which substantial numbers of residues have been deleted from the C-terminal region form compact states with many properties of molten globules *(29,49)*, even at neutral pH. An A state conformation of staph nuclease has been observed *(60)*.

5.6. Apomyoglobin

Apomyoglobin forms a molten globule that can be detected at low pH in the presence of moderate salt concentrations *(36,38)*, and even in the pH 4–5 region at low salt. Interestingly, unlike the native state, the stability of the intermediate state is relatively insensitive to the effect of mutations *(50)*. The structure of the molten globule, formed at pH 2 as a function of different anions shows different amounts of secondary structure with different anions *(38)* as well as different degrees of compactness (A. L. Fink, K. A. Oberg, and S. Seshadri, unpublished observations).

5.7. Carbonic Anhydrase

A molten globule intermediate of carbonic anhydrase has been reported under both equilibrium and refolding kinetics conditions *(22,51)*. When carbonic anhydrase is refolded from GuHCl, the transiently formed molten globule aggregates *(52)*. Polyethylene glycol can bind to the molten globule and decrease aggregation *(53)*. Mitaku et al. *(54)* showed that the hydrophobic fluorescent dye pyrene bound preferentially to an intermediate present at low concentrations of denaturant: At pH 7, 30°C, the intermediate is maximally populated at 1.5*M* GuHCl. The intermediate was characterized by significant secondary structure, and the absence of tertiary structure. A phase diagram as a function of GuHCl and temperature was constructed, showing that at higher temperatures the intermediate occupied a much larger fraction of conformational space. Rodionova et al. *(55)* showed that the intermediate also bound ANS.

5.8. Ubiquitin

2D-NMR studies of ubiquitin in the A state, induced by methanol at pH 2.0, have been reported recently by Harding et al. *(28)*. Solutions of the A state were prepared by dissolving ubiquitin in 40 or 60% (v/v) aqueous methanol and adjusting the pH to 2 with HCl or DCl. It should be noted that native ubiquitin is very stable and remains native at pH 2 at temperatures close to 80°C. The A state NMR spectrum is considerably different from those of the native and unfolded proteins.

5.9. Other Examples

Dryden and Weir *(56)* have reported evidence that interleukin-2 forms a molten globule at low pH (<2), and in the 3–4*M* GuHCl range at pH 4.6. Molten globule intermediates have been reported for human stefin B under equilibrium conditions of low pH (pH 4, 0.6*M* GuHCl, 25°C), denaturant (1.7*M* GuHCl, pH 8, 25°C) and high temperature (68–75°C, pH 7.5) *(57)*. Immunoglobulin has been reported to form a molten globule-like intermediate at pH 2, 40 m*M* potassium phosphate *(58)*.

It appears that reduction of the disulfide bonds in large proteins with several disulfides may often give rise to molten globules. For example, ovotransferrin has 15 intrachain disulfides; under native conditions, but with reduced disulfides, the molecule adopts a molten globule conformation *(31)*. Reduced serum albumin behaves similarly *(59)*.

There have been fewer studies of molten globules formed under conditions of thermal denaturation, perhaps in part because of the propensity of thermally denatured proteins to aggregate (which is, as noted, a common property of molten globules). There are several reports indicating that thermally denatured proteins retain significant residual structure, however, it is less clear if these species are molten globules or just compact unfolded states, since there has been little detailed characterization of them. As noted above, hsp70 proteins form a molten globule intermediate on raising the temperature to 45°C, pH 7, in the absence of nucleotides *(44)*.

6. Structural Models for the Molten Globule

The main difference between the native and molten globule states appears to be disruption of the hydrophobic interactions binding the structural units together in the native state, with the consequent exposure of nonpolar surfaces to solvent. Ptitsyn's model for the molten globule has been described above. Another model that would account for most of the experimental observations of molten globules is the following.

The native conformation is assumed to consist of a number of structural units, which we will call building blocks, that are tightly packed together via tertiary interactions. These building blocks may consist of regions of secondary structure, or other types of subdomains or autonomous folding units, and will have significant intrinsic stability. Such building blocks would minimally consist of a helix or two-stranded beta-sheet, and maximally of a domain (a large building block could, in turn, be composed of smaller building block units). The transition from the native state to the molten globule state, which corresponds experimentally to the loss of tertiary structure, but retention of secondary structure, would correspond to the separation of these building block units, with concomitant solvent penetration, and with expansion of the molecule as a whole. A range of possible conformations for the molten globule is anticipated, depending on the extent of the loss of side chain contacts between building blocks. Once the contacts between building blocks are lost the molecule will become a collection of structural units (the building blocks) linked by flexible polypeptide chain links. It is likely that these conformations will be quite mobile, in the sense that even with loss of some of the building block contacts there may still be transient contacts between the structural units.

In a small molecule the separation of these building blocks leads to their decreased stability owing to loss of the stabilizing influence of the tertiary interactions, and thus may also result in significant loss, or possibly reorganization, of their secondary structure. The molten globule state is thus pictured as an ensemble of substates ranging from those that are quite compact, with substantial contacts between all, or most of, the building block units, to those with relatively few contacts between the building blocks that are thus quite expanded. Whereas the building block units themselves may retain a rather rigid, native-like structure, there will be substantial mobility between them, as dictated by the freely rotating connecting links. Depending on the destabilizing forces of the external environment, the predominant substates may be quite compact and relatively native-like, or more expanded *(7,60)*.

7. "Kinetic" Molten Globules

Using stopped-flow CD measurements, Kuwajima and coworkers demonstrated that an intermediate was formed very rapidly in the refolding of α-lactalbumin which had very similar far-UV CD properties to the A state *(61)*. Subsequently similar rapidly formed intermediates with substantial secondary structure but little tertiary structure have been demonstrated for several other proteins. In addition the propensity of molten globules to bind ANS has been used to demonstrate the formation and breakdown of molten globule-like intermediates during refolding using stopped-flow fluorescence techniques. On the basis of the apparent generality of this phenomenon, Ptitsyn, Pain, and coworkers *(11)* have proposed the molten globule as a common folding intermediate.

Transient molten globule-like intermediates can be detected by their affinity for the hydrophobic dye ANS *(11,32,34)*. Typical experimental conditions are (final concentrations) [ANS] = 30–75 μ*M*, and [protein] = 5 μ*M*. The choice of protein determines the particular experimental conditions in terms of pH, temperature, and denaturant. The rapidity with which the ANS-binding species are formed, typically in the 50 ms to 1 s range *(11)*, means that a fluorescence stopped-flow system is usually required. Interestingly, this timescale is two to three orders of magnitude longer than that for the initial formation of secondary structure. Excitation is in the vicinity of 380–390 nm, emission in the 480–490 nm region. Typically the protein is unfolded in a fairly high concentration of urea or GuHCl and diluted to a level at which the protein is native. The rate of refolding is proportional to the concentration of denaturant present during refolding, so some control of the folding rate is possible. Examples include carbonic anhydrase, α-lactalbumin, β-lactoglobulin, and phosphoglycerate kinase *(11)* and tryptophan synthase *(62)*.

A typical experiment involving the observation of a transient molten globule-like intermediate is that involving the refolding of carbonic anhydrase in which formation of the molten globule intermediate can be observed by unfolding the protein in 8.5*M* urea and then diluting the urea solution to a final concentration of 4.1–4.3*M* urea, at 23°C *(63)*. The presence of the urea decreases the rate of refolding. The protein concentration can be as high as 4–5 mg/mL. The intermediate can be detected by ANS binding (via fluorescence, excitation at 390 nm, emission at 500 nm), intrinsic tryptophan fluorescence, esr of a spin-labeled derivative, and by the change in intensity of high-field NMR signals *(63)*. Transient molten globule-like intermediates have been observed using far-UV stopped-flow CD for several additional proteins, including cytochrome c and β-lactoglobulin *(10)*, lysozyme, and α-lactalbumin *(61)*, tryptophan synthase subunits *(62,64)*, staphylococcal nuclease *(65)*, and dihydrofolate reductase *(66)*. Because of the propensity of ANS to promote aggregation of molten globules it is important that controls to check for this be performed when ANS is used to monitor transient intermediates (as well as under equilibrium conditions).

8. Possible Physiological Roles for the Molten Globule

Bychkova et al. *(67)* and more recently van der Goot et al. *(68)* have reviewed evidence that the molten globule state is involved in membrane translocation. The A state of α-lactalbumin has been shown to induce fusion of phospholipid vesicles *(69)*. Low pH (as in lysosomes) has been shown to play an important role in membrane penetration by toxins, such as the diphtheria toxin. It has been shown that diphtheria toxin forms a molten globule in the vicinity of pH 5 and that fusion with membranes correlates with formation of the molten globule conformation *(70–72)*. Similarly, it has been shown that membrane insertion of colicin A correlates with its transition to the molten globule *(73,74)*.

Transient molten globules are likely to play a significant role in the folding of proteins *in vivo*. It is expected that when an unfolded protein is placed under native-like conditions it will very rapidly collapse to a compact state such as the molten globule *(4)*. In addition, molten globule-like conformations would be expected to be found for nascent polypeptides bound to hsp70 chaperones *(75)* and perhaps to hsp60 chaperones *(76)*. In fact, it is possible that molten globules are the substrates for some chaperones.

References

1. Ptitsyn, O. B. (1987) *Prot. Chem.* **6,** 273–293.
2. Ptitsyn, O. B. (1992) in *Protein Folding* (Creighton, T. E., ed.), W. H. Freeman, New York, pp. 243–300.
3. Kuwajima, K. (1989) *Proteins: Struct. Fun. Genet.* **6,** 87–103.
4. Dill, K. A. and Shortle, D. (1991) *Ann. Rev. Biochem.* **60,** 795–825.

5. Christensen, H. and Pain, R. H. (1991) *Eur. Biophys. J.* **19,** 221–230.
6. Baldwin, R. L. (1991) *Chemtracts, Biochem. Mol. Biol.* **2,** 379–389.
7. Palleros, D. R., Shi, L., Reid, K., and Fink, A. L. (1993) *Biochemistry* **32,** 4314–4321.
8. Alonso, D. O. V., Dill, K. A., and Stigter, D. (1991) *Biopolymers* **31,** 1631–1649.
9. Finklestein, A. V. and Shakhnovich, E. I. (1989) *Biopolymers* **28,** 1681–1694.
10. Kuwajima, K., Yamaya, H., Miwa, S., Sugai, S., and Nagamura, T. (1987) *FEBS Lett.* **221,** 115–118.
11. Ptitsyn, O. B., Pain, R. H., Semisotnov, G. V., Zerovnik, E., and Razgulyaev, O. I. (1990) *FEBS Lett.* **262,** 20–24.
12. Aune, K. C., Salahuddin, A., Zarlengo, M. H., and Tanford, C. (1967) *J. Biol. Chem.* **242,** 4486–4489.
13. Privalov, P. L., Tiktopoulo, E. I., Venyaminov, S. Y., Griko, Y. V., Makhatadze, G. I., and Khechinashvili, N. N. (1989) *J. Mol. Biol.* **205,** 737–750.
14. Xie, D., Bhakuni, V., and Freire, E. (1991) *Biochemistry* **30,** 10,673–10,678.
15. Yutani, K., Ogasahara, K., and Kuwajima, K. (1992) *J. Mol. Biol.* **228,** 347–350.
16. Ohgushi, M. and Wada, A. (1983) *FEBS Lett.* **164,** 21–24.
17. Dolgikh, D. A., Gilmanshin, R. I., Brazhnikov, E. V., Bychkova, V. E., Semisotnov, G. V., Venyaminov, S. Y., and Ptitsyn, O. B. (1981) *FEBS Letts.* **136,** 311–315.
18. Stigter, D. and Dill, K. A. (1990) *Biochemistry* **29,** 1262–1271.
19. Chan, H. S. and Dill, K. A. (1990) *Proc. Natl. Acad. Sci. USA* **87,** 6388–6392.
20. Hamada, D., Hoshino, M., Kataoka, M., Fink, A. L., and Goto, Y. (1993) *Biochemistry* **32,** 10,351–10,358.
21. Goto, Y. and Fink, A. L. (1990) *J. Mol. Biol.* **214,** 803–805.
22. Dolgikh, D., Abaturov, L., Bolotina, I., Brazhnikov, E., Bychkova, V., Gilmanshin, R., Lebedev, Y., Semisotnov, G., Tiktopulo, E., and Ptitsyn, O. B. (1985) *Eur. Biophys. J.* **13,** 109–121.
23. Ikeguchi, M., Kuwajima, K., and Sugai, S. (1986) *J. Biochem.* **99,** 1191–1201.
24. Baum, J., Dobson, C. M., Evans, P. A., and Hanley, C. (1989) *Biochemistry* **28,** 7–13.
25. Hughson, F. M., Wright, P. E., and Baldwin, R. L. (1990) *Science* **249,** 1544–1548.
26. Jeng, M.-F., Englander, S. W., Elöve, G. A., Wand, A. J., and Roder, H. (1990) *Biochemistry* **29,** 10,433–10,437.
27. Bycroft, M., Matouschek, A., Kellis, J. T., Jr., Serrano, L., and Fersht, A. R. (1990) *Nature* **346,** 488–490.
28. Harding, M. M., Williams, D. H., and Woolfson, D. N. (1991) *Biochemistry* **30,** 3120–3128.
29. Flanagan, J. M., Kataoka, M., Shortle, D., and Engelman, D. M. (1992) *Proc. Natl. Acad. Sci.* **89,** 748–752.
30. Chaffotte, A., Guillou, Y., Delepierre, M., Hinz, H.-J., and Goldberg, M. E. (1991) *Biochemistry* **30,** 8067–8074.
31. Hirose, M. and Yamashita, H. (1991) *J. Biol. Chem.* **266,** 1463–1468.
32. Semisotnov, G. V., Rodionova, N. A., Razgulyaev, O. I., Uversky, V. N., Gripas, A. F., and Gilmanshin, R. I. (1991) *Biopolymers* **31,** 119–128.
33. Tandon, S. and Horowitz, P. M. (1989) *J. Biol. Chem.* **264,** 9859–9866.
34. Goto, Y. and Fink, A. L. (1989) *Biochemistry* **28,** 945–952.

35. Fink, A. L., Calciano, L. J., Goto, Y., and Palleros, D. R. (1990) *Current Research in Protein Chemistry,* (Villafranca, J., ed.), Academic, San Diego, pp. 417– 424.
36. Goto, Y., Calciano, L. J., and Fink, A. L. (1990) *Proc. Natl. Acad. Sci. USA* **87,** 573–577.
37. Mitaku, S., Ishido, S., Hirano, Y., Itoh, H., Kataoka, R., and Saito, N. (1991) *Biophys. Chem.* **45,** 217–222.
38. Goto, Y., Takahashi, N., and Fink, A. L. (1990) *Biochemistry* **29,** 3480–3488.
39. Ikeguchi, M. and Sugai, S. (1989) *Int. J. Peptide Protein Res.* **33,** 289–297.
40. Jeng, M.-F. and Englander, S. W. (1991) *J. Mol. Biol.* **221,** 1045–1061.
41. Kuroda, Y., Kidokoro, S.-I., and Wada, A. (1992) *J. Mol. Biol.* **223,** 1139–1153.
42. Kataoka, M., Hagihara, Y., Mihara, K., and Goto, Y. (1993) *J. Mol. Biol.* **229,** 591–596.
43. Palleros, D. R., Welch, W. J., and Fink, A. L. (1991) *Proc. Natl. Acad. Sci.* **88,** 5719–5723.
44. Palleros, D. R., Reid, K., and Fink, A. L. (1992) *J. Biol. Chem.* **266,** 5279–5285.
45. Creighton, T. E. and Pain, R. H. (1986) *J. Mol. Biol.* **137,** 431–436.
46. Zerovnik, E. and Pain, R. H. (1986) *Protein Eng.* **1,** 248.
47. Mitchinson, C. and Pain, R. H. (1985) *J. Mol. Biol.* **184,** 331–342.
48. Calciano, L. J., Escobar, W. A., Millhauser, G., Miick, S., Rubaloff, J., Todd, P., and Fink, A. L. (1993) *Biochemistry* **32,** 5644–5649.
49. Shortle, D. and Meeker, A. K. (1989) *Biochemistry* **28,** 936–944.
50. Dolgikh, D., Kolomiets, A., Bolotina, E., and Ptitsyn, O. B. (1984) *FEBS Lett.* **165,** 88–92.
51. Hughson, F. M., Barrick, D., and Baldwin, R. L. (1991) *Biochemistry* **30,** 4113–4118.
52. Cleland, J. L. and Wang, D. I. C. (1990) *Biochemistry* **29,** 11,072–11,078.
53. Cleland, J. L. and Randolph, T. W. (1992) *J. Biol. Chem.* **267,** 3147–3153.
54. Mitaku, S., Ishido, S., Hirano, Y., Itoh, H., Kataoka, R., and Saitô, N. (1991) *Biophys. Chem.* **40,** 217–222.
55. Rodionova, N. A., Semisotnov, G. V., Kutyshenko, V. P., Uverskii, V. N., Bolotina, I. A., Bychkova, V. E., and Ptitsyn, O. B. (1989) *Mol. Biol. (Mosc.)* **23,** 683–692.
56. Dryden, D. and Weir, M. P. (1991) *Biochim. Biophys. Acta* **1078,** 94–100.
57. Zerovnik, E., Jerala, R., Kroon-Zitko, L., Pain, R. H., and Turk, V. (1992) *J. Biol. Chem.* **267,** 9041–9046.
58. Buchner, J., Renner, M., Lilie, H., Hinz, H.-J., Jaenicke, R., Kiefhaber, T., and Rudolph, R. (1991) *Biochemistry* **30,** 6922–6929.
59. Lee, J. Y. and Hirose, M. (1992) *J. Biol. Chem.* **2867,** 14,753–14,758.
60. Fink, A. L., Calciano, L. J., Goto, Y., Nishimura, M., and Swedberg, S. A. (1993) *Prot. Sci.* **2,** 1155–1160.
61. Kuwajima, K., Hiraoka, Y., Ikeguchi, M., and Sugai, S. (1985) *Biochemistry* **24,** 874–881.
62. Chaffotte, A. F., Cadieux, C., Guillou, Y., and Goldberg, M. E. (1992) *Biochemistry* **31,** 4303–4308.
63. Semisotnov, G. V., Rodionova, N. A., Kutyshenko, V. P., Ebert, B., Blanck, J., and Ptitsyn, O. B. (1987) *FEBS Lett.* **224,** 9–13.

64. Goldberg, M. E., Semisotnov, G. V., Friguet, B., Kuwajima, K., Ptitsyn, O. B., and Sugai, S. (1990) *FEBS Lett.* **263,** 51–56.
65. Sugawara, T., Kuwajima, K., and Sugai, S. (1991) *Biochemistry* **30,** 2698–2706.
66. Kuwajima, K., Garvey, E. P., Finn, B. E., Matthews, C. R., and Sugai, S. (1991) *Biochemistry* **30,** 7693–7703.
67. Bychkova, V. E., Pain, R. H., and Ptitsyn, O. B. (1988) *FEBS Lett.* **238,** 231–234.
68. van der Goot, F. G., Lakey, J. H., and Pattus, F. (1992) *Trends Cell Biol.* **2,** 343–348.
69. Kim, J. and Kim, H. (1986) *J. Amer. Chem. Soc.* **25,** 7867–7874.
70. Zhao, J.-M. and London, E. (1986) *Proc. Natl. Acad. Sci. USA* **83,** 2002–2006.
71. Blewitt, M. G., Chung, L. A., and London, E. (1985) *Biochemistry* **24,** 5458–5464.
72. Jiang, J. X., Abrams, F. S., and London, E. (1991) *Biochemistry* **30,** 3857–3864.
73. van der Goot, F. G., González-Mañas, J. M., Lakey, J. H., and Pattus, F. (1991) *Nature* **354,** 408–410.
74. Lakey, J. H., González-Mañas, J. M., van der Goot, F. G., and Pattus, F. (1992) *FEBS Lett.* **307,** 26–29.
75. Beckman, R. P., Mizzen, L. A., and Welch, W. J. (1990) *Science* **248,** 850–854.
76. Martin, J., Langer, T., Boteva, R., Schramel, A., Hoewich, A. L., and Hartl, F. U. (1991) *Nature* **352,** 36–42.

CHAPTER 16

Chaperonin-Assisted Protein Folding of the Enzyme Rhodanese by GroEL/GroES

Paul M. Horowitz

1. Introduction

Recent interest has focused on the influence of protein factors in the folding of other proteins both in vivo and in vitro *(1)*. Among these factors are catalysts of folding-related processes, such as peptidyl-prolyl cis-trans isomerase *(2)* or protein disulfide isomerase *(3)*. Other proteins, called "molecular chaperones," have been described that apparently influence protein folding and interactions by forming noncovalent associations with folding intermediates *(4)*. These chaperones include the heat shock proteins of the hsp10, hsp60, and hsp70 classes *(5–13)*. The number indicates the approximate molecular weight of the subunit making up the particular hsp. The molecular chaperones are among the most thoroughly reviewed topics in the field of protein folding and structure. The recent review by Gething and Sambrook may serve as an *entre* to relevant issues and literature *(1)*.

The functions of these chaperone proteins have been suggested to follow from their ability to stabilize folding intermediates and release them in a controlled, ordered, and often energy-dependent fashion. Thus, they have been proposed to be involved in protein folding, the assembly of oligomeric proteins, assisting the passage of proteins across biological membranes, and preventing irreversible protein inactivation or aggregation that would follow conformational perturbations, as induced, for example, by high temperature.

From: *Methods in Molecular Biology, Vol. 40: Protein Stability and Folding: Theory and Practice*
Edited by: B. A. Shirley

One class of chaperonins is exemplified by the GroEL (cpn60) protein of *E. coli*, which is homologous to a mitochondrial matrix protein, hsp60, that is a required component of the specific pathway for import, folding, and maturation of mitochondrial matrix proteins *(14,15)*. The GroEL tetradecamer, composed of 14 identical 60 kDa subunits, has been reported to assist the refolding of a number of proteins *(10,11,16)* in a process that requires MgATP, K^+, and a second protein, GroES (cpn 10) in addition to GroEL. Some proteins do not require GroES. The mitochondrial homolog of the bacterial GroEL, hsp60, has been reported to be active as a single heptameric ring *(19)*. Although the yields of correctly folded proteins are increased by the GroEL/GroES system, the effects on the rates of folding depend on the specific protein. For example, the GroEL/GroES system slows the rate of folding of the enzyme rhodanese while increasing the yield of active enzyme. On the other hand, the rate of folding of the enzyme Rubisco is greatly enhanced compared with the spontaneous process. Evidence has been provided that GroEL can interact with high affinity with a very broad spectrum of proteins, with specificity for their unfolded or partially folded states, and with ATP-induced release of the bound proteins *(17)*.

The mechanisms by which GroEL/GroES influence protein folding are currently being actively investigated. Most of the mechanisms suggested for the action of the chaperonins GroEL/GroES include the following elements *(14)*. The 14 subunit GroEL oligomer is assembled into a structure resembling a double doughnut composed of two stacked, seven-member rings. Electron micrographs suggest that the GroEL oligomer is in the form of a cylinder containing a central space with a diameter of approx 60 Å *(18)*. This cylindrical structure can bind nonnative, partially folded conformers of the target proteins that are suggested to be in the form of folding intermediates that have been called "molten globules" *(20,21)* or "compact folding intermediates" *(22)*. These structures are somewhat ill defined, but they are generally considered to be compact, have a high degree of secondary structure, contain reduced levels of tertiary interactions, and display increased exposure of hydrophobic surfaces as evidenced by the binding of hydrophobic probes such as 1,8 ANS (*see* Chapter 3 and 15). Even though interactive surfaces on the passenger protein may be exposed, they can be protected sterically by being part of a macromolecular complex. Trapped structures of this type that can be formed with other polypeptide chains may also allow assembly at these interactive surfaces prior to release of the bound polypeptide, thus providing the appropriate sequencing of interactions. This may be applicable to the orderly assembly of oligomeric proteins. For the protein rhodanese that will be discussed in this chapter, the release step requires additional factors including ATP and a second oligomeric protein, GroES *(23)*.

The detailed molecular mechanism by which the GroEL/GroES system exerts its influence still is under investigation actively in a number of laboratories. It has been suggested that ATP dependent steps, possibly including hydrolysis, lead to alternating release and rebinding of the target protein that allows for the progressive transition of the protein to its thermodynamically most stable state. The role of GroES in this model would be to modulate and coordinate the release and rebinding steps. Thus, GroEL/GroES would influence folding by offering kinetic guidance rather than "thermodynamic" influence. It has recently been reported, based on electron microscopy and biochemical studies, that GroES appears to bind asymmetrically to GroEL on one end of the cylinder, and that the protein to be folded is bound in the central space on the opposite end of the GroEL cylinder *(24)*.

There is considerable current interest in defining the structural motif(s) that are recognized by GroEL when it binds the proteins that are to be folded. It is generally believed that GroEL binds proteins and stabilizes interactive folding intermediates as molten globules, thereby preventing misfolding and/or aggregation. Based on NMR studies of peptide binding, it has been suggested that amino-terminal α-helices form at least part of the recognition site in the proteins whose folding is influenced by GroEL *(24)*. However, GroEL binds peptides weakly, whereas proteins are bound very tightly. This is expected based on experiments that suggest that there are multiple interaction sites for a target protein on GroEL.

This chapter will detail the use of the proteins GroEL/GroES in the folding of the mitochondrial enzyme rhodanese. The assay can be adapted for use with other proteins or used directly to test various features of chaperonin assisted refolding. The *E. coli* GroEL is of interest because it is promiscuous in its interactions, as noted above.

The enzyme rhodanese has several advantages for assays of chaperonin assisted refolding. The protein is found in the matrix of all mammalian mitochondria *(25)*. This 33 kDa protein is synthesized in the cytoplasm. Rhodanese is monomeric and contains no cofactors, prosthetic groups, or required side chain modifications for its enzymatic activity. Thus, only its amino acid sequence is needed to understand its structure, function, folding, and transport. The mature mitochondrial protein has not been processed and, with the exception of the initiating methionine, it retains all the sequence present at synthesis. Thus, all the sequence information required for interaction with the mitochondrial targeting, import, and folding machinery is present in the isolated protein. In addition, the active protein is monomeric under a wide variety of conditions, and a high resolution structure is available *(26,27)*. In short, there are fewer complications with rhodanese than are present in many other systems used to study chaperonins.

GroEL can interact with partially folded rhodanese and arrest its spontaneous folding. Subsequent or simultaneous addition of GroES and the other components leads to the folding and reactivation of enzymatically active rhodanese.

We will describe the procedure by which the assisted refolding of rhodanese can be performed. This procedure can be suitably modified for attempting the refolding of other proteins. The interactions of other proteins with GroEL, for example, could be conveniently monitored by the ability of the bound proteins to prevent GroEL from arresting the spontaneous folding of rhodanese.

2. Materials

The following materials and solutions are used for the assay of chaperonin assisted refolding of rhodanese. The volume is noted of each component to give a final incubation volume of 250 µL.

1. 0.1*M* Tris-HCl, pH 7.8: 125 µL.
2. 0.25*M* $MgCl_2$: 10 µL.
3. 0.25*M* KCl: 10 µL.
4. GroEL; 50 µ*M* protomer; 12.5 µL to give final concentration of 2.5 µ*M* protomer (*see* Note 1). GroEL and cpn 10 are purified from lysates of cells carrying a multicopy plasmid pGroESL or pOF39 *(28,30)*, by procedures that have been previously described *(31)*.
5. cpn 10; 50 µ*M* protomer; 12.5 µL to give final concentration of 2.5 µ*M* protomer.
6. 7*M* β-mercaptoethanol: 7 µL.
7. 1*M* Sodium thiosulfate: 12.5 µL.
8. 0.1*M* ATP: 5 µL.

9. Water, to give a final vol of 250 μL after addition of rhodanese.
10. Rhodanese; unfolded at 0.3 mg/mL diluted in the assay to 3.6 μg/mL. Some investigators use commercially available rhodanese (type II, cat.# R 1756, Sigma, St. Louis, MO). Alternatively, rhodanese can be prepared as described in the literature *(29,30)* (*see* Note 2).
11. Rhodanese assay reagents as follows *(31)*:
    a. 0.15*M* $Na_2S_2O_3$.
    b. 0.15*M* KCN.
    c. 0.12*M* KH_2PO_4.
    d. 18% formaldehyde: dilute reagent (approx 37%) 1:1 with water.
    e. Concentrated ferric nitrate reagent (2X Stock): Make this solution in the hood! Add nitric acid to stirring water, slowly! The final solution will contain: 400 g $FeNO_3 \cdot 9H_2O$; 800 mL 65% HNO_3; in a final vol of 3 L. Filter the solution if it is turbid. Dilute this stock with an equal volume of water to make the assay reagent.

3. Methods

A typical GroEL/GroES assisted folding assay for rhodanese is based on regaining activity from a sample of rhodanese that has been unfolded for 45 min at 25°C in buffer containing 1 m*M* β-mercaptoethanol and 8*M* urea (*see* Notes 3 and 4).

1. Initiate refolding by diluting 3 μL of unfolded rhodanese into a final vol of 250 μL of a buffer containing the following components at the indicated final concentrations: 2.5 μ*M* protomer GroEL, 2.5 μ*M* protomer GroES, 200 m*M* β-mercaptoethanol, 50 m*M* Tris-HCl, pH 7.8, 50 m*M* sodium thiosulfate, 10 m*M* $MgCl_2$, 10 m*M* KCl, and 2 m*M* ATP at 25°C (*see* Notes 5 and 6).
2. Mix the samples by vortexing. Take care to mix the samples thoroughly and consistently (*see* Note 7).
3. Prepare the rhodanese assay mixture by combining $Na_2S_2O_3$, KCN, and KH_2PO_4 (reagents 11 a, b, and c, above) in a 1:1:1 volume ratio. The final pH of the mixture should be 8.6.
4. Measure the regain of rhodanese activity by withdrawing a 25-μL aliquot from the incubation mixture and adding it to 1 mL of the standard assay mixture.
5. Incubate the mixture for 15 min and then add 0.5 mL of the 18% formaldehyde to terminate the reaction.
6. Add 1.5 mL of the ferric nitrate reagent. Mix.
7. Determine the enzyme activity from the absorbance at 460 nm of the colored complex formed between the product, thiocyanate, and ferric ion in the standard assay for rhodanese activity *(31)*. Use an appropriate reagent blank.

The reactivation will be complete in about 45 min with a $t_{1/2}$ of approx 10 min. Although recoveries are somewhat variable, it has been possible to regain 80% of the activity based on the activity of an equal quantity of nondenatured enzyme that has been incubated along with the experimental samples (*see* Note 2).

4. Notes

1. The concentrations of GroEL and GroES are often determined from their extinction coefficients. In our laboratory we use the values of 12,200/*M*/cm for GroEL and 3440/*M*/cm for GroES. There have been reports that GroEL contains tryptophan, although the DNA sequence for the protein shows that no tryptophan residues are encoded. Part of the discrepancy may result from the fact that GroEL has been shown to bind a very diverse group of peptides with a small stoichiometry relative to the 14-mer with a molecular weight of 840 kDa. Thus, SDS gels would reveal prominent bands corresponding to GroEL, whereas the low occupancy of GroEL sites, together with the small content of any particular bound protein, would make it difficult to detect a particular bound passenger protein as a distinct component. The aggregate of bound peptides could contribute to the absorbance and the reported tryptophan content.
2. Commercially available rhodanese has been used in a number of reports and is apparently good enough for some studies; the specific activities of the commercial enzyme are quoted as 100–200 U/mg. The enzyme is fairly homogeneous by SDS gel electrophoresis. The available purifications use conventional procedures, and they can typically yield more than 200 mg of homogenous rhodanese at a specific activity of >500 U/mg.
3. Appropriate controls should be used to assess the effects of carryover of denaturant, the linearity of the enzyme assay with time and protein concentration, and the spontaneity of folding for the protein of interest.
4. Several other bases can be used to construct assays for individual proteins. These follow from the basic principles underlying the interactions with the GroEL/GroES system. For example, GroEL interactions with rhodanese can be measured using light scattering. GroEL can interact with rhodanese folding intermediate(s) and prevent aggregation that competes with successful folding. Therefore, GroEL will lead to a decrease in the light scattering caused by aggregation when denatured rhodanese is diluted into buffer. This effect only depends on the interaction with GroEL, and it is, therefore, independent of ATP, GroES, or any of the other components. Subsequent addition of the missing components to the GroEL–rhodanese complex will permit the folding and reactivation of the protein.
5. Some experimental variables may be important for the particular protein

being folded and not for the GroEL/GroES system itself. For example, in the assay described here, β-mercaptoethanol and thiosulfate are required to get high levels of reactivation, because rhodanese is sensitive to oxidation and is stabilized by its substrate. We routinely use this assay, although these components have not been reported in some investigations.

6. A number of experimental parameters relate to the GroEL/GroES system itself or the interactions of the chaperonins with particular proteins. For example, the assay described here is sensitive to temperature, and the refolding yield is very low at low temperatures. Although all proteins that are favorably influenced by these chaperonins require GroEL and ATP, not all require GroES. Under the conditions investigated so far, GroES is required for rhodanese reactivation. Although it is not absolutely needed for the folding of dihydrofolate reductase, the rates of refolding are influenced by the presence of GroES.
7. As noted above, samples should be mixed in a consistent way, and the sample volumes and the type of tubes that are used should be standardized. We have noted some dependence of the recoveries on the volume of the samples and on the vigor of mixing. Surface effects and/or oxidation may be responsible for some variability.
8. In reporting results or comparing them with published reports, it is important to be clear about the basis on which the refolding yield is compared. In some reports, the yield is based on maximum attainable recovery in the particular assay, so that although the absolute recoveries are low, the percentage recoveries are high. Care must be exercised to note whether published values give relative recovery compared with theoretical recoveries or relative to controls. This difference can lead to some ambiguity in comparing results among laboratories.

Conditions recently have been developed that permit rhodanese refolding in the absence of chaperonins. Under these conditions, GroEL can interact with a rhodanese folding intermediate and arrest the spontaneous folding. This arrest phenomenon can be used to assess the influence of a number of factors on the interaction of rhodanese with GroEL.

The limited stoichiometry for the interactions involving GroEL and test proteins can be used to develop competition assays. Thus, for example, the arrest of rhodanese spontaneous folding by GroEL can be measured in the presence of a protein whose interaction is being monitored. Interactions with the test protein will interfere with rhodanese interactions and spontaneous folding of rhodanese will occur. Careful controls are required to be sure that specific interactions are being monitored.

References

1. Gething, M.-J. and Sambrook, J. (1992) *Nature* **355,** 33–45.
2. Lang, K., Schmid, F. X., and Fischer, G. (1987) *Nature* **329,** 268–270.
3. Freedman, R. B. (1989) *Cell* **57,** 1069–1072.
4. Ellis, J. (1987) *Nature* **328,** 378–379.
5. Laskey, R. A., Honda, B. M., Mills, A. D., and Finch, J. T. (1978) *Nature* **275,** 416–420.
6. Bochkareva, E. S., Lissin, N. M., and Girshovich, A. S. (1988) *Nature* **336,** 254–257.
7. Chirico, W. J., Waters, M. G., and Blobel, G. (1988) *Nature* **332,** 805–810.
8. Hemmingsen, S. M., Woolford, C., van der Vies, S. M., Tilly, K., Dennis, D. T., Georgopoulos, C. P., Hendrix, R. W., and Ellis, R. J. (1988) *Nature* **333,** 330–334.
9. Flynn, G. C., Chappell, T. G., and Rothman, J. E. (1989) *Science* **245,** 385–390.
10. Goloubinoff, P. I., Gatenby, A. A., and Lorimer, G. H. (1989) *Nature* **337,** 44–47.
11. Goloubinoff, P. I., Christeller, J. T., Gatenby, A. A., and Lorimer, G. H. (1989) *Nature* **342,** 884–889.
12. Buchner, J., Schmidt, M., Fuchs, M., Jaenicke, R., Rudolph, R., Schmid, F. X., and Kiefhaber, T. (1991) *Biochemistry* **30,** 1586–1591.
13. Martin, M., Langer, T., Boteva, R., Schramel, A., Horwich, A. L., and Hartl, F. U. (1991) *Nature* **352,** 36–42.
14. Ostermann, J., Horwich, A. L., Neupert, W., and Hartl, F. U. (1989) *Nature* **341,** 125–130.
15. Georgopoulos, C. and Ang, D. (1990) *Semin. Cell Biol.* **1,** 19–25.
16. Viitanen, P. V., Lubben, T. H., Reed, J., Goloubinoff, P., O'Keefe, D. P., and Lorimer, G. H. (1990) *Biochemistry* **29,** 5665–5671.
17. Viitanen, P. V., Gatenby, A. A., and Lorimer, G. H. (1992) *Protein Sci.* **1,** 363.
18. Langer, T., Pfeifer, G., Martin, J., Baumeister, W., and Hartl, F.-U. (1992) *EMBO J.* **11,** 4757–4765.
19. Viitanen, P. V., Lorimer, G. H., Seetharam, R., Gupta, R. S., Oppenheim, J., Thomas, J. O., and Xowan, N. J. (1992) *J. Biol. Chem.* **267,** 695–698.
20. Kuwajima, K. (1989) *Protein Struc. Funct. Genet.* **6,** 87–103.
21. Ptitsyn, O. B. (1991) *J. Prot. Chem.* **6,** 272–293.
22. Creighton, T. E. (1990) *Biochem. J.* **270,** 1–16.
23. Mendoza, J. A., Rogers, E., Lorimer, G. H., and Horowitz, P. M. (1991) *J. Biol. Chem.* **266,** 13,044–13,049.
24. Landry, S. J. and Gierasch, L. M. (1991) *Biochemistry* **30,** 7359–7362.
25. Westley, J. (1973) *Adv. Enzymol. Relat. Areas Mol. Biol.* **39,** 327–368.
26. Ploegman, J. H., Drent, G. H., Kalk, K. H., Hol, W. G. J., Heinrikson, R. L., Keim, P., Weng, L., and Russell, J. (1978) *Nature* **273,** 1245–1249.
27. Hol, W. G. J., Lijk, L. J., and Kalk, K. H. (1983) *Fundament. Appl. Toxicol.* **3,** 370–376.
28. Fayet, O., Siegelhoffer, T., and Georgopoulos, C. (1989) *J. Bacteriol.* **171,** 1379–1385.
29. Horowitz, P. M. (1978) *Anal. Biochem.* **86,** 751–753.
30. Kurzban, G. P. and Horowitz, P. M. (1991) *Prot. Express. Purif.* **2,** 379–384.
31. Sorbo, B. (1953) *Acta Chem. Scand.* **7,** 1129–1133.

Index